Wildlife Ecology and Management

or

2

7

Wildlife Ecology and Management

The Late GRAEME CAUGHLEY PhD
CSIRO Division of Wildlife and Rangelands Research
Lyneham, Canberra, Australia

ANTHONY R.E. SINCLAIR PhD
Department of Zoology
University of British Columbia, Vancouver, Canada

BOSTON
Blackwell Scientific Publications
OXFORD LONDON EDINBURGH
MELBOURNE PARIS BERLIN VIENNA

For Graeme, friend and colleague.

© 1994 by
Blackwell Scientific Publications
Editorial offices:
238 Main Street, Cambridge
Massachusetts 02142, USA
Osney Mead, Oxford OX2 0EL, England
25 John Street, London WC1N 2BL
England
23 Ainslie Place, Edinburgh EH3 6AJ
Scotland
54 University Street, Carlton
Victoria 3053, Australia

Other Editorial Offices:
Librairie Arnette SA
1, rue de Lille
75007 Paris
France

Blackwell Wissenschafts-Verlag GmbH
Düsseldorfer Str. 38
D-10707 Berlin
Germany

Blackwell MZV
Feldgasse 13
A-1238 Wien
Austria

First published 1994

Set by Excel Typesetters Co., Hong Kong
Printed and bound in the United States of
America by Edwards Brothers, Ann Arbor,
Michigan

94 95 96 97 98   5 4 3 2 1

DISTRIBUTORS

USA
Blackwell Scientific Publications, Inc.
238 Main Street
Cambridge, Massachusetts 02142
(Orders: Tel: 617 876-7000
800 759-6102)

Canada
Oxford University Press
70 Wynford Drive
Don Mills
Ontario M3C 1J9
(Orders: Tel: 416 441-2941
800 268-4178)

Australia
Blackwell Scientific Publications Pty Ltd
54 University Street
Carlton, Victoria 3053
(Orders: Tel: 03 347-5552)

Outside North America and Australia
Marston Book Services Ltd
PO Box 87
Oxford OX2 0DT
(Orders: Tel: 0865 791155
FAX: 0865 791927
Telex: 837515)

Library of Congress
Cataloging-in-Publication Data

Caughley, Graeme.
Wildlife ecology and management/
Graeme Caughley, A.R.E. Sinclair.
p.   cm.
Includes bibliographical references
(p. 305) and index.
ISBN 0−86542−144−7
1. Wildlife management.
2. Animal ecology.
I. Sinclair, A. R. E.
(Anthony Ronald Entrican) II. Title.
SK355.C38 1994
639.9—dc20

Contents

Preface

Our objective in writing this book is modest. We seek to provide a text that can be used in undergraduate courses in wildlife management. If parts of it are of interest to people who have advanced beyond that stage then that will be a bonus.

This book is structured as three interlocking parts. There is first a chapter on how wildlife management objectives are formulated.

The book then launches into a number of chapters providing a contracted overview of general ecology, as distinct from that portion of applied ecology that is called wildlife management. We have observed that many courses offered in wildlife management do not stipulate a solid grounding in ecology as a prerequisite. The chapters on general ecology (see Chapters 2–11) are there to remedy that defect. We view wildlife management as applied ecology. You will have trouble applying it unless you know some. In particular you will need a knowledge of the theory of population dynamics and of the relationship between populations and their resources if you are to make sensible judgments on the likely consequences of this or that management action.

If you already have an adequate grounding in general ecology (test questions: what is an intrinsic rate of increase?; what is a functional response?), you should move directly to the part dealing with wildlife management (see Chapter 12 onwards). These chapters cover census, how to test hypotheses, and the three aspects of wildlife management: conservation, sustained yield, and control.

We are particularly grateful to Anne Gunn, who critically read previous versions of the majority of chapters and added numerous improvements, and to David Grice who acted as text manager, coordinated the work of the two authors separated by the width of the Pacific Ocean, and prepared many of the figures. These two swung into action to complete the task after Graeme Caughley fell ill. We thank them profusely and sincerely.

We would also like to thank the following people for their contructive comments: Chris Collet for help with Chapter 3, Steve Cork for help with Chapter 7, Roger Pech for help with Chapter 8, David Hik for help with Chapter 9, Dave Spratt and Grant Singleton for help with Chapter 11, and Nick Nicholls for help with Chapter 14.

Inge Newman, Barbara Staples, and Robin Barker provided invaluable detective work when searching for references. We also thank Fleur Sheard who prepared the drawings. For the remaining figures we thank David Grice, Juliet Davies, and Frank Knight.

Graeme Caughley died February 16 1994 as this book went to press.

A.R.E.S.

1 Introduction

This chapter explains what wildlife management is and how it should operate. You are faced with the difference between value judgments and technical judgments and how these relate to goals and policies compared to options and actions. We take you through the various steps involved in deciding what to do and how it should be done. We describe decision analysis and matrices and how they help evaluate feasible management options.

1.1 What is wildlife management?

Wildlife is a word whose meaning expands and contracts with the viewpoint of the user. Sometimes it is used to include all wild animals and plants. More often it is restricted to terrestrial and aquatic vertebrates. In the discipline of wildlife management it designates free-ranging birds and mammals and that is the way it is used here. Until about 25 years ago wildlife was synonymous with game, those birds and mammals that were hunted for sport. The management of such species is still an integral part of wildlife management but increasingly it embraces other aspects such as conservation of endangered species.

Wildlife management may be defined for present purposes as the *management of wildlife populations*. This may be too restrictive for some who would argue that many of the problems of management deal with people and therefore that education, extension, park management, law enforcement, and land evaluation are legitimate aspects of wildlife management and ought to be included within its definition. They have a point, but expansion of the definition to take in all these aspects diverts attention from the core around which management activities are organized: the manipulation or protection of a population to achieve a goal. Obviously, people must be informed as to what is being done, they must be educated to an understanding of why it is necessary, and their behavior may have to be regulated with respect to that goal. But the most important task is to choose the right goal and to know enough about the animals and their habitat to assure its attainment. Hence wildlife management is restricted here to its literal meaning, thereby emphasizing the core at the expense of the periphery of the field.

1.1.1 Kinds of management

Wildlife management implies stewardship, that is the looking after of a population. A population is a group of coexisting individuals of the

1

same species. Wildlife management may be either *manipulative* or *custodial*. Manipulative management does something to a population, either changing its numbers by direct means or influencing numbers by the indirect means of altering food supply, habitat, density of predators, or prevalence of disease. Manipulative management is appropriate when a population is to be harvested, or when it slides to an unacceptably low density or increases to an unacceptably high level.

Custodial management on the other hand is preventative or protective. It is aimed at minimizing external influences on the population and its habitat. It is not aimed necessarily at stabilizing the system but at allowing free rein to the ecological processes that determine the dynamics of the system. Such management may be appropriate in a national park where one of the stated goals is to protect ecological processes and it may be appropriate for conservation of a threatened species where the threat is of external origin rather than being intrinsic to the system.

1.2 **Goals of management**

A wildlife population may be managed in one of four ways:
1   make it increase;
2   make it decrease;
3   harvest it for a continuing yield;
4   leave it alone but keep an eye on it.
These are the only options available to the manager.

Three decisions are needed: (i) what is the desired goal; (ii) which management option is therefore appropriate; and (iii) by what action is the management option best achieved? The first decision requires a judgment of value, the others technical judgments.

1.2.1 *Who makes the decisions?*

It is not the function of the wildlife manager to make the necessary value judgments in determining the goal any more than it is within the competence of a general to declare war. Managers may have strong personal feelings as to what they would like, but so might many others in the community at large. Managers are not necessarily provided with heightened aesthetic judgment just because they work on wildlife. They should have no more influence on the decision than does any other interested person.

However, when it comes to deciding which management options are feasible (once the goal is set) and how goals can best be attained, wildlife managers have the advantage of their professional knowledge. Now they are dealing with testable facts. They should know whether current knowledge is sufficient to allow an immediate technical decision or whether research is needed first. They can advise that a stated goal is unattainable, or that it will cost too much, or that it will cause unintended side-effects. They can consider alternative routes to a goal and advise on the time, money, and effort each would require. These are all technical judgments, not value judgments. It is the task of the wildlife manager to make them and then to carry them through.

Since value judgments and technical judgments tend to get confused with each other it is important to distinguish between them. By its essence a value judgment is neither right nor wrong. Let us take a hypothetical example. The black rat (*Rattus rattus*) is generally unloved. It destroys stored food, it is implicated in the spread of bubonic plague and several other diseases, and it has been known to bite babies. Suppose a potent poison specific to this species were discovered, thereby opening up the option of removing it from the face of the earth. Many would argue for doing just that, and swiftly. Others would argue that there are strong moral objections to exterminating a species, however repugnant or inconvenient that species might be. Most of us would have a strong opinion one way or the other but there is no way of characterizing either competing opinion as right or wrong. That dichotomy is meaningless. A value judgment can be characterized as hardheaded or sentimental (these are also value judgments), or it may be demonstrated as inconsistent with other values a person holds, but it cannot be declared right or wrong. In contrast, technical judgments can be classified as right or wrong according to whether they succeed in achieving the stated goal.

1.2.2 Decision analysis

In deciding what objective (goal) is appropriate we consider a range of influences, some dealing with the benefits of getting it right and others with the penalties of getting it wrong. Social, political, biological, and economic considerations are each examined and given due weight. Some people are good at this and others less so. In all cases, however, there is a real advantage, both to those making the final decision and to those tendering advice, to have the steps of reasoning laid out before them as a decision is approached.

At its simplest, this need mean no more than the people helping to make the decision spelling out the reasons underpinning their advice. But with more complex problems it helps to be more formal and organized, mapping out on paper the path to the decision through the facts, influences, and values that shape it. That process should be explicit and systematic. Different people will assign different values (weights) to various possible outcomes and, particularly if mediation by a third party is required, an explicit statement of those weights allows a more informed decision. It helps also to determine which disagreements are arguments about facts and which are arguments about judgments of value.

Table 1.1 is an objective/action matrix in which possible objectives are ranged against feasible actions. The objectives are not mutually exclusive. The matrix comes from the response of the Department of Agriculture of Malaysia to the attack of an insect pest on rice (Norton 1988). It allows the departmental entomologists and administrators to view the full context within which a decision must be made. Each of the listed objectives is of some importance to the department. The next step would be to rank those objectives and then to score the

Table 1.1 Possible objectives and management actions for public pest management. The initial problem is to assess how each action is likely to meet each objective. (After Norton 1988.) The manager must check off the objectives that can be met by each of the actions for a particular management problem. Which actions are finally chosen may be decided by financial, social, or political constraints, but the manager can see which objectives can be met with such constraints

Actions	Objectives					
	Improve farmers' ability to control pest	Improve farmers' incentives	Strengthen political support	Keep department's cost low	Reduce damage	Reduce future pest outbreaks
Short term						
1 Warn and advise farmers						
2 Advise and provide credit						
3 Advise and subsidize pesticides						
4 Advise, subsidize, and supervise spraying						
5 Mass treat and charge farmers						
6 Mass treat at department's cost						
Medium term						
7 Intensive pest surveillance						
8 Implement area-wide biological control						
9 Training courses for farmers						

management actions most appropriate to each. The final outcome is the choice of one or more management actions that best meet the most important objective or objectives. Such very simple aids to organizing our thoughts are often the difference between success and failure.

Another such aid is the feasibility/action matrix. Table 1.2 is Bomford's (1988) analysis of management actions to reduce the damage wrought by ducks on the rice crops of the Riverina region of Australia. The feasibility criteria are here ranked so that if a management action fails according to one criterion there is no point in considering it against further criteria. Note how this example effortlessly identifies areas of ignorance that would have to be attended to before a rational decision is possible.

Our third example of decision aids is the pay-off matrix (Table 1.3). It expresses the state of nature (level of pest damage in this example) as rows and the options for management action as columns (Norton

Table 1.2 A matrix to examine possible management actions against criteria of feasibility. (After Bomford 1988)

Control options	Feasibility criteria					
	Technically possible	Practically feasible	Economically desirable	Environmentally acceptable	Politically advantageous	Socially acceptable
1 Grow another crop	1	0				
2 Grow decoy crop	1	1	?	1	1	1
3 Predators and diseases	0					
4 Sowing date	1	1	?	1	1	1
5 Sowing technique	1	1	?	1	1	1
6 Field modifications	1	1	?	1	1	1
7 Drain or clear daytime refuges	?	0				
8 Shoot	1	1	?	1	?	1
9 Prevent access, netting	1	1	0	1	1	1
10 Decoy birds or free feeding	?	1	?	1	1	
11 Repellants	1	0				
12 Deterrents	1	1	?	1	1	1
13 Poisons	1	1	?	0		
14 Resowing or transplanting seedlings	1	?	1	?	1	1

1, Yes; 0, no; ?, no information.

Table 1.3 A pay-off matrix for pest control. (After Norton 1988)

	Actions			
	Do nothing	Pest control strategies		
State of nature	(0)	(1)	(2)	(3)
Level of pest attack				
Low (L)	Outcome L, 0	Outcome L, 1	Outcome L, 2	Outcome L, 3
Medium (M)	Outcome M, 0	Outcome M, 1	Outcome M, 2	Outcome M, 3
High (H)	Outcome H, 0	Outcome H, 1	Outcome H, 2	Outcome H, 3

1988). The problem is to assess the probable outcome of each combination of level of damage and the action mounted to alleviate it. Note that the column associated with doing nothing gives the level of damage that will be sustained in the absence of action. It is the control against which the net benefit of management must be assessed. The cells of this matrix are best filled in with net revenue values (benefit minus cost) rather than with benefit/cost ratios because it is the absolute rather than relative gain that shapes the decision.

1.3 Hierarchies of decision

Before we begin manipulating a wildlife population and its environment we must ask ourselves why we are doing it and what is it supposed to achieve. In management theory, that decision is usually divided into hierarchical components.

At the bottom, but here addressed first, is the management action. It might be to eliminate feral pigs (*Sus scrofa*) on Lord Howe Island off the coast of Australia. The management action must be legitimized by a technical objective, e.g., to halt the decline of the Lord Howe Island woodhen (*Tricholimnas sylvestris*) on Lord Howe Island. Above that is the policy goal, a statement of the desired endpoint of the exercise, which in this example might be to secure the continued viability of all indigenous species within the nation's National Park system.

In theory the decisions flow from the general (the policy goal) to the special (the management action), but in practice that does not work because each is dependent on the others, in both directions. Nothing is achieved by specifying "halt a species decline" as a technical objective unless there is available a set of management actions that will secure that objective. Obviously a management action cannot be specified to cure a problem of unknown cause. All three levels of decision must be considered together such that the end product is a feasible option.

A feasible option is identified by answering the following questions.

1   Where do we want to go?
2   Can we get there?
3   Will we know when we have arrived?
4   How do we get there?
5   What disadvantages or penalties accrue?
6   What benefits are gained?
7   Will the benefits exceed the penalties?

The process is iterative. There is no point in persevering with the policy goal thrown up by the first question if the answer to the second is negative. The first choice of destination is therefore replaced by another, and the process repeated.

Question 3 is particularly important. It requires formulating stopping rules. That does not mean necessarily that management action ceases on attainment of the objective, rather that management action is altered at that point. The initial action is designed to move the system towards the state specified by the technical objective; the subsequent action is designed to hold the system in that state. If we cannot determine when the objective has been attained, either for reasons of logic (ambiguous or abstract statement of the objective) or for technical reasons (inability to measure the state of the system), the option is not feasible.

1.4 Policy goals

Policies are usually couched in broad terms that provide no more than a general guide for the manager. The specific decisions are made when the technical objectives are formulated. However, there are two types of policy goals that the manager must know about in case they clash with the choosing of those objectives.

1.4.1 *The nonpolicy*

Nonpolicies stipulate goals that are not clearly defined. They are usually formulated in that way on purpose so that the administering agency is not tied down to a rigidly dictated course of action. Policies are usually formulated by the administering agency whether or not they are given legislative sanction. If the agency has not developed a policy it may fill the gap with a nonpolicy that commits it to no specified action. For example, take the goal of "protecting intrinsic natural values." It reads well but is entirely devoid of objective meaning.

1.4.2 *The nonfeasible policy*

In contrast to the relatively benign nonpolicy, the nonfeasible policy can be damaging. Although it may give each interest group at least something of what they desire, sometimes the logical consequence is that two or more technical objectives are mutually incompatible.

An example is provided by the International Convention for the Regulation of Whaling of 1946 which was "to provide for the proper conservation of whale stocks" and "thus make possible the orderly development of the whaling industry." This pleased both those people concerned about conservation of whales and those wishing to harvest whales. Unfortunately, the goal is a nonsense because, for reasons that are elaborated in Chapter 16, species with a low intrinsic rate of increase are not suitable for sustainable harvesting. The two halves of the policy goal contradict each other. The history of whaling since 1948, in which the blue (*Balaenoptera musculus*), the fin (*B. physalus*), the sei (*B. borealis*), the Brydes (*B. edeni*), the humpback (*Megaptera novaeangliae*), and the sperm (*Physeter macrocephalus*) were reduced to the level of economic extinction, is a direct consequence of choosing a policy goal that was not feasible.

Another form of the nonfeasible policy is that, in contrast with the nonpolicy, the policy is so specific that it actually determines technical objectives and sometimes even management actions. If these are unattainable in practice, the policy goal itself is also unattainable. An example is provided by the now defunct policy to exterminate deer in New Zealand. It was always an impossibility.

1.5 **Feasible options**

Objectives must be attainable. It is the wildlife manager's task to produce the attainable technical objectives by which the policy goal is defined. In contrast to the goal, which may be described in somewhat abstract terms, a technical objective must be stated in concrete terms and rooted in geographic and ecological fact. It must be attainable in fact and it should be attainable within a specified time. A technical objective should therefore be accompanied by a schedule.

1.5.1 *Criteria of failure*

It follows as a corollary that there must be an easy way of recognizing the failure to attain an objective. The most common is to measure the outcome against that specified by the technical objective. Another is to compare the outcome with a set of *criteria of failure*, set before the management action is begun. These two are not the same. Comparison of outcome with objective can produce assessments like "not quite" or "not yet." Not so with criteria of failure. They take the form: "the operation will be judged unsuccessful, and will therefore be terminated, if outcome x has not been attained by time t."

1.6 Summary

We view wildlife management as simply the management of wildlife populations. Four management options are available: (i) to make the population increase; (ii) to make it decrease; (iii) to take from it a sustained yield; or (iv) to do nothing but keep an eye on it. We have first to decide our goal for the population, and that will be largely a value judgment. To help us steer through social, political, and economic influences we use a decision analysis to reveal those influences and their effect on goals and policies. A series of questions about the selected option must be posed and answered to ensure that it is feasible and that its success or failure can be determined.

2 Biomes

2.1 Introduction

This chapter provides a brief overview of the main ecological divisions in the world and will supply a background of natural history for the chapters that follow.

The earth, or biosphere, can for convenience be divided up into major regions. On land these regions are characterized by a similarity of geography, land form, and major floral and faunal groupings. Thus, we can talk about the tundra – high latitude, cold, usually flat or rolling relief, and with low-growing shrubs like willows and mat-forming herbs. Tropical lowland forest is very different – moist, warm regions near the equator dominated by dense forest. Regions with similar characteristics are called *biomes*. They are divided further into units of greater similarity, called *ecosystems*, based on environment and groupings of plants and animals.

Ecosystems comprise the abiotic environment and the biotic groupings of plant and animal species called *communities*. Each of the species in a community has a characteristic density (or range of densities) and it is the interaction of these various populations that gives a particular community its special features. Populations have their own features, e.g., age and sex ratios, and these are affected by both the environment in which the animals live and the particular adaptations of the individuals, their morphology, physiology, and behavior. Thus, in the study of wildlife ecology and management we need to understand both the large-scale spatial and temporal events occurring in biomes and ecosystems, and the smaller scale characteristics of individuals and populations.

Habitat is the suite of resources (food, shelter) and environmental conditions (abiotic variables such as temperature and biotic variables such as competitors and predators) that determine the presence, survival, and reproduction of a population. In Chapter 6 we will examine the relationships between populations and their resources. In Chapters 8, 9, and 10 we examine how some components of their habitat, such as predators and competitors, impinge on the populations and their role in wildlife management.

We will now review the main features of the various biomes and some of the wildlife forms that inhabit them. Although biomes are characterized by many different properties, they can be summarized conveniently according to mean annual temperature and rainfall (Fig.

9

2.1). Biomes are groupings of ecosystems with similar environment and vegetation structure (physiognomy). There are six major terrestrial biomes distinguished by their physiognomic characteristics: forests, woodlands, shrublands, grasslands, semidesert scrub, and deserts. We include one group of marine biomes. Walter (1973) provides more detailed descriptions.

2.2 Forest biomes

2.2.1 Boreal forest

Taiga in Eurasia or *boreal forest* in North America starts where 10°C mean daily temperature is exceeded for >30 days/year. Tundra takes over where this temperature is exceeded for <30 days. The boreal forest is dominated by several species of conifer of the genera *Pinus* (pine), *Picea* (spruce), *Abies* (fir), and *Larix* (larch), although only white spruce (*Picea glauca*) spans the whole of North America. Eastern Asia also has many species of conifer, but, in contrast, only two species, Norway spruce (*P. abies*) and Scots pine (*Pinus sylvestris*) predominate in Europe.

In dense boreal forest the shrub layer is almost absent and mosses dominate the herb layer. In openings of the forest, and in wetter areas where trees are absent, there is a sparse shrub layer of willows (*Salix*),

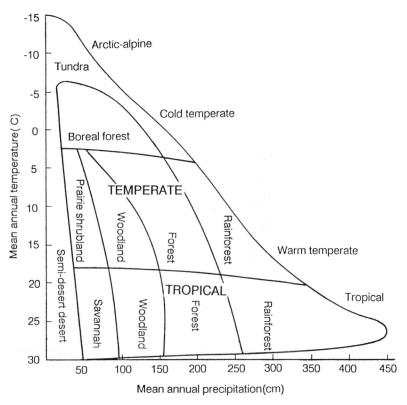

Fig. 2.1 World biome types in relation to rainfall and temperature. (After Whittaker 1975.)

birches (*Betula*), and alders (*Alnus*). Soils are acid, low in nutrients, and with a thick humus layer that takes a long time to decompose.

Boreal forest is the main habitat of the snowshoe hares (*Lepus canadensis*) and their main predators the lynx (*Lynx canadensis*), great horned owls (*Bubo virginianus*), and goshawks (*Accipiter gentilis*). Among other birds, ravens, swallows, chickadees, woodpeckers, and forest grouse are common. During the Pleistocene this was also the habitat for browsing mammoths (*Mammuthus*), woolly rhino (*Coelodonta*), and giant ground sloths (*Megalonyx*).

2.2.2 Temperate forest

Temperate forests can be divided into deciduous forests, rainforests, and evergreen forests. Deciduous trees drop their leaves as an adaptation to winter. Leaves being delicate structures are likely to be damaged by freezing. Thus, nutrients are withdrawn from the leaf and stored in the roots. The dead leaf is then shed. Because trees need to regrow their leaves in spring they require a growing season of 4–6 months with moderate summer rainfall and mild winters. They avoid the extremes of the wet maritime and cold continental climates. These forests are found mostly in the midlatitudes of the northern hemisphere, particularly in western Europe, eastern North America, and eastern Asia. There is a large variety of tree species with oak (*Quercus*), beech (*Fagus*), maple (*Acer*), and elm (*Ulmus*) being common. Forests of western Europe are not as rich in tree species because of extinctions during the last ice age.

Small mammals such as voles (*Microtus*), mice (*Clethrionomys*), and shrews (*Sorex*), are numerous although with relatively few species. Large mammals are represented by deer (*Odocoileus, Cervus*) and bison (*Bison*). The majority of the insectivorous bird species, such as thrushes (*Turdus*) and old and new world warblers (*Phyloscopus, Dendroica*), migrate to the tropics or southern hemisphere.

Temperate rainforests occur along the western coasts of North America, Chile, New Zealand, and southern Australia, in maritime climates with high year-round rainfall. They are known for their large trees (60–90 m high) such as redwoods (*Sequoia sempervirens*) in California, Douglas fir (*Pseudotsuga douglasi*) in British Columbia, eucalypts (*Eucalyptus regnans*) in Australia, and podocarps (*Podocarpus*) in New Zealand. Diversity is often low for both plant and animal species. Large vines, lianes, palms, and epiphytes are rare, but tree ferns in the southern hemisphere are common.

Temperate evergreen forests vary around the world. In this biome are included the dry sclerophyll forests of eastern Australia dominated by *Eucalyptus* species with their tough, elongated leaves; the dry pine forests of western North America including Monterey (*Pinus radiata*) and ponderosa pine (*P. ponderosa*); and the dry forests of southeast Asia. Canopies are open and the understory vegetation is sparse and often adapted to dry conditions. Evergreen forests in New Zealand are quite different for they occur in wet regions. Although close to

Australia, these forests must have been separated by continental drift before the Australian flora developed, for there are no eucalypts or acacias. The forests are dominated by evergreen conifers, notably the kauri (*Agathis*) in the warmer north, and several species of *Podocarpus* and *Dacrydium* in the south. There are five evergreen species of southern beech (*Nothofagus*). New Zealand forests are noted for their endemic birds and the absence of indigenous terrestrial mammals.

2.2.3 *Tropical forest*

Daily temperatures in tropical forests remain similar year round (24–25°C), and day length varies less than an hour. Seasons are determined by rainfall, there being some months with less rain than others – 10 cm of rain is a dry month. In Malaysia, Indonesia, and some parts of the Amazon basin (Rio Negro) all months have >20 cm of rain, some receiving >45 cm. In Africa and India there is a short dry season. High temperatures in these forests cause high transpiration rates and plants have adaptations to overcome water loss through thickening of the cuticle, producing leathery leaves. Examples are the rubber tree (*Ficus elastica*) and *Philodendron*. Leaves in the shade are large; those in the light are smaller.

In contrast to the relative paucity of species in temperate forests there is a high diversity of plants and animals in tropical rainforests. The most extensive rainforest is found in the Amazon basin of South America, but other forests are found in central and west Africa, southeast Asia, Indonesia, and northern Australia. One can find more than 200 species of trees in small areas. Leaf shapes are similar between species of tree. The canopy is high and closed, and 70% of plant species are trees. Most of the other plant species are also concentrated in the canopy: associated climbing lianes and epiphytes such as orchids form part of the canopy. The lack of light results in a relatively sparse understory. The roots of the large trees do not reach far into the soil because it is permanently wet. These giant trees therefore develop buttress roots reaching 9 m up the trunks to support them. Individual trees have a periodicity of growth and flowering, but two individuals of the same species can be out of phase. Periodicities of growth differ between species; they are not related to the annual cycle and vary between 2 and 32 months.

Most of the animal species are adapted to the canopy. The greatest diversity of primates occurs in these forests, and in South America other mammals such as sloths (*Brachydura*) are also adapted to feeding in the canopy. The diversity of bird species is high, the highest being in the Amazon forests. Feeding and breeding of many bird and bat species are adapted to the flowering periodicities of their preferred feeding trees.

2.3 Woodland biomes

Tropical broadleaf woodlands are an extension of the tropical forests in drier seasonal climates and low nutrient soils. As an adaptation to this climate trees have large leaves which they shed during the dry season. A few species, such as *Balanites* in Africa and *Eucalyptus* in Australia, have small xeromorphic leaves which are retained throughout the year. Trees often flower at the end of the dry season before leaf formation. The dense herb layer leads to frequent fires in the dry season so that shrubs and trees have evolved fire resistance.

Typical of this biome are the extensive *Colophospermum* and *Brachystegia* woodlands of southern Africa and *Isoberlinia* woodlands in west Africa. The canopy varies from 3 to 10 m and is relatively open. Soils and grasses are low in nutrients; ungulate species are also at low density, although some, e.g., roan (*Hippotragus equinus*) and sable antelope (*H. niger*), are adapted to this habitat. Similar vegetation occurs in Brazil, India, and southeast Asia. The Indian and Asian woodlands are the centers of radiation for the cattle group (*Bos*) – gaur, banteng, kouprey, and yak.

As in the tropics, temperate woodlands occur in drier environments than the forests. This biome covers a heterogeneous collection of small conifer and deciduous tree habitats in the Mediterranean and Mexico, but none of them is very extensive.

2.4 Shrublands

The best known of these types is what is called the Mediterranean vegetation – a scrub adapted to the dry conditions of a Mediterranean climate which consists of dry hot summers and cool wet winters. Similar types are found in South Africa, southern Australia, central Chile, and southern California. Shrubs are low with sclerophyllous leaves. Many are adapted to annual fires, regrowing from the root stock.

Typical trees and shrubs of the Mediterranean are various oaks (*Quercus*), holly (*Ilex*), the evergreen pines and junipers (*Juniperus*), and olive (*Olea*); in California *Quercus*, *Cupressus*, and chaparral shrubs (*Ceanothus*); in Chile various cacti (*Trichocereus*); in South Africa *Elytropappas* and the major radiation of *Protea*; and in Australia the mallee scrub made up of *Eucalyptus* shrubs as well as "grass trees" (*Xanthorrhoea*, *Kingia*), cycads (*Macrozamia*), the evergreen *Casuarina*, and several members of the Proteacae. All are adapted to a period of slow growth and the prevention of water loss by closing stomas during the summer drought. The leaves are hard and leathery, characteristic of sclerophyllous vegetation. In isolated areas, such as southwest Australia or South Africa, plants show a high degree of speciation and many of the species are endemic. There are several small mammals and passerine birds adapted to the regime of summer drought, but their diversity is usually low. For example, in California chaparral there is the wrentit (*Chamaea fasciata*) and kangaroo rat (*Dipodomys venustus*). In the Mediterranean there is the Sardinian warbler (*Sylvia melanocephala*), in South Africa the Cape sugarbird (*Promerops cafer*) on proteas, and in Australia the western spinebill (*Acanthorhynchus superciliosus*) feeding on banksias.

2.5 Grassland biomes

2.5.1 Tropical savanna

Tropical savanna comprises grassland with scattered trees. Often the trees are sparse, as on the open plains of East Africa, or quite dense with up to 30% canopy cover (proportion of ground under tree canopy) as in some of the *Acacia* savannas of Africa and Australia. Although temperature is fairly constant, rainfall is highly seasonal and falls in the range of 50–100 cm. Grasses are mostly perennial, 2–20 cm in height, and are usually burned each dry season. Most savannas in Africa are maintained by fire rather than by soil moisture; examples of the latter (edaphic grassland) are seen in the flood plains of the larger rivers such as the Zambesi and Nile, or shallow lake beds of Africa, and the llanos of the Orinoco in Venezuela.

The African savannas support a wide range of large mammal species, some communities having as many as 15 ungulate and seven large carnivore species as well as many rodent and lagomorph herbivores, mongooses, civets, and other small carnivores. Among birds, pipits, larks, sandgrouse, guinea fowl, and ostrich (*Struthio camelus*) are common types. The Australian savannas support an array of macropod herbivores (kangaroos) but no large carnivores, although the three that used to occur have become extinct on the mainland in the last 30 000 years. Small carnivorous marsupials are represented by dasyurids. In the birds, finches, parrots, and emus (*Dromaeus novae-hollandiae*) are common. South American wet savannas, e.g., in Venezuela and Brazil, have a range of large rodents such as capybaras (*Hydrochoeris hydrochoeris*) and coypus (*Myocastor coypus*) which are ecologically similar to the ungulates in Africa, but the drier pampas has very few large herbivores. There may be historical reasons for their absence: in the Tertiary there were many endemic herbivores belonging to the Notoungulate group which have since died out. Of the birds, pipits, buntings, and tinamous are characteristic.

2.5.2 Temperate grasslands

Temperate grasslands are similar to the tropical savannas in that they support perennial grasses and are often maintained by fire. They are seasonal in both precipitation (rain or snow) and temperature. They occur in dry climates in the centers of the North American and Asian continents. In South America we see this vegetation as the pampas of Argentina. Temperate grasslands experience cold winters with low snowfall, spring rains, and a summer drought. Like tropical savannas they support large herds of ungulates – bison and pronghorn (*Antilocapra*) on the American prairies, saiga (*Saiga*) and horses on the Asian steppes – and carnivores such as wolves (*Canis lupus*). Nonetheless the number of species is low. Birds are represented by larks, pipits, buntings, grouse, buzzards (*Buteo*), and falcons.

2.5.3 Tundra

Arctic tundras occur north of the tree line in both North America and Eurasia. There is a maximum of 188 days with mean temperature above 0°C, but sometimes as few as 55 days. The growing season spans the four summer months and is determined locally by when the snow melts. Exposed areas have longer growth whereas those under snow drifts have shorter seasons, and so a mosaic of vegetation is maintained. Plant communities consist of a complex mixture of sedges, grasses, lichens, mosses, and dwarf shrubs.

In the Arctic, soils are frozen in permafrost except for a shallow layer at the surface which thaws in summer. Lemmings (*Lemmus*) feed on the vegetation year-round, being protected under the snow in winter. Geese nest in large numbers and impose a heavy grazing impact in summer. Ptarmigan (*Lagopus*) are another abundant bird group. Because of the permafrost the ground snow does not drain easily in the summer and much of the tundra is swampy; these swamps provide ideal breeding grounds for mosquitos, which form dense swarms in late summer. This abundance of insects and other invertebrates, combined with the almost constant daylight, provides good breeding conditions for insectivorous birds – many shorebirds (plovers, sandpipers) and passerines, e.g., snow bunting (*Plectrophenax nivalis*), migrate to this biome to breed. Large mammals include muskox (*Ovibos moschatus*) and caribou (*Rangifer tarandus*), small mammals such as arctic hare (*Lepus arcticus*) are sometimes numerous, and wolves, arctic foxes (*Alopex lagopus*), and snowy owls (*Nyctea scandiaca*) are common predators.

2.5.4 *Alpine*

In contrast to the tundra where precipitation is low and drainage poor, many alpine areas have high precipitation, good drainage, and a high degree of fragmentation. In temperate regions this leads to relatively high growth.

Alpine meadows have a similar vegetation structure to that of the tundra but because they are confined to mountain tops they are often found in small scattered patches. Fewer bird and mammal species use these areas for breeding in comparison to the tundra. Birds which specialize here are rosy finches (*Leucosticte*) and horned larks (*Eremophila alpestris*) in North America, and accentors (*Prunella*) in Europe and Asia. In North America the characteristic mammals are marmots (*Marmota*), pikas (*Ochotona*, a small lagomorph), and voles (*Microtus, Phenacomys*). Elk (*Cervus elephus*), moose (*Alces alces*), caribou (*Rangifer tarandus*), and bears (*Ursus*) use the meadows in summer. In Asia the Himalayan alpine zone is the center of evolutionary radiation for the goats and sheep. These species form the prey of snow leopard (*Panthera uncia*). Pikas (*Ochotona*) have also diversified here.

Alpine meadows on the tropical mountains of Africa produce some extraordinary adaptations in the vegetation. The weather is extreme: it freezes every night and becomes relatively hot every day. Several plant types (*Senecio, Lobelia*) show gigantism – plant genera that are small herbs in temperate regions become large trees in this environment. The leaves are fleshy and store water. Few animal species are adapted to these conditions but one is the hill chat (*Pinarochroa sordida*).

2.5.5 *Summary*

It is no accident that all the large herds of ungulates occur in the grassland biomes; caribou in the Canadian tundra, saiga on the Asian steppes, bison on the American prairies, and various antelopes on the African savannas. They have in common the ability to migrate in response to the seasonal climate and changing vegetation, and so place themselves in the areas of highest food production at the time. They avoid thereby many of their predators who cannot migrate to the same extent. These two abilities – to find temporary food patches and to avoid predators – allow a higher density of animals than if the population did not migrate in a similar area; the large herds are not just a consequence of the extensive area of these biomes.

2.6 Semidesert scrub

Warm semidesert scrub is most extensive in a band surrounding the Sahara and extending through Arabia, Iran, and to India. The Somali horn of Africa and the Namibian zone of southwest Africa have, in prehistory, been joined to the Sahara. The vegetation is scattered thorn bush (*Acacia*) and succulents with a sparse herb layer. Several of the antelopes in the Somali–Sahara area are browsers with convergent adaptations of long necks and the ability to stand up on their back legs (dama gazelle (*Gazella dama*), dibatag (*Ammodorcas clarkei*), gerenuk (*Litocranius walleri*)). In both Asia and Africa, the main arid-adapted small mammals are the gerbils (*Gerbillus*, *Tatera*) and jerboas (*Jaculus*, *Allactaga*).

North American semidesert scrub surrounds the Sonoran and Mojave deserts. Creosote bush (*Larrea divaricata*) is common and there is a wide variety of other spiny and succulent plants such as prickly pear (*Opuntia*). A number of arid-adapted small mammals such as pocket mice (*Perognathus*) and kangaroo rats (*Dipodomys*) live on seeds. Ground-feeding birds such as doves, new world sparrows, and juncos are characteristic. The equivalent Australian vegetation is dominated by shrubs of the family Chenopodiaceae. Small mammals include hopping mice (*Notomys*) and the marsupial jerboa pouched mouse (*Antechinomys*). However, most of the mammals and birds are derived from the temperate woodlands and are recent invaders. These areas are known for the large flocks and nomadic movements of Australian finches (*Ploceidae*) and budgerigars (*Melopsittacus undulatus*) following the unpredictable pattern of rainfall.

At higher latitudes in the rain shadow of the Rocky Mountains and the Himalayas, a cool semidesert vegetation is characterized by low aromatic shrubs such as sagebrush (*Artemisia*) and perennial tussock grasses. Small mammals and birds are similar to those in the warm semideserts. Ground squirrels (*Spermophilus*) are common in this type of vegetation in North America.

2.7 Deserts

Deserts tend to occupy the midlatitudes and extend from the west towards the middle of continents – the Sahara in Africa, the Gobi in Asia, and the deserts of Australia, southern California, and Arizona are examples. They receive on average < 25 cm of rain/year. Smaller ones include the Namib desert of southern Africa, the Sonoran and others in southwest USA, and the Atacama of Chile. Below 2 cm annual rainfall there is no vegetation, and from 2 to 10 cm it is very sparse: plants have typically xeric adaptations – many species lie dormant as seeds for periods of several years, but germinate, flower, and set seed again in quick succession after a rain storm. At this time the desert comes to life as insects breed and nomadic birds move in to take advantage of the high seedset. Few large mammals are adapted to this environment but the addax (*Addax nasomaculatus*) in the Sahara, the camel (*Camelus*) in Asia, and the red kangaroo (*Macropus rufus*) of Australia are examples.

2.8 **Marine biomes**

Marine biomes can be divided into open ocean (pelagic), sea floor (benthos), and continental shelf.

2.8.1 *Pelagic*

The surface layers of the pelagic biome receive light and so support phytoplankton, small single-cell algae, and diatoms. These support zooplankton, a mixture of small crustaceans, molluscs, worms, and many other forms that are fed upon by fish. Small fish are transparent as a way of avoiding predation. Larger species such as tuna (*Thunnus*) are fast swimming and move in large shoals. The essential chemicals for growth (nutrients) in these waters are not high and so the amount of plant and animal material is also low.

In the deep pelagic zone there is no light and the animals have to survive on the dead material that sinks from the surface layers. These are called heterotrophic systems because they depend on food from outside sources rather than on plants which trap their own light and make carbohydrates (these are called autotrophic systems). One still finds crustacea, colonial protozoans (foraminifera, radiolaria), and fish, many of which cannot see or have extraordinary adaptations to lure other fish within catching distance. This biome also contains the giant squids.

2.8.2 *Benthos*

The deep ocean benthos is one of the most extreme of all environments: cold, dark, and pressurized. Nevertheless, a diversity of animals live in the bottom mud. Some are attached to the mud (sea anemones, sponges, and brachiopods), others are burrowers, and yet others crawl over the surface.

2.8.3 *Continental shelf*

The continental shelf and the surface waters above it are the richest in nutrients, plankton, and animal life. Dense algal forests can grow because light reaches the sea bottom and these in turn support communities of inshore fish. The higher density of marine invertebrates and fish in these environments supports larger mammal predators such as seals, sealions, and some whales, but this occurs mostly in temperate regions. Tropical continental shelves are less productive and support fewer mammals: the dugong (*Dugong dugon*) and manatees (*Trichechus*) which graze on submarine vegetation.

Cold-water currents high in nutrients well up at the edge of the continental shelf. Upwellings occur particularly in arctic and antarctic waters, but there are some in the tropics such as the Humboldt current off Peru. The upwellings are rich in plankton, and a wealth of fish, sea birds, and whales feed on them.

Coral reefs are a special biome forming a rim around oceanic islands. Although not usually associated with a continental shelf, they have similar ecological characteristics.

2.9 Summary

The world can be divided into broad ecological divisions, each of which has a characteristic vegetation and wildlife. The forest biomes are diverse, being subdivided into boreal, temperate, tropical, woodland, and shrubland. Grassland biomes include tropical savanna, temperate grassland, alpine grassland, and tundra. The deserts constitute a further biome. Each of these can be divided further into ecosystems and communities based on groupings of plants and animals. Within these larger groupings each animal species selects its habitat.

3 Animals as individuals

3.1 Introduction

In order to manage populations we need to know something about the characteristics of its members. We seek a knowledge of their morphologic and physiologic adaptations to environment, their behavior, particularly with respect to dispersal, reproduction, and use of habitat, and the genetic variability among them.

In this chapter we begin broadly by outlining the mechanisms by which these adaptations come about: the evolutionary process of speciation, convergence, and radiation. We then focus in on the methods by which the genetic constitution of individuals, or groups of individuals, can be determined and the importance of such information in wildlife management.

3.2 Adaptation

To understand why a population of a species lives where it does, i.e., to explain its *distribution* in nature, we should know how an individual is *adapted* to its environment, what types of *environment* it encounters, and what resources are available. When we talk about the adaptations of individuals we mean the way in which an animal fits into its environment and uses its resources. The adaptive characters that describe an individual – its physical attributes (morphology), physiology, and behavior – are determined first by the processes of natural selection and second by its history over evolutionary time, its phylogeny.

The physical environment – temperature, humidity, and other features that we call the *abiotic* environment – together with the effects of other species that form the food, competitors, and predators (the *biotic* environment), acts through natural selection to produce a suite of adaptations which are called life-history traits.

3.3 The theory of natural selection

The term *evolution* refers simply to change in a population over time. It does not necessarily mean speciation, i.e., the process of forming species (although this may be an outcome) and it does not imply a mechanism of change. The idea of evolution was already being talked about in Europe in the early 1800s, albeit as a radical concept. Charles Darwin described a mechanism for this change in his book *On the Origin of Species* in 1859. It was called *natural selection* and proposed jointly by Darwin and A.R. Wallace in 1858. Darwin based his theory on three observations.

19

1 Populations increase geometrically through reproduction.

2 All individuals are different – the genetic mechanism for this being demonstrated later by Mendel (1959).

3 Populations remain constant (at least within broad limits) owing to a lack of resources. This point came from Malthus (1798) in his essay on populations. From these observations there follow two postulates.

(a) There is competition for resources between individuals.

(b) Those individuals that are most capable of obtaining the resources, and can survive and reproduce, are those that will leave the most progeny. The next generation will contain a greater proportion of those types.

The selection comes from the relative success of the different types in leaving progeny. The process of natural selection is *the relative reproductive success in the long term* where "reproductive success" includes births, survival, and reproduction of the offspring, "long term" means over several generations, and "relative" means in comparison with other members of the population.

This is the theory of natural selection at its simplest. It carries the following corollaries.

1 Natural selection results in adaptation to the environment because the types leaving more progeny are by definition better at surviving and reproducing in that environment. The most successful types are the fittest individuals.

2 Since no population has all possible varieties, natural selection cannot produce perfect adaptation – only the best among those available, and these best may be quite imperfectly adapted.

3 Natural selection results in adaptation to past and present conditions, not to future conditions. It cannot anticipate future conditions or select for individuals preadapted to them. If the changing conditions suit a currently rare individual type, it is through chance alone and does not indicate predetermined design.

4 Natural selection acts only on the inherited components of an individual, i.e., the *genes*. For these purposes genes are those elements of the chromosome that segregate independently, and therefore may include several DNA groups if they are linked. Natural selection cannot maintain either whole *phenotypes* or whole *genotypes*. The genotype is the total complement of genes in the individual. The phenotype is the individual organism, which is a product of the genotype interacting with the environment during development. Phenotypic variation is reflective of genotypic and environmental variation.

5 A favorable gene can have both advantageous and disadvantageous effects within the same individual due to *pleiotropy* and *polygenic effects*. Pleiotropy describes a gene affecting more than one character in the individual, and some effects may benefit while others are disadvantageous. Polygenic effects implies that a character is affected by several genes, some good some bad. All that is required is that the beneficial effects outweigh the detrimental ones.

6   Natural selection does not guard against the extinction of species. Many adaptations do indeed promote the continued existence of a species but there are also many that result in extreme specialization to unusual environments, restricted habitats, or isolated areas. These species are vulnerable to environmental change. On the island of Mauritius in the Indian Ocean the endemic tree *Calvaria major* has not germinated for the past several centuries. Temple (1977) showed by feeding the seeds to domestic turkeys that they will germinate only after passing through the alimentary tract of a bird. He suggested that the dodo (*Raphus cucullatus*), a large flightless pigeon that became extinct about 1735, shortly after people arrived, originally ate these seeds. The tree seems to have become specialized to these particular conditions for germination of these seeds. The close dependence of the tree on the dodo makes it vulnerable to extinction. Similarly, the extinction of many species of the Hawaiian honeycreepers has resulted in the extinction or near extinction of all species in the plant genus *Hibiscadelphus*. The honeycreepers, with their long curved bills (see Fig. 3.1), were the pollinators of the curved tubular flowers of the *Hibiscadelphus* (Diamond and Case 1986).

3.4 **Examples of adaptation**

3.4.1 *Convergence*

Convergence occurs when organisms of different ancestry (i.e., from different phyletic groups) adapt to similar environments and thus develop similar characteristics. One of the classic examples is the placental mammals and the marsupials which have evolved similar morphology and behavior even though they are quite unrelated (Fig. 3.2).

The rock ringtail possum (*Pseudcheirus dahli*), a marsupial of northern Australia, lives in the crevices of large rock piles. Bruce's hyrax (*Heterohyrax brucei*), from the very different placental order Hyracoidea confined to Africa and Arabia, has precisely the same homesite. In North America the hoary marmot (*Marmota caligata*), a rodent of similar size, lives in rock piles on the mountains and feeds on surrounding vegetation. The three species have converged in form and ecology.

There are many examples of convergence in birds. The yellow-throated longclaw (*Macronyx croceus*), a member of the pipit family Motacillidae, lives in the dry open grasslands of eastern Africa. It is brown, yellow below, with a black chest band. It sits on bushes and sings constantly. In North America the western meadowlark (*Sturnella neglecta*) is similar in appearance, behavior, and habitat, but it belongs to the entirely different new world family Icteridae. Penguins (Spheniscidae) of the southern hemisphere are the ecological equivalents of the unrelated Alcidae (auks, murres, puffins, guillemots) of the northern hemisphere.

3.4.2 *Adaptive radiation*

Adaptive radiation is the name given to the divergence of a single lineage to provide a variety of forms. The best known example is that

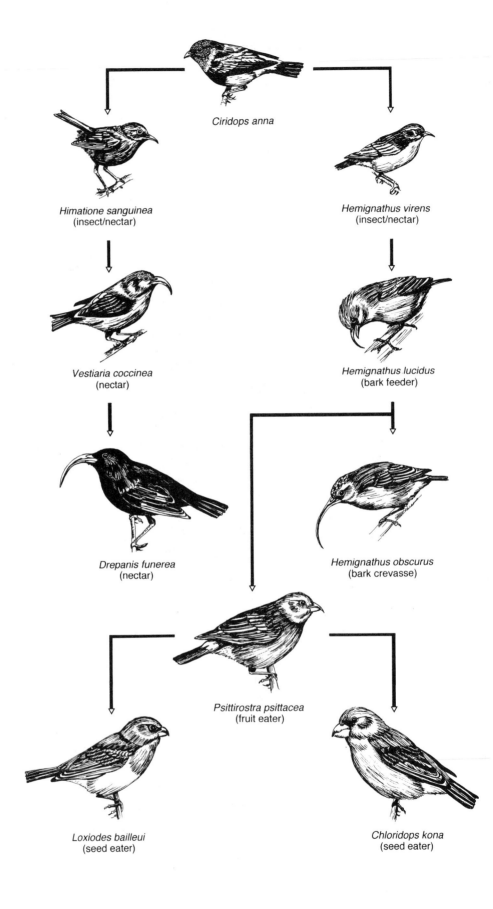

Ciridops anna

Himatione sanguinea
(insect/nectar)

Hemignathus virens
(insect/nectar)

Vestiaria coccinea
(nectar)

Hemignathus lucidus
(bark feeder)

Drepanis funerea
(nectar)

Hemignathus obscurus
(bark crevasse)

Psittirostra psittacea
(fruit eater)

Loxiodes bailleui
(seed eater)

Chloridops kona
(seed eater)

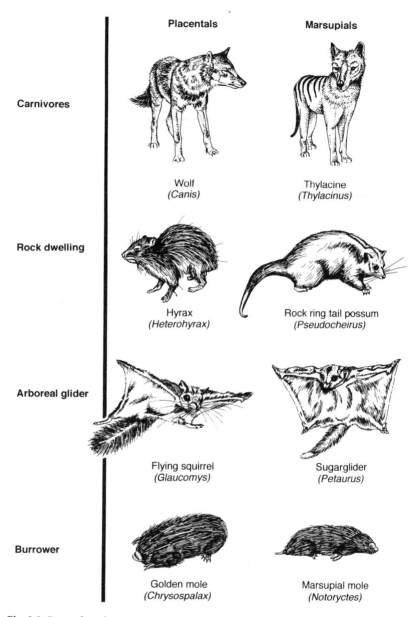

Fig. 3.2 Examples of convergent evolution between placental and marsupial mammals.

Fig. 3.1 Adaptive radiation of the Hawaiian honeycreepers (Drepanidini). Only a few of the species are illustrated here. There are four main functional groups: insect feeders in foliage, nectar feeders, bark feeders, and seed eaters. (After Futuyma 1986, with nomenclature from Pratt *et al.* 1987.)

of Darwin's finches on the Galapagos Islands (Lack 1947; Grant 1986). They were probably founded by a finch species from the Americas. Subsequently they diverged into types that feed on insects and others that feed on seeds; some species live on the ground, others in trees, and still others among cacti.

Another example of adaptive radiation is seen in the endemic honeycreepers (Drepanididae) of Hawaii (Fig. 3.1). Many of these species have become extinct in the past 150 years. They appear to have evolved from a thin-billed insect eater. From this type, one group of species developed into long-billed nectar feeders while another group evolved into long-billed bark-crevice feeders. Yet another group developed thick bills and feeds on fruit and seeds. This one family has filled the niches normally filled on continents by many families of birds.

3.5 **The effects of history**

3.5.1 *Phylogenetic constraints*

In discussing how animals fit into their environment we have considered the process of adaptation through natural selection. We have noted that adaptation is not perfect because conditions change and animals are constantly trying to catch up. Furthermore, the organisms are limited by the evolutionary pathways that their ancestors have followed: both birds and mammals have evolved from reptiles but selection on mammals today cannot produce feathers. The potential for growing feathers was lost a long time ago.

Natural selection is constrained in what it can produce by what is currently available. The giant panda (*Ailuropoda melanoleuca*) of China is a large herbivore that eats bamboo shoots almost exclusively. These bamboos provide low-quality food. Most large mammal herbivores, such as horses, deer, and kangaroos, have long intestines and special fermentation mechanisms in the gut to allow maximum digestive efficiency (see Section 7.6); in contrast, carnivores, such as cats, and omnivores, such as bears, have relatively short guts. Current thinking is that the giant panda evolved from bears in the Miocene about 20 million years ago (O'Brien *et al.* 1985a) and changed to a herbivorous diet. Because of its carnivore ancestry it has a short gut and cannot now make the evolutionary jump to the longer, more complex, digestive system of herbivores. Hence the giant panda has one of the least efficient digestive systems known for a terrestrial vertebrate: only 18% of the food is digested compared with 50–70% for horses, antelopes, and deer. The giant panda compensates with other adaptations, in particular eating a very large amount of food and spending most of the day doing so. This prolonged feeding in turn leads to behavioral adaptations: pandas are solitary and spend little time in social and mating activities.

3.5.2 *Geographic constraints*

Movement of the continents
Earlier this century Wegener (1924), a meteorologist, proposed that the continents were at one time joined together and subsequently drifted apart. Wegener's idea was generally rejected. The discovery in the

1960s that the earth's surface is made up of plates, and that these move, proved that Wegener was essentially correct. Volcanic activity and earthquakes along midoceanic ridges produce prodigious amounts of submarine basalt, and this spreads the sea floor. The continents that float on these basaltic plates are thereby forced apart.

Some 150 million years ago there were two great landmasses, Laurasia in the north and Gondwana in the south. Figure 3.3 shows how Gondwanaland split apart. The process began about 115 million years ago, with Africa and India breaking away first. These and Madagascar separated 65 million years ago, while South America, Antarctica, and Australia were still joined. Australia finally separated from Antarctica much later, about 40 million years ago.

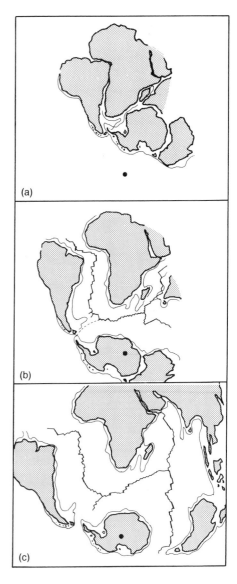

Fig. 3.3 Gondwanaland at different time periods before the present. The thin line around each continent is the limit of water < 1000 m deep. The dot indicates the South Pole. (a) At 150 million years BP the southern continents were joined. (b) At 65 million years BP Africa had separated from the other continents, which were still joined. (c) The present-day distribution of continents. (After Norton and Sclater 1979.)

These historical movements explain some of the more peculiar distributions of animal groups, e.g., why marsupials are found today only in Australia, New Guinea, and the Americas.

Figure 3.4 shows the distribution of the large flightless ratites (ostriches, rheas, emus, and their extinct relatives). Similar distributions of tree-ducks, penguins, and parrots attest to the breakup of the southern continent.

The joining of North and South America in the Pliocene provides another example of historical events determining the nature of faunas. South America originally had a remarkably diverse mammalian fauna,

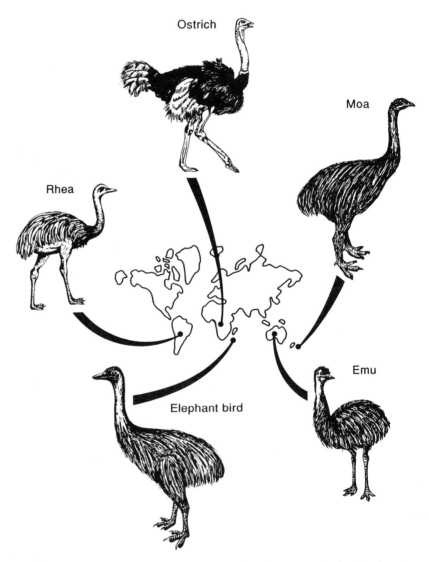

Fig. 3.4 Present-day evidence of Gondwanaland can be seen in the distribution of ratite birds on the southern continents. The moa in New Zealand and elephant bird of Madagascar became extinct only in the last few centuries. (After Diamond 1983.)

resembling the radiation of the ungulate fauna of Africa today. It included a wide range of marsupial carnivores, such as big sabertooth types (*Thylacosmilus*) and hyena types of the family Borhyaenidae, and smaller mongoose types represented by the Didelphids (the group that includes the opossum *Didelphis marsupialis*). These carnivores fed on the herbivorous notoungulates, a huge placental group now entirely extinct.

After the two continents joined, a few South American forms moved north, e.g., the armadillos and sloths, but most died out as a result of competition and predation from North American invaders. The present-day deer, camels, bears, cats, and wolves of South America are all derived from northern forms.

The ice ages: historical effects of climate

During the Pleistocene (the last 2 million years) the earth went through a series of cold and warm periods. Ice-caps developed over Canada, the northern palearctic (Europe and Asia), and on the main mountain chains such as the Alps, Rockies, Andes, and Southern Alps of New Zealand. Sea levels dropped as much as 100 m and "land bridges" were formed across the Bering Strait between Asia and North America and across the English Channel between Britain and France. The cold and warm periods in temperate regions were paralleled by dry and wet periods in the tropics.

The ice ages had a significant influence on the present-day distribution of animals. The Beringian land bridge across the present-day Bering Strait allowed an earlier invasion of North America by mammoths, mastodons, and sabertooth cats (*Smilodon*), and later invasions by more modern forms such as beaver (*Castor*), sheep (*Ovis*), muskox (*Ovibos*), caribou (*Rangifer*), elk (*Cervus*), moose (*Alces*), Bison (*Bison*), brown bear (*Ursus arctos*), and wolf (*Canis lupus*). There was a smaller reverse migration from North America into Asia of horses and camels, both of which subsequently became extinct in North America. Typical American mammals are deer (*Odocoileus*), mountain goat (*Oreamnos*), and pronghorn (*Antilocapra*). Most of the others are Eurasian forms.

During the last glaciation (12 000 years ago) a few areas within the northern ice-sheets were free of ice and some animals survived and evolved in these "refuge" areas. The northern end of Vancouver Island in Canada was a refuge for elk and marmots, which differentiated into new races.

The climatic fluctuations causing the ice ages also caused the expansion and retreat of the tropical forests of South America and Africa. The South American forests contain the highest diversity of bird species anywhere. The centers of endemism within these forest match fairly closely the forest refuge patches left by ice ages (Fig. 3.5). In general, the ice ages have accounted for many present-day distributions of mammals and birds.

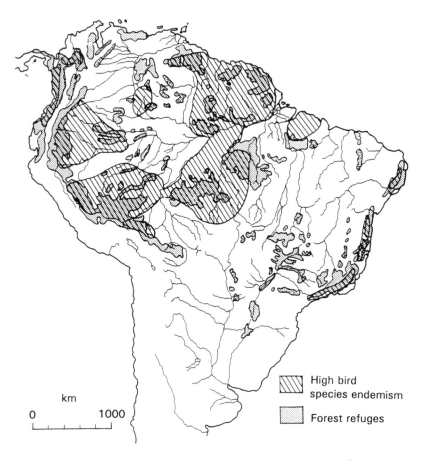

Fig. 3.5 Patches of high species endemism in birds of the Amazon rainforest (hatched) coincide with the areas that were forest refuges in the past during periods of maximum aridity (stippled). (After Brown 1987.)

The invasion of people

There is one other historical influence that determined the distribution of the larger mammals and birds: the spread of people over the world. They spread into Eurasia from Africa some 200 000 years before present (BP), reaching Australia some 35 000 years BP, North America during the last ice age at 12 000 years BP, and New Zealand, Madagascar, Hawaii, and Easter Island only about 1000 years ago.

Although there is considerable debate on the effects of these human migrations, one school of thought, discussed in Martin and Klein (1984) and Anderson (1989), holds that the arrival of people resulted in the extinction of large mammals either directly through hunting or indirectly through habitat change. Thus, in North and South America mammoths and giant ground sloths disappeared, in New Zealand the large ratites (moas) were hunted to extinction, in Madagascar both giant ratites (elephant birds, *Aepyornis*) and giant lemurs (*Mega-ladapsis*) vanished, and in Polynesia a variety of birds such as the giant flightless galliform (*Sylviornis neocaledoniae*), twice the size of a turkey, became extinct with the arrival of people. Another school of thought holds that rapid climate change caused their extinction (Guthrie 1990). For example, the giant Irish Elk (*Megaloceros giganteus*) is thought to have died out at the end of the ice age, coincident with change in climate.

Knowledge of past events allows us to answer such questions as why Africa has a wide diversity of large mammals whereas North America and Europe do not. When we ask questions concerning the distribution of a species, e.g., why the white-tailed deer (*Odocoileus virginianus*) is found in South America or why the nine-banded armadillo (*Dasypus novemcinctus*) is found in Texas, we need to know about their individual adaptations of habitat selection, diet, and behavior, but we need also to know about the movement of continents and the effects of the ice ages.

A new evolutionary force now affects animal communities: intensive agriculture and industrialization. This is a post-Pleistocene development which has altered many habitats through pollution and large-scale clearing for agriculture and industry (Morrison *et al.* 1992).

3.6 The abiotic environment

The abiotic environment includes the sets of conditions that determine where an animal can live and reproduce. *Conditions* are those factors such as temperature and rainfall that affect an animal but which are not themselves influenced by the population. Because environments are not constant, animals are adapted to a range of conditions, and usually the less constant the conditions the wider the range (Stevens 1989). The limits of adaptation are called the *tolerance limits* for the animal, and we need to specify whether we mean the limits for reproduction or for occupation. The latter are usually broader. Section 5.4 discusses the conditions, and the adaptations to these by individuals, that determine the distribution of a population and the position of the range boundary.

3.7 Genetic characteristics of individuals

Individuals differ from each other genetically as well as physically and behaviorally in sexually reproducing species. As an example of genetic differences leading to differing behavior, females of 13-lined ground squirrels (*Spermophilus tridecemlineatus*) mate with several males at the start of the breeding season (Schwagmeyer and Woontner 1985). What advantage is this to the males? Do they all stand a chance of producing some offspring; does the first male to mate contribute to all or most of the conceptions; or does the last male to do so score most of the conceptions? It turns out that the first male to mate contributes to 75% of the conceptions (Foltz and Schwagmeyer 1989) and so being first is clearly an advantage. There is some advantage in being second because those males contribute to the other 25%, but subsequent males receive no benefit. What advantage is this mating with several males to the female? Since males are intolerant of juveniles that are not their own offspring, is this a tactic of females aimed at ensuring the cooperation of all surrounding males?

Waterbuck (*Kobus defassa*) in Africa defend territories through which female herds pass while grazing. The male mates with any estrous females when the female herd is in his territory. He also has to defend his territory against other territorial males and bachelor males that have no territory. In some areas a territorial male allows one other male into his territory (Wirtz 1982). What advantage is there to the territory holder in allowing the second male in? One suggestion is that the second male helps to defend the territory and so allows more opportunities for the primary male to mate. In return the second male may be able to "steal" some matings when the primary male is occupied elsewhere.

So far we do not have the answer to most of these questions. To obtain the answers, we must identify the individual parents of offspring. Recent genetic techniques have allowed us to do this.

3.7.1 *Methods for studying genetic variability*

Allozyme gel electrophoresis

Until recently the standard techniques for detecting genetic variability within and between individuals was to measure differences in amino-acid composition of allozymes or proteins encoded by different alleles of locus. Blood or tissue homogenate from individuals is placed on a gel matrix, such as cellulose acetate, and an electrical charge applied. The proteins migrate along the gel at rates dependent on their total electrical charge. Changes in amino-acid composition, the result of mutation, are often reflected as changes in electrical charge. The electrical current is switched off and the gel stained for a particular protein after a set time. Differences between individuals are evident as different configurations of the protein bands on the gel.

The method has been used to measure differences between races and between species by assessing the variability in many proteins from several individuals in each population. Phylogenetic trees have been constructed by this method.

The technique has several limitations. First, some proteins with different mutations can move at the same rate, thereby appearing to be the same. This problem becomes greater the more distant the relationship between individuals or species. Second, much of the genetic variability is not evident at the protein level because of the redundant nature of the genetic code. Other techniques assess the genetic diversity present in the individual's DNA itself. We examine these next.

Restriction fragment length polymorphisms

Mutations in the DNA can be detected by presence or absence of a cleavage site revealed by an enzyme called a restriction endonuclease. Restriction enzymes cleave or cut the DNA at particular recognition sites located at random along the DNA molecule. Total DNA is isolated from a tissue sample, challenged with restriction enzymes, and then electrophoresed on an agarose gel matrix. The shorter the DNA fragment, the greater the movement along the gel under application of an electrical current. The DNA is then transferred by capillary action from the gel matrix to a nylon membrane. The nylon membrane can be probed in this fashion with a radioactive copy of a target gene previously isolated and characterized. The radioactive probe will bind to homologous DNA sequences on the nylon membrane and, by using X-ray film to resolve the position of the radioactivity, the location and size of the DNA fragments encoding the target gene can be determined. DNA mutations can remove or add restriction sites which, when cleaved, produce fragments of different length. Differences between individuals are then evident from the distribution of fragment lengths.

This technique reveals a much greater range of genetic variability than does protein allozyme electrophoresis. Restriction fragment length polymorphism (RFLP) analysis of DNA from individual humans is used extensively to detect genetic disorders such as thalassemia (Weatherall 1985).

An example of the use of the RFLP technique is provided by studies of the reproductive behavior of lesser snowgeese (*Anser caerulescens*) around Hudson's Bay in northern Canada. These geese have two color phases, blue and white, with blue being genetically dominant. Field workers had observed that some goslings in family broods were blue when both parents were white, suggesting that either other females had laid eggs in the nest (*intraspecific brood parasitism*) or the female had mated with another male (*extrapair fertilization*). Using probes that could identify RFLPs, Quinn et al. (1987) showed that some goslings were not related to either parent (so demonstrating brood parasitism) and some not related to the male adult (suggesting extrapair fertilization) – and these were not just the blue goslings. There is a much higher degree of unrelated offspring raised by these "monogamous" birds than had previously been suspected.

DNA fingerprinting

DNA fingerprinting has been used primarily by forensic scientists to discriminate between individual humans. In 1984 Alec Jeffreys of Leicester University discovered particular sequences within human DNA that vary greatly between unrelated individuals (Jeffreys et al. 1985). These sequences are highly repetitive, occurring many times along each chromosome. The length of each repeated sequence, the number of repeats, and their exact location within the chromosomal DNA molecule differ between individuals. The different patterns of repeated sequences are resolved as bands on X-ray film produced by probing with a radioactive copy of the repeat sequence. An individual derives half of its bands from each biological parent, and the pattern of bands is thought to be unique to that individual. The complex pattern of variable bands produced by this method is called a genetic bar-code and is, in essence, an analysis of the multiple allelic states at the multiple loci detected by the repeat sequence probe (Collet 1991).

DNA fingerprinting has since been adopted by ecologists as a means of determining genetic identity and familial relationships, and has given rise to the field of molecular ecology. DNA fingerprinting has also been applied in the classification of organisms into genus, species, and subspecies, to studies of evolution, to monitoring captive breeding programs to avoid excessive inbreeding and inheritance of recessive traits, to monitoring breeding markers for valuable traits associated with high levels of meat and milk production, identification of the origin of natural migratory populations of animals such as fish and turtles, and in wildlife law enforcement.

Although Jeffreys et al. (1985) developed their probes for humans, the same two probes have now been used successfully on dogs, cats, house sparrows, and many other species to identify genetic relatedness (Burke and Bruford 1987; Wetton et al. 1987). New probes are being developed to allow even more precise identification of genetic sites (Burke 1989). This technique was used in the polygamous red-winged blackbird (*Agelaius phoeniceus*) to show that half the nests contained offspring from extrapair fertilization (Gibbs et al. 1990).

Mitochondrial DNA techniques

Mitochondria in cells have their own DNA (mtDNA) whose strands are relatively short (1.6×10^4 base pairs compared with 10^9 base pairs for the nuclear DNA) and which mutates at about six times the rate of nuclear DNA. Regions of mtDNA can be monitored for mutations with radioactive probes, in the same way as nuclear DNA. Genetic variability accumulates rapidly and large differences between populations are thereby often evident. mtDNA is inherited by matrilineal descent only, thus permitting an assessment of novel sources of variability. For example, there are areas in Texas, California, Montana, and Alberta where the ranges of white-tailed and mule deer (*Odocoileus hemionus*) overlap. Examination of male deer in Texas shows that although they look like mule deer their mtDNA resembles that of white-tailed deer. This suggests that at some point in the past these populations were derived from matings between male mule deer and female white-tailed deer; the female mtDNA was retained although the other characters are those of mule deer (Carr et al. 1986; Derr 1991). The process of taking some feature of one species into the genome of another is called *introgression*. In other areas of overlap, such as in Montana, there is little introgression (Cronin et al. 1988).

Polymerase chain reaction

Taxonomy, population genetics, and molecular ecology are advancing rapidly as a result of the development of a technique called the *polymerase chain reaction* (PCR). This allows millions of copies of a particular target sequence of DNA to be produced so that DNA sequencing, originally difficult, can now be used easily to identify individuals or groups of organisms.

The DNA of most interest is that which is highly variable between individuals. It is found in two regions, the major histocompatibility complex (the region of the chromosome containing genes for the immune system) and the region of the mitochondrial DNA involved with replication. Once a target region is chosen, a short piece of DNA is synthesized to act as a primer (a piece onto which new DNA is attached). We start with a mixture of the original double-stranded DNA, the primers, free nucleotides, and a heat-stable DNA polymerase. The mixture is heated and the strands of the double DNA are separated. Upon cooling, the primers attach themselves at one end of the target DNA and serve as starting points from which the polymerase builds the copy. A new cycle of heating starts the process again and is repeated for 30 rounds to make over a million copies of the selected DNA region.

This technique has been used on single cells from hair, sperm, and tissue cultures collected from live animals. It has been used also on degraded DNA from long dead animals such as the quagga (*Equus quagga*, an extinct South African zebra), a 13 000-year-old giant ground sloth and a 36 000-year-old extinct bison (Collet 1991). It was used to show that genetic variation in humpback whales (*Megaptera novaeangliae*) was not reduced when the population went through low numbers from commercial harvesting (Baker *et al.* 1993).

3.7.2 Individual variability between geographic regions

Populations whose individuals interbreed freely (a panmictic population) will have similar genetics throughout their range. Individuals should be adapted to the average conditions of the range, ideally those of the center. At the edge of the range where conditions are extreme the individuals should be less well adapted; they cannot develop local adaptations to these edge conditions because any tendency towards genetic divergence is swamped by gene flow from the center. However, if gene flow through a population is slow relative to the rate of local genetic adaptation, individuals at one end of the range can differ from those at the other end, with intermediate forms between. This gradual trend in appearance or behavior is called a *cline*. A good example of this is provided by *Cervus elephus* (Fig. 3.6). In Europe it is called red deer and is a relatively small dark animal. Males produce a deep-throated roar during the rut. At the other end of its range in North America the same species is called elk. Here it is larger, more tan colored, and the mating call of males is a high-pitched whistle or "bugle." Forms intermediate in both morphology and behavior occur across Asia.

The range of some species extends around the world. Where the two ends meet the animals have diverged sufficiently such that individuals no longer interbreed and behave as separate species. These are called *ring species*. The classic example is the black-backed gull–

Fig. 3.6 The distribution of *Cervus elephus* shows a cline from the small red deer of Scotland to the large elk of North America. The numbers and their associated shaded areas indicate different races in the cline. (After Whitehead 1972.)

herring gull pair. In Europe the herring gull (*Larus argentatus*) is light gray on the upper surface of its wings and back, but these parts gradually become darker through Asia and North America so that they are entirely black on the eastern seaboard of the USA. This form has crossed the Atlantic and lives in Europe as the black-backed gull (*L. fuscus*) without interbreeding with the herring gull; it behaves as a separate species.

Populations isolated by geographic barriers are called *allopatric*. In isolation they can become genetically different through adaptation to their different areas. If the populations then meet and overlap in range (become *sympatric*) they may not interbreed if they have diverged too far, in which case they have become separate species; or they may interbreed and form a zone of hybrids, an area of higher genetic variability with many intermediate forms. The parent types would then be called races or subspecies. An example is seen in waterbuck in Africa which has a northern form with a white rump patch and a southern form with a thin white ring on the rump. They overlap in a narrow zone in Kenya where they interbreed, producing various rump patterns.

3.7.3 *Individual variability within geographic regions*

So far we have considered genetic variability within and between populations. There is another form of variability which we call a *polymorphism*. The formal definition of this by Ford (1940) is "the occurrence together in the same habitat of two or more discontinuous forms of a species in such proportions that the rarest of them cannot merely be maintained by recurrent mutation or immigration." This means that the different morphs, often quite visibly different, live together in the same habitat instead of living geographically apart as described above for subspecies. For example, the lesser snowgoose has two color morphs, blue and white, which occur together in the same population. This species is polymorphic for color. The common guillemot or murre (*Uria aalge*) has a "bridled" morph with a ring of white feathers around the eye. The normal morph has no white marks. Along the coast of Europe the frequency of the bridled morph increases with latitude and humidity from 0.5% to over 50% (Southern 1951).

Where the frequencies of the morphs in these populations have remained relatively constant, as in the bridled guillemot, the state is called a *stable polymorphism*. There are also cases where one morph displaces another, perhaps because of changing conditions or because two races, originally separate, have recently become sympatric. Such may have happened to lesser snowgeese where the two color morphs used to spend the winter in separate areas but now share their wintering grounds in the central USA (Cooke 1988). The temporary state in which one morph is replacing another is called a *transient polymorphism*.

The selective advantages accruing to the various morphs are usually unknown. Mechanisms that could maintain the polymorphism include the following.

1 *Heterozygote advantage*. The heterozygote has a selective advantage over both homozygotes. Often the rare allele is a genetic dominant, lethal or disadvantageous in homozygous form.

2 *Frequency dependent selection*. The rarest morph has a selective advantage over the others. For example, this could occur where predators have a search image for the common morph and so overlook the rare one. A search image occurs when a predator concentrates it's hunting on one prey species to the exclusion of others.

3 *Alternating selection*. Different morphs are advantageous under different environmental conditions. Some morphs may be adapted to wet and cool conditions, others to hot and dry conditions, such that they are advantaged seasonally.

3.8 **Applied aspects**

3.8.1 *Adaptation*

Why do we need to know about adaptation from the point of view of wildlife management? Many species are becoming rare through loss of their habitat. To rectify this by preserving habitats we need to know their physiologic and behavioral adaptations and constraints. For example, to improve the breeding ponds for ducks in the Canadian prairies we need to know the tolerances they have for levels of alkalinity and salinity. Recently hatched ducklings of mallard (*Anas platyrhynchos*) and blue-winged teal (*A. discors*) require fresh water for survival because their salt glands are not completely functional until the ducklings are 6 days old. Growth of mallards in the first month of life is slowed when they live in moderately saline water. Although dabbling ducks often nest on islands in lakes of high salt content, female mallards lead their ducklings to freshwater lakes, and gadwall (*A. strepera*) ducklings use freshwater seepage zones (Swanson *et al.* 1984).

The social organization of animals is an adaptation to habitat, food supply, avoidance of predators, and courtship. Jarman (1974) compares the social organization of the African antelopes (see Section 7.8). At one extreme is the tiny dikdik (*Rhynchotragus kirkii*), which lives in pairs jointly defending a territory in thick scrub. They avoid predators by staying very still when predators are around, and they run only at the last moment. They rely on concealment. Their food is high-protein shoots and buds and flowers on bushes. Being small (5 kg), dikdik do not eat large amounts of food. However, the food is sparsely distributed and so dikdik are also dispersed – they cannot live in large groups. Equally, this sparse food supply should be defended to prevent others from eating it, and so they are territorial. Since females are widely dispersed (because of the food they eat), a male can obtain a mate only by keeping a female in his territory, and so a monogamous pair-bond develops.

On the other end of the scale is the African buffalo (*Syncerus caffer*) (500 kg) which lives in herds of several hundred animals, often in open savanna, and eats abundant but low-quality grass. By living in herds they obtain protection from predators – only lions (*Panthera leo*) are big enough to attack them. Because grass tends to be closely and uniformly distributed, buffalo are able to live in herds and still find enough to eat. It is not worth their while defending a patch of grass because there is plenty more beyond. Thus, the mating system has evolved into a dominance hierarchy among males in which the dominant males obtain most of the matings – a polygynous system. These comparisons suggest that an adaptation to one thing, say food type, leads to complementary adaptations to habitat utilization, antipredator responses, and social behavior.

3.9 Summary

The characteristics of individual animals are shaped by the process of evolution through the associated process of speciation. Geographic barriers, earth movements, and the migration of climatic zones split up the distribution of species, the separated components then adapting to their own disparate environments. Evolution of higher order taxa lead to convergence on one hand and radiation on the other.

These large processes determine the detailed characteristics of the individuals of a population, their morphologic and behavioral traits differing within populations and among populations according to genetic programming. Methods developed comparatively recently (electrophoresis, RFLP analysis, DNA fingerprinting, and mtDNA techniques) allow us to determine more accurately the genetic constitution of individuals and the genetic differences among races, species, and higher taxa.

4   **Animals in populations**

4.1 **Introduction**

In this chapter we deal with the those internal workings of a population that result in a change of population size. The speed of that change is measured as rate of increase. Any such change alerts us to a change in the fecundity rate, the mortality rate, the age distribution, or more than one of these. Each of these parameters will be considered in turn and the relationships between them explained.

This chapter has two quite distinct functions. The first is to arm the reader with the theory of population dynamics. The second is to indicate which parts of that theory are immediately applicable to wildlife management and which are necessary for a background understanding only. The first function may appear to load a manager with unnecessary mental baggage but without such knowledge mistakes are more than just possible, they are likely. A knowledge of atomic theory is not needed to mix a medicine, but without that knowledge a pharmacist will, sooner or later, make a critical mistake.

4.2 **Rate of increase**

If a population comprising 100 animals on (say) January 1 contained 200 animals on the following January 1 then obviously it has doubled over 1 year. What will be its size on the next January 1 if it continues to grow at the same rate? The answer is not 300, as it would be if the growth increment (net number of animals added over the year) remained constant each year, but 400, because it is the growth rate (net number of animals added, divided by numbers present at the beginning of the interval) that remains constant. Thus the growth of a population is analogous to the growth of a sum of money deposited at interest with a bank. In both cases the growth increment each year is determined by the rate of growth and by the amount of money or the number of animals that are there to start with. Both grow according to the rules of compound interest and all calculations must therefore be governed by that branch of arithmetic.

Populations decrease as well as increase. The population of 100 animals on January 1 might have declined to 50 by the following January 1, in which case we say that the population has halved. If its decline continues at the same rate it will be down to 25 on the next January 1. Halving and doubling are the same process operating with equal force, the only difference being that the process is running in opposite directions. The terms by which we measure the magnitude of the process should reflect that equivalence. It is poorly achieved by simply giving the multiplier of the growth, 2 for a doubling and 0.5 for a halving, and it becomes even more confusing when these are given as percentages. We need a metric that gives exactly the same figure for a halving as for a doubling, but with the sign reversed. That would make it obvious that a decrease is simply a negative rate of increase.

It is achieved by expressing the rate of increase, positive or negative, as an exponential rate according to the following equation:

$$N_{t+1} = N_t e^r$$

in which N_t is population size at time t, N_{t+1} is the population size a unit of time later, e is the base of natural logs taking the value 2.7182817 and r is the exponential rate of increase. The *finite rate of increase* is the ratio of the two censuses:

$$e^r = N_{t+1}/N_t$$

(often symbolized by the Greek letter λ) and therefore the *exponential rate of increase* is:

$$r = \log_e(N_{t+1}/N_t) = \log_e e^r$$

We will try this out on a doubling and halving. With a doubling:

$$e^r = 200/100 = 2$$

and so:

$$r = \log_e e^r = 0.693$$

With a halving:

$$e^r = 50/100 = 0.5$$

and so:

$$r = \log_e e^r = -0.693$$

Thus a halving and a doubling both provide the same exponential rate of increase, 0.693, which in the case of a halving has the sign reversed (i.e., −0.693). It makes the point again that a rate of decrease is simply a negative rate of increase.

The finite rate of increase (i.e., the growth multiplier e^r) and the exponential rate of increase r must each have a unit attached to them. In our example the unit was 1 year, and so we can say that the population is multiplied by e^r per year. The exponential rate r is actually the growth multiplier of $\log_e$ numbers per year. That is something of a mouthful, and so we say that the population increased at an exponential rate r on a yearly basis. Note that e^r and r are simply different ways of presenting the same rate of change. They do not contain independent information.

The exponential rate of increase, unlike the finite rate, can be changed from one unit of time to another by simple multiplication and division. If $r = -0.693$ on a yearly basis then $r = -0.693/365 = -0.0019$ on a daily basis. That simplicity is not available for e^r.

The equations given above were simplified to embrace only one unit of time. They can be generalized to:

$$N_t = N_0 e^{rt}$$

where N_0 is the population size at the beginning of the period of interest and N_t is the population size t units of time later. The average exponential rate of increase over the period is:

$$r = (\log_e(N_t/N_0))/t$$

which can also be written as:

$$r = (\log_e N_t - \log_e N_0)/t$$

It would be a waste of data to use only the population estimates at the beginning and end of the period to estimate the average rate of increase between those two dates. If intermediate estimates are available these can and should be included in the calculation to increase its precision. The appropriate technique is to take natural logarithms of the population estimates and then fit a linear regression to the data points, each comprising $\log_e N$ and t. A linear regression takes the form $y = a + bx$ in which y is the dependent variable (in this case logged population size) and x the independent variable (in this case time measured in units of choice). Our equation thus becomes:

$$\log_e N = a + bt$$

in which a is the fitted value of $\log_e N$ when time $t = 0$ and b is the increase in $\log_e N$ over one interval of time. But that is the definition of r, and so $r = b$. The equation for the linear regression may thus be rewritten:

$$\log_e N = a + rt$$

It can be converted back to the notation used in the example where rate of increase was measured between only two points by designating the start of the period as time 0:

$$\log_e N_t = \log_e N_0 + rt$$

which, with a little rearranging, converts to:

$$r = (\log_e N_t - \log_e N_0)/t$$

as before. Figure 4.1 shows such use of linear regression to estimate a population's rate of increase, and yields $r = 0.14$. The 5.73 estimates the log of numbers in year 0.

4.2.1 *Intrinsic rate of increase*

The rate of increase of a population of vertebrates usually fluctuates gently for most of the time, around a mean of zero. If conditions suddenly become more favorable the population increases, the environ-

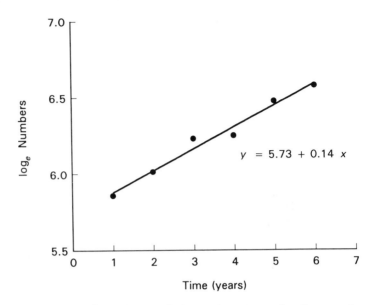

Fig. 4.1 Calculating the mean rate of increase from a run of caribou population estimates. (After Bergerud 1971.)

mental favorability being reflected in a rise of fecundity and a decline in mortality. The environmental change might have been an increase in food supply, perhaps a flush of plant growth occasioned by a mild winter and a wet spring. The rate at which the population increases is then determined by two things: on one hand the amount of food available and on the other the intrinsic ability of the species to convert that extra energy into enhanced fecundity and diminished mortality. Thus it depends on an environmental effect and an intrinsic effect, but neither is without limit. From the viewpoint of the animal both are constrained. There comes a point at which the animal has all the food it can eat, any further food having no additional effect on its reproductive rate and probability of survival. Similarly, an animal's reproductive rate is constrained at the upper limit by its physiology. Litters can be only so big and the interval between successive litters cannot be reduced below the gestation period. The potential rate of increase can never be very high, irrespective of how favorable the environmental conditions are, if the period of gestation is long (e.g., 22 months for the African elephant (*Loxodonta africana*)). All species therefore have a maximum rate of increase which is called their *intrinsic rate of increase* (Fisher 1930) and symbolized r_m. It is a particularly important parameter in estimating sustainable yield (see Section 16.3).

Populations do not attain that maximum very often. It requires a very high availability of food and a low density of animals such that there is negligible competition for that food. These conditions are most closely approached when a population is in the early stage of active growth subsequent to the release of a nucleus of individuals into an area from which they were formerly absent. Figure 4.2 gives

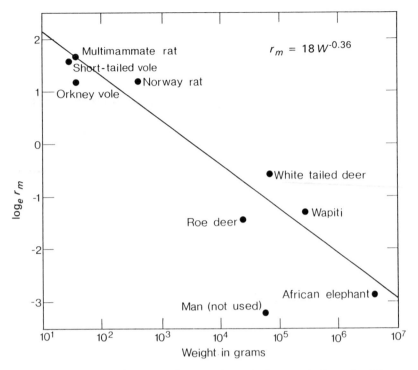

Fig. 4.2 Intrinsic rate of increase of mammals graphed against bodyweight. (After Caughley and Krebs 1983.)

intrinsic rates of increase of several mammals, most of the data being gathered in that way. Alternatively the rate could be estimated from the initial stages of growth of a population recovering from over-hunting. That would work for blue whales (*Balaenoptera musculus*), for example, which are presently recovering from intense overharvesting between about 1925 and 1955 (Cherfas 1988).

Intrinsic rate of increase r_m tends to vary with body size. The relationship has been calculated (Caughley and Krebs 1983) for herbivorous mammals as:

$$r_m = 1.5W^{-0.36}$$

where W is mean adult live weight in kilograms. Table 4.1 gives r_m

Table 4.1 Expected intrinsic rates of increase r_m on a yearly basis for herbivorous mammals as estimated from mean adult live weight

Weight (kg)	r_m
1	1.50
10	0.65
100	0.29
1000	0.08

calculated by that equation for a range of bodyweights. In the absence of other data it provides an approximation that can be used to make a first estimate of sustained yield (see Section 16.3).

4.3 Fecundity rate

A population's rate of increase is determined by its size, by how many animals are born, and by how many die during a year. Hence birth rate is an important component of population dynamics which can be measured in a number of ways. Of these the most useful is fecundity rate.

We measure fecundity rate as the number of female live births per female per unit of time (usually 1 year). That figure is often broken down into age classes to give a fecundity schedule as in Table 4.2 and each value is symbolized m_x, female births per female in the age interval x to $x + 1$.

4.4 Mortality rate

The number of animals that die over a year is another important determinant of rate of increase, and again it can be measured in a number of ways. We measure it as the mortality rate, the number of animals that die during a unit of time (usually 1 year) divided by the number alive at the beginning of the time unit. As with fecundity, the rate is often given for each interval of age.

The pattern of mortality with age is summarized as a life table which has a number of columns as in Table 4.3. The first is the age

Table 4.2 A fecundity schedule calculated for chamois. (After Caughley 1970)

Age (years) (x)	Sampled number (f_x)	Number pregnant or lactating (B_x)	Female births per female ($B_x/2f_x$) (m_x)
0	–	–	0.000
1	60	2	0.017
2	36	14	0.194
3	70	52	0.371
4	48	45	0.469
5	26	19	0.365
6	19	16	0.421
7	6	5	0.417
>7	10	7	0.350

Table 4.3 Construction of a partial life table

Age (years) (x)	Survival frequency (f_x)	Survivorship (l_x)	Mortality (d_x)	Mortality rate (q_x)
0	1200	1.00	0.58	0.58
1	500	0.42	0.17	0.40
2	300	0.25	0.08	0.32
3	200	0.17	–	–
⋮	⋮	⋮	⋮	⋮

interval labeled by the age at the beginning of the interval and symbolized x. The second is survivorship l_x, the probability at birth of surviving to age x. The third is mortality d_x, the probability at birth of dying in the age interval x, $x + 1$. And the fourth, the most useful, is mortality rate q_x, the probability of an animal age x dying before the age of $x + 1$.

Probabilities are estimated from proportions. The probability of a bird surviving to age x can be estimated, for example, by banding 1200 fledglings and recording the number still alive 1 year later, 2 years later, 3 years later, and so on. Let us say those frequencies were 500, 300, and 200. Survivorship at age 0 (i.e., at birth) is 1200/1200 = 1, by age 1 year it has dropped to 500/1200 = 0.42, further to 300/1200 = 0.25 at age 2 years, and further still to 200/1200 = 0.17 at age 3 years.

No further data are needed to fill in the other columns corresponding to these values of l_x because each is a mathematical manipulation of the l_x column. Mortality d_x is calculated as $l_x - l_{x+1}$ (1 − 0.42 = 0.58 for $x = 0$ and 0.42 − 0.25 = 0.17 for $x = 1$). Mortality rate q_x is calculated as $(l_x - l_{x+1})/l_x$ or d_x/l_x (0.58/1 = 0.58 for $x = 1$ and 0.17/0.42 = 0.40 for $x = 2$). Table 4.3 shows the table fully constructed up to age 2 years, that for age 3 years being partial because data for age 4 years are needed to complete it. The subsequent rows would be filled in each year as the data became available.

So, constructing a life table is easy and even fun when the appropriate data are available. But pause for a moment to contemplate the difficulty of obtaining those data. Banding 1200 fledglings, or whatever number, poses no more than a problem in logistics. The difficulty comes in estimating what proportion of those birds are still alive at the end of the year. How does one go about it? It is virtually impossible unless they are in an aviary.

There are, however, a couple of approximation methods available. If one can age a sample of the living population, or alternatively establish the ages at death of a sample of deaths from that population, an approximate life table can, in some circumstances, be constructed from those age frequencies.

4.4.1 Indirect estimation of life tables

If certain conditions (see the end of this section) are met, the age distribution of the living population can be used as a surrogate for the survival frequency f_x of Table 4.3 to produce an approximate life table. Many of the bovids can be aged from annual growth rings on the horns; some species of deer, seals, and possums from growth layers in the teeth; and fish from growth lines on the scales. An unbiased sample of such a population, taken in the middle of the season of births to ensure that the ages are integral, yields an age distribution of the living population that may be amenable to life-table analysis.

Table 4.4 shows such an analysis. Reindeer (*Rangifer tarandus*) from a shot sample were assumed to be collected randomly from 2 years onward. Frequencies for calves and yearlings were taken from

Table 4.4 Construction of a life table from the age composition of a living population. Male reindeer on South Georgia uncorrected for rate of increase. (After Leader-Williams 1988)

Age (years) (x)	Survival frequency (f_x)	Survivorship (l_x)	Mortality (d_x)	Mortality rate (q_x)
0	78	1.000	0.487	0.487
1	40	0.513	0.269	0.525
2	19	0.244	0.064	0.263
3	14	0.179	0.051	0.286
4	10	0.128	0.641	0.500
5	5	0.064	0.064	1.000
	166			

composition counts and added in *pro rata*. A "sample" of 166 male reindeer is assumed to reflect the age composition of the living population, and if a zero rate of increase over a period of time is also assumed it would reflect survival frequencies by age. The survivorship series is then constructed by dividing each age frequency by 166, the d_x series as $l_x - l_{x+1}$, and the q_x series as d_x/l_x. Elaborations to cope with nonzero rates of increase are given in Caughley (1977b).

Table 4.5 Construction of a life table from a pick-up sample. The table is not corrected for rate of increase

Age (years) (x)	Mortality frequency (f_x)	Mortality corrected (fd_x)	Mortality (d_x)	Survivorship (l_x)	Mortality rate (q_x)
0	–	485	0.485	1.000	0.485
1	–	129	0.129	0.515	0.250
2	2	4	0.004	0.387	0.010
3	5	11	0.011	0.383	0.029
4	5	11	0.011	0.372	0.030
5	6	13	0.013	0.361	0.036
6	18	38	0.038	0.348	0.109
7	17	36	0.036	0.310	0.116
8	20	42	0.042	0.274	0.153
9	17	36	0.036	0.232	0.155
10	15	32	0.032	0.196	0.163
11	16	34	0.034	0.164	0.207
12	18	38	0.038	0.130	0.292
13	15	32	0.032	0.092	0.348
14	14	29	0.029	0.060	0.483
15	8	17	0.017	0.031	0.548
16	5	10	0.010	0.014	0.714
17	1	2	0.002	0.004	0.500
18	0	0	0.000	0.002	0.000
19	1	2	0.002	0.002	1.000
	183	1001	1.001		

In similar fashion an unbiased sample of ages at death, as might be obtained by a picked-up collection of skulls, may in some circumstances be treated as a multiple of the d_x series. Table 4.5 gives an example from African buffalo (Sinclair 1977). Sinclair counted only those skulls aged 2 years or older because skulls from younger animals disintegrate quickly. These age frequencies are given in the second column of the table and total 183 skulls. The third column corrects for the missing younger frequencies: Sinclair had independent evidence that the mortality rate over the first year of life was 48.5% and that 12.9% of the original cohort died in the second year. Hence, if the original cohort is taken as 1000, 485 of these would die in the first year of life and 129 in the second year. These values are tabled. They account for 614 of the original cohort, leaving 386 to die at older ages. The age frequencies of the 183 animals in the second column are thus each multiplied by 386/183 to complete the third column. The fourth column, d_x, is formed by dividing the fd_x frequencies by 1000 so that they sum to unity. Survivorship at age 0 (i.e., birth) is then set at one and the subsequent l_x values calculated by subtracting the corresponding d_x from each. Mortality rates q_x are calculated as before, as $q_x = d_x/l_x$.

No allowance has been made here for the rate of population increase. Krebs (1989) shows for this same example how to correct the life table for an observed 8% rate of increase.

The reliability of such a life table depends on how closely the data meet the underlying assumptions of the analysis.

1   The sample is an unbiased representation of the living age distribution in the first case or of the true frequency of ages at death in the second. The exercise would have to control the usual biases implicit in hunting activities if the sample of the living age distribution were obtained by shooting. One would be unlikely to use a sample obtained by sporting hunters for example. The first age class is usually underestimated in a picked-up sample of ages at death because the skulls of young animals disintegrate much faster than do those of adults, thereby significantly biasing the table.

2   Age-specific fecundity and mortality must have remained essentially unchanged for a couple of generations.

3   Whether the sample is of the living population or of the ages at death, the population from which it came must have a rate of increase very close to zero.

Note that these approximate life tables are constructed on the assumption, true or false as it may be, that rate of increase is zero and has been so for the length of one, and preferably two, generations. Such a life table cannot then be used to deduce anything about the dynamics of the population. That limits the usefulness of such exercises in wildlife management. The resultant table has next to no management application.

4.5 **Relationship between parameters**

We restrict the following discussion to females for simplicity, but the points made apply also to the male segment of the population.

Remember that l_x is survivorship to age x, m_x is production of daughters per female at age x, and r is the exponential rate at which the population increases. Then:

$$\Sigma l_x m_x e^{-rx} = 1$$

which is the basic equation of population dynamics. If the survivorship and fecundity schedules hold constant, the population's age distribution will converge to the constant form of:

$$S_x = l_x e^{-rx}$$

which is called the *stable age distribution*. S_x is the number of females in a particular age class divided by the number of females in the first age class. The basic equation may thus be written $\Sigma S_x m_x = 1$. In the special case of rate of increase being zero the stable age distribution, now called the stationary age distribution, is $S_x = l_x$ by virtue of $e^{-0x} = 1$. That is the justification for using such an age distribution to construct a life table. The *stationary age distribution* is the special case of the stable age distribution that occurs when $r = 0$.

It has been argued that, since fecundity and mortality schedules seldom remain constant for long, the stable age distribution is little more than a mathematical abstraction. In fact it is a robust configuration that is attained very quickly after mortality and fecundity patterns stabilize. Likewise, a population of large mammals whose mortality and fecundity schedules fluctuate little from year to year will have an age distribution close to the stable form. That does not hold for small mammals, fish, and birds whose age-specific fecundity and mortality is much more labile, nor for those large mammals that live in a fluctuating environment.

4.6 **Patterns of population growth**

The way a population grows is determined by its relationship with one or more resources. To anticipate Chapter 6, a *resource* is something an animal needs to survive and reproduce. For example, the most important resource for a population of parrots may be the nesting holes in old trees. That may limit how many parrots can live in an area. However, another population of the same species may have more nesting holes available than it could ever saturate because its most important resource comprises seeds which are in short supply. Yet another species of parrot may be limited neither by food nor by nesting sites but simply by a shortage of space. Some species are antisocial and each pair needs a large area for its exclusive use. For example, pairs of Major Mitchell's parrot (*Cacatua leadbeateri*) in Australia will not breed closer together than 1 km (Rowley and Chapman 1991). They are limited by space, and in this context we will consider space as a resource, just as food is a resource.

Many tasks in wildlife management require the modification of population growth, either to slow it down (as in control problems) or to speed it up (conservation problems). It is thus necessary to know what sort of process one is modifying. These can be complex, but we start with simple patterns by way of illustration.

4.6.1 Population growth: no resource limiting

When resources are unlimited, animal populations grow at their intrinsic rate of increase r_m determined by innate physiology interacting with those components of the environment that are not resources: temperature, slope of the land, and such like. Note that the intrinsic rate of increase is not a constant for a species. Emperor penguins released in the tropics, although having all the resources they need, will not increase at the same rate as if they were on the coast of Antarctica. The temperature is too high for them in the tropics. Thus the intrinsic rate of increase varies for a species according to the physical environment.

Where no resource is limiting the population increases (or decreases) at the rate r_m for that species in that environment. Thus numbers conform to the formula:

$$N_t = N_0 e^{rt}$$

where $r = r_m$. The trajectory of population growth will be a steepening curve if population size N is graphed against time t, or a straight line if the logarithm (any base) of N is graphed against t. Where the log base chosen is e, the slope (b) of that straight line (increase in $\log_e N$ over one unit of time) will be $r = r_m$. Figure 4.3 shows the relationship.

4.6.2 Population growth limited by a nonconsumable resource

The size of a population may ultimately be limited by a shortage of space, as when individuals need an empty zone of fixed size around them (spacing behavior), or by a shortage of a nonconsumable resource

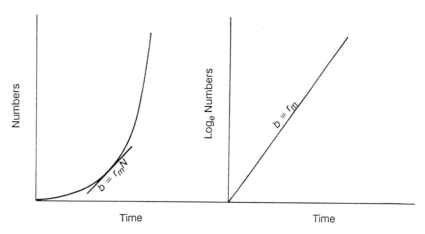

Fig. 4.3 Relationship between population size and time when population growth is exponential because no resource is limiting. (After Caughley 1977b.)

such as nesting sites. The resource is not used consumptively but preemptively (see Section 6.5). There are only winners or losers, no sharing of the burden of misery. The expected pattern of population growth is an increase at or near r_m followed by a rapid leveling off. Thus the process of population growth shifts from one state to another rather abruptly after a threshold density is passed. The simplest way of describing such a process is by a pair of equations:

$$dN/dt = r_m N \quad \text{(for } N < K)$$
$$dN/dt = 0 \quad \text{(for } N = K)$$

Figure 4.4 sketches the growth of a population obeying these rules.

4.6.3 *Population growth limited by a consumable resource*

Where individuals consume a resource rather than preempt it, the pattern of population growth must necessarily differ from that of the previous example. How it differs depends on the rate at which the resource is renewed, by whether that rate is influenced by the absolute level of the resource, and by whether the animals are able to get at the "capital" (the absolute level) or whether they have access only to the "interest" (renewal rate). The pattern of population growth can be complicated but here we look at a few possibilities to note the spread of outcomes that are possible.

When the resource is gone its gone
Consider a population of mice started by a couple of individuals dispersing into a grain store. The capital supply of food is immense but it is not renewed. Since the rate at which mice survive and multiply is critically dependent on how much they eat, and since each will eat as much as it can each day while the food lasts, the population will increase at r_m until the food is exhausted. Then the population crashes to extinction.

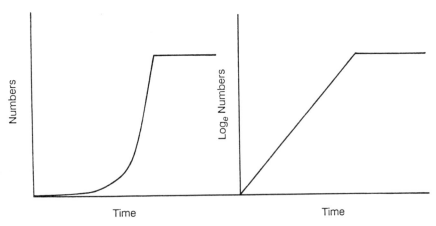

Fig. 4.4 Relationship between population size and time when population growth is limited by a nonconsumable resource such as nesting sites.

Now consider a similar but subtly different case: mice erupting in response to an abnormal seed fall. Southern beech (*Nothofagus*) flowers only about once every 10 years but when it does there is suddenly an abundance of food. A carpet of seed is spread through the forest and the mice build up in numbers as they attack it. With no more seed falling, the carpet of seed is progressively thinned out. An individual alive in the early stages of the bonanza will eat as much as it is able; one present in the latter stages will be able to find far fewer seeds in a day's foraging. The rate of increase will slow progressively as the seed carpet thins.

The process of population growth will pass through three states. First, the population increases at rate r_m while there is enough seed to satiate each mouse each day; then the growth rate will tail off progressively as the seed is thinned out below that threshold; and finally the population trajectory will turn over and the population will decline to extinction. Figure 4.5 shows the expected trend.

Reindeer introduced in 1944 onto St Matthew Island to the west of Alaska followed a similar trend (Klein 1968), driven by the same process. Their major food is lichen, of which there was much on St Matthew Island. Unfortunately for the reindeer, although lichen can attain a high biomass in the absence of grazing it has a very low rate of renewal. On the timescale of this natural experiment that rate of renewal was essentially zero. The reindeer increased rapidly while chewing into the standing crop of food. When it was exhausted the population plunged to extinction.

Consumable resource: reactive relationship
The most simple case of a renewable resource being utilized by a population of animals is that where the animals have no influence on the rate of renewal of the resource, where they consume the "interest"

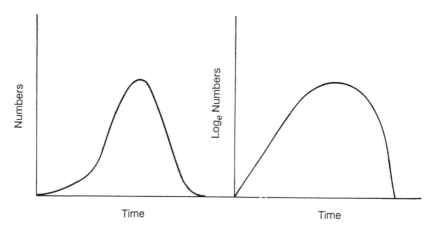

Fig. 4.5 Expected trend of a population reacting to a resource that is consumed but not renewed.

but cannot touch the "capital." The animals thus have no influence over the amount of the resource available to their next generation. We might envisage, for example, an animal feeding on fruit that drops to the floor of a rainforest. Suppose that fruit-fall is 100 fruit per hectare per day and that each animal needs to eat 12 of these per day to keep it healthy enough to allow just its replacement in the next generation. However, its satiating intake is 25 per day. Thus a population starting from low density will increase at rate r_m, each animal eating 25 fruit per day, until the density of animals reaches 4/ha. Thereafter the rate of increase slows because the 100 fruit per hectare per day is insufficient to feed all animals to their satiating level. Intake per individual is reduced progressively as numbers rise until it averages 12 fruit per day. At that point the rate of population increase settles on zero.

If rate of increase declines linearly with declining food intake per individual the pattern of population growth is exponential at rate r_m while intake is at the level of satiation, and logistic thereafter. Symbolizing:

i = satiating intake per individual per day = 25 fruit,
g = production of the resource per hectare per day = 100 fruit,
b = maintenance intake per individual per day = 12 fruit,
N = number of individuals per hectare.

The proportion of the fruit channeled into maintenance and replacement is bN/g, leaving $1 - bN/g$ that can be utilized for generating population increase. The population finally stabilizes at a level of $N = g/b = 100/12 = 8.33$ individuals per hectare:

$$dN/dt = r_m N \quad (\text{for } g/N \geqslant i)$$
$$dN/dt = r_m N(1 - bN/g) \quad (\text{for } g/N < i)$$

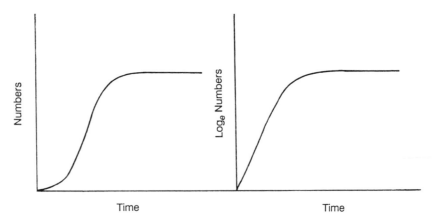

Fig. 4.6 Expected trend of a population reacting to a resource where the animals have access to the "interest" but not the "capital." They hence have no effect on the rate of renewal of the resource. These conditions lead to logistic growth.

If we modify the second equation by substituting K for g/b, the equilibrium number of animals per hectare, it becomes:

$$dN/dt = r_mN(1 - N/K) \quad \text{(for } g/N < i)$$

which is the *logistic equation*, the simplest and most favored theoretical depiction of limited population growth used in wildlife management. Figure 4.6 shows the relationship.

The logistic equation is applicable only in the special and unusual circumstances of a population's limiting resource being produced at a rate independent of the number of animals using it. That will hold only when the animals use the "interest" of the resource but cannot get at the "capital" generating the interest. One might wonder why the equation is employed so often to model systems that do not meet these requirements.

The major reason is that the equation in its form:

$$dN/dt = r_mN(1 - N/K)$$

is stripped of information about the resource. At face value it implies that dN/dt is a function only of population density N, the intrinsic rate of increase r_m, and the maximum sustainable number K. Rate of increase r in the logistic equation is a linear declining function of N:

$$r = 1/N \times dN/dt = r_m(1 - N/K)$$
$$= r_m - (r_m/K)N$$

which is exactly in the form of the equation $y = a + bx$ which describes a straight line. The line relating r to N by the logistic equation has a y-intercept of r_m and a slope of $-r_m/K$. It implies therefore that one need know nothing about resources or the population's relationship with them to predict how the population will grow, but as we have seen that holds only when the relationship between the animals and their limiting resource is of a particular and unusual form.

So beware of the logistic equation. A population's rate of increase is not necessarily a simple function of its density, unless it is determined by spacing behavior.

Having said that, we note that there are some instances in which the logistic equation has a valid use in wildlife management.

1 Plants cannot influence the rate of renewal of their resources, water and sunlight, and therefore cannot influence the availability of these resources to the next generation. The growth of many plant populations should therefore be close to logistic and there are several field studies indicating just that (Noy-Meir 1975). Theoretically and evidentially the logistic is the first choice to model the growth of a population of plants.

2 The relationship between a wildlife population and its limiting resource sometimes comes close to validating logistic assumptions. Wildebeest (*Connochaetes taurinus*) in the northern Serengeti of

Tanzania migrate in the dry season into a supply of food that represents the year's growth of grass (Sinclair *et al.* 1985). They consume it and move on. They have little or no measurable effect on the capital supply of food, which is represented by an invulnerable root system and meristematic tissue. Growth of a wildebeest population living in this way could be close to logistic, and logistic growth is close to what is observed (Sinclair *et al.* 1985). The limiting resource for the North American elk (*Cervus elaphus*) of Yellowstone National Park appears to be grass on its wintering grounds. The elk eat that grass only after it has hayed off. This consumption has no effect on the amount of grass available at the beginning of the next winter. Growth of that elk population would be expected to approximate a logistic, and so it does (Houston 1982).

3   When a species has an intrinsic rate of increase >0.1 on a yearly basis, its trajectory of population growth tends to proceed to an asymptote without overshooting too much. The logistic then provides a reasonable approximation to the growth curve even when the mechanism of growth has little in common with the logistic assumptions. In such circumstances the logistic can be used to calculate a first approximation to sustained yield (see Section 16.5).

4   The logistic pattern of growth can sometimes be used in strategic modeling even when it is not entirely appropriate. We might ask how species diversity changes with the introduction of a new species of predator or a new herbivore. Or we might seek the general pattern of spread of a gene through a population. Questions like these require answers that deal with kind rather than with number or degree. Those answers generally will be robust to the details of population growth and so the logistic equation can be used confidently as a representative example of a whole family of models where growth is ultimately limited by the shortage of a resource.

Consumable resource: interactive relationship

The previous section dealt with population growth powered by the utilization of a renewable resource and where the relationship between the dynamics of the resource and that of the animals is reactive. The essence of a reactive system is that the dynamics of the population react to the amount of the resource on offer but the population is incapable of influencing the rate at which that resource is renewed. The reaction is in one direction. This section deals with population growth where the relationship between the resource and the population is interactive. Such a system differs from a reactive system in that rate of increase of the animals reacts to the level of the resource as before but the rate of renewal of the resource also reacts to the number of animals using it. The reaction runs in both directions to produce an interaction. The animals use both the "capital" and the "interest" of the resource, there being no distinction between the two.

The interactive relationship between a population of animals and the resource upon which it depends is probably the most common in nature. All prey–predator systems and most plant–herbivore systems fall neatly into this category. Theoretical work by May (1972) in particular has shown that, irrespective of the details of the interactive relationship, only two general outcomes are likely in a constant environment (nonfluctuating conditions): an equilibrium where the animal population and the resource reaches an accommodation with each, the two settling on fixed densities (a point equilibrium), or both fluctuate in what is called a *stable limit cycle*. The latter should not be confused with Lotka–Volterra cycles beloved of ecological textbooks. Those are neutrally stable, their behavior duplicating that of a frictionless pendulum whose amplitude is determined by initial conditions. If it is bumped the amplitude changes and holds indefinitely to the new value. The biological ground rules needed for the production of neutrally stable cycles are ecologically impossible and so we do not investigate them further.

The stable limit cycle and the point equilibrium are the product of precisely the same processes. Whether one or other occurs is determined only by the values of the constants of the equations describing the interaction. Thus exactly the same process can produce outcomes that differ in kind.

Although the stable limit cycle is theoretically a likely outcome of interaction between a population and its limiting resource (i.e., the resource it depends upon), such an outcome is conspicuously rare in nature. In fact no-one has yet identified unambiguously such a stable limit cycle, although they have been invoked tentatively to describe the relationship between elephants and trees in Africa (Caughley 1976b) and they are periodically given to account for the interlinked cycles between snowshoe hare and lynx in the boreal forests of Canada.

The point equilibrium is clearly the most common underlying outcome for wildlife, but that bald statement should not be taken as indicating that wildlife populations and their consumable resources are locked into a fixed ratio, one to the other. The equilibrium is the expected outcome in a constant environment but, unfortunately, there is no such thing. We expect both the level of the limiting resource and hence the density of the animals to fluctuate in the vicinity of the theoretical equilibrium point under the influence of environmental perturbations, particularly perturbations in weather. The equilibrium point may never be occupied, or it may be occupied fleetingly only when the standing state of the system passes through it. The equilibrium point is not even the mean position of the system during year-to-year fluctuation.

Thus we might conclude that a theoretical equilibrium between animals and resources, manifested only in a constant environment which is itself only of theoretical interest, is of little practical importance in wildlife management. This is not necessarily so because the essence of a system characterized by a theoretical equilibrium point is that the dynamics of the system are centripetal. The density of the animals and the level of the resource, through interaction between the two, will tend to come back towards the equilibrium point. The further one or other is displaced from that point the more powerful is the pull back towards it.

The mechanism enforcing centripetality is called *negative feedback*. It is the process manipulated by the wildlife manager to produce a sustained yield (Chapter 16), and it is the process that in turn manipulates the wildlife manager who is trying to control nuisance wildlife (Chapter 17).

4.7 **Summary**

The dynamic behavior of a population – whether it increases, decreases, or remains stable – is determined by its age-specific mortality and fecundity rates interacting with its age distribution. When age-specific rates of fecundity and mortality remain constant, the population's age distribution assumes a stable form even though its size may be changing. The form of the growth curve sketched by a population increasing from low numbers is determined by the relationship between the population and the dynamics of its resources. The overworked logistic curve is but one possibility. In almost all cases, however, the endpoint of population growth is a relatively steady density reflecting centripetal dynamics controlled by a negative feedback. It is this force that is manipulated by management to generate a sustained yield.

5 Dispersal, dispersion, and distribution

5.1 Introduction

This chapter explores some of the reasons why populations are found where they are. We describe the finer-scaled pattern as the dispersion and the broader scale as the distribution. We then offer examples of how different factors such as temperature and seasonality limit the distribution of wildlife.

Dispersal is the movement an individual animal makes from its place of birth to the place where it reproduces. Dispersal is not to be confused with *migration* (movement backward and forward between summer and winter home ranges) or with *local movement* (movement within a home range). The terms immigration and emigration are used in mark–recapture studies to describe movement into and out of a study area of arbitrary size and location. Migration is used by population geneticists to mean "the movement of alleles between semi-isolated subpopulations, a process that by definition involves gene flow between subpopulations" (Chepko-Sade *et al.* 1987). Although these uses differ from their ecological uses the difference is usually obvious from context and causes little confusion.

Dispersion is the pattern of spatial distribution taken up by the animals of an area. Dispersions may be fixed if the animals are sessile but more commonly they change with time under the influence of a changing dispersion of resources. A dispersion at a given time may be changed by dispersal, or local movement, or both.

The *distribution* of a population or species is the area occupied by that population or species. It is depicted as the line drawn around the dispersion. The distribution can be subdivided into gross range and breeding range, and can be mapped at different scales.

5.2 Dispersal

Dispersal is an action performed by an individual. An animal disperses or it remains within its maternal home range. If it disperses, it may move only that distance sufficient to bring it to the nearest unoccupied and suitable area within which to establish its own home range, or it might move a considerable distance, crossing many areas that look suitable enough, before settling down.

The mechanism of dispersal may also vary. The individual may be pushed out of the maternal home range by a parent or it may move without any prompting save for that supplied by its genes. The young of some species never meet their parents (e.g., frogs, reptiles, the mound-building birds of the family Megapodidae) and so must provide their own motivation.

The urge to disperse, or at least the manifestation of that urge, clearly differs markedly between individuals of a population. Figure 5.1 shows a sample of distances dispersed by juvenile kangaroo rats (*Dipodomys spectabilis*) (Jones 1987), a solitary, nocturnal, grain-eating desert rodent. Females average 29 m and males 66 m, but the majority of individuals do not disperse at all. Jones (1987) reported that adults of this species do not disperse much: 70% of adult males and 61% of adult females remain in one mound for the rest of their lives. Juvenile females of red deer (*Cervus elaphus*) seldom disperse but adopt home ranges that overlap those of their mothers. In contrast the males leave the natal home range between the ages of 2 and 3 years, mostly joining stag groups in the vicinity (Clutton-Brock *et al.* 1982).

5.3 **Dispersion**

Dispersions may be random, clumped, or spaced. The most common is a *clumped dispersion* (sometimes called a *contagious dispersion*). If the area is divided into quadrats and the frequency distribution of animals per quadrat recorded, the variance of that distribution will equal its mean if the animals are randomly distributed (a Poisson distribution); the variance will be greater than the mean if the animals are clumped at that scale; and the variance will be less than the mean if the animals space themselves. Most animals are clumped into herds, flocks, or schools. A few are evenly spaced when on mating territories as in many birds, territorial antelopes, or salmon in streams.

Scale is important when dispersions are considered because two or more orders of dispersion may be imposed upon each other – randomly distributed clumps of animals, for example. In these circumstances a quadrat in a grid of small quadrats will include either part of a group or it will miss a group; its count will be of many animals or of no animals. When the grid comprises large quadrats, an average quadrat

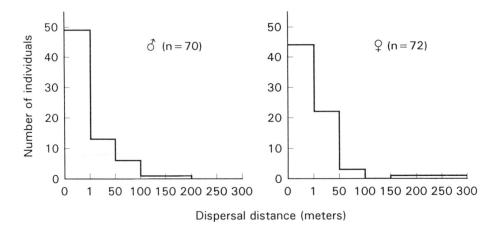

Fig. 5.1 Frequency distributions of distances dispersed by juvenile rat kangaroos. (After Jones 1987.)

will contain several groups of animals and the variation in counts between quadrats will be less marked. The dispersion is the same whether the quadrats used to sample it are large or small, but in this case the clumping as measured by the variance/mean ratio will appear to be more intense when quadrats are small.

An alternative to characterizing dispersion in terms of the frequency distribution of quadrats containing 0, 1, 2, etc., animals per quadrat is instead to record the frequency distribution of nearest-neighbor distances or of the distances between randomly chosen points and the nearest animal to each. The problem of quadrat size does not arise because no quadrats are involved, but no neat measure is presently available for distributions of distances that clearly differentiates classes of dispersions, one from the other, given the wealth of possible dispersions.

In fact a lot of time is spent trying to make something "scientific" out of dispersions when a simple statement of how the animals tend to distribute themselves, and why, would suffice. Some individuals of some species are gregarious while those of others are solitary. That statement is not illuminated by variance/mean ratios. Some species have tight habitat preferences, their dispersion reflecting where that habitat is to be found. Others are more catholic in their requirements and will therefore be distributed more evenly across the landscape. The ecology of the dispersion is important; the mathematics of it less so.

When designing surveys to count wildlife (see Chapter 12), attention must be paid to its dispersion and sampling units allocated accordingly. We explore this practical aspect of dispersion more fully in Section 12.4.

5.4 Distribution

Krebs (1985) considered that "the simplest ecological question one can ask is simply: Why are organisms of a particular species present in some places and absent in others?"

Figure 5.2 shows three hypothetical distributions, not as a map but as a plot within a range of mean annual temperature and rainfall. For species A, temperature and rainfall act independently of each other in setting limits to distribution. A single mean temperature and a single mean annual rainfall is all one needs to predict whether or not the species will be in a given area.

The distribution of species B is also determined by temperature and rainfall but this time in an asymmetric interactive manner. Distribution is determined absolutely by an upper and lower limiting temperature but it is demarcated within those bounds by rainfall, whose effect varies with temperature. High rainfall is tolerated only in hot areas and low rainfall only in colder areas where evaporation is reduced.

The distribution of species C is controlled by a symmetric interaction of rainfall and temperature. The species' tolerance of high temperatures increases with increasing annual rainfall and the tolerance of

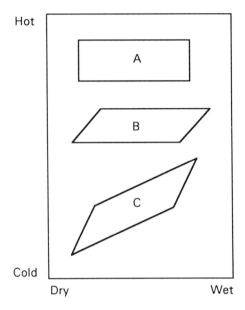

Fig. 5.2 Three hypothetical envelopes of adaptability of a species to temperature and moisture: A, the two factors act independently; B, the level of one factor influences the effect of the other; and C, the effect of each factor varies according to the level of the other. (After Caughley *et al.* 1988.)

the species to rainfall increases with temperature. This is a two-way interaction.

A known range of tolerance to one or more factors such as temperature and rainfall does not translate directly into a map of distribution because the factors may interact as in examples B and C of Fig. 5.2, the level of one factor determining the effect of another (see Section 3.6). Whether distribution is determined by one or several factors depends critically on the geographic dispersion of the levels of each factor. Is the interval of adaptability to one factor nested geographically within the interval of adaptability to another as in Fig. 5.3(a) and (b), or do they overlap as in Fig. 5.3(c)?

5.4.1 *Determinants of the range boundary*

The edge of the range marks that point at which, on average, an individual's contribution to the next generation (i.e., its fitness) is less than unity. This is the same as saying that $r_m < 0$ or that the individual does not quite replace itself in the next generation. For purposes of considering range boundaries we need to classify the environmental factors that affect an individual's fitness (in this sense the chances of it surviving and reproducing) and hence the intrinsic rate of increase r_m in different regions of the distribution of the species. The following classification divides environmental factors into three primary groups.

1 A beneficial or deleterious environmental factor whose level cannot be affected by the animals.

(a) A component of climate *per se* (e.g., a gradient of mean annual temperature, Section 5.4.3).

(b) A specific substrate (e.g., a particular rock type that is present or absent, Section 5.4.2).

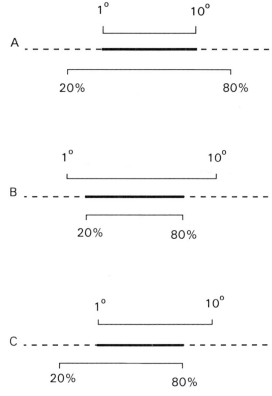

Fig. 5.3 A hypothetical species cannot persist where temperature is outside the range 1–10°C or where relative humidity falls outside the range 20–80%. Which factor defines the range of the species is determined by whether, in the field, the range of adaptability to temperature lies within that defined by humidity (a), or the reverse (b), or whether they overlap such that the position of the range boundary is determined in some places by humidity and in others by temperature (c). (After Caughley *et al.* 1988.)

(c) A source of food or water whose absolute amount cannot be reduced by the activities of the animals but which nonetheless may influence their dispersion by being distributed heterogeneously or sparsely. The animals face what Andrewartha and Birch (1984) call an *extrinsic shortage* because it is not of their making.

2 A beneficial environmental factor whose use by one animal reduces the amount available to others.

(a) A resource used consumptively. All individuals have access to the resource and each individual's activities influence the level of the resource available to it and other individuals. Sinclair's (1977) account of the interaction of grass and buffalo (*Syncerus caffer*) and Houston's (1982) account of the ecology of elk (*Cervus elaphus*) describe such a relationship. In Andrewartha and Birch's (1984) parlance a shortage of this resource, caused as it is by the animals themselves, is an "intrinsic shortage."

(b) A resource used preemptively. Individuals are either winners or losers. An example is provided by nesting holes used by parrots.

3 A deleterious environmental factor whose effect on an animal may vary with the density of the population to which that animal belongs.

(a) A facultative parasite or pathogen.

(b) An obligate parasite or pathogen.

(c) A facultative predator.

(d) An obligate predator.

5.4.2 Population characteristics near the range boundary

The characteristics of a population will change as the edge of the range is approached. Of the many ways in which a population may be described we concentrate on three: its intrinsic rate of increase, its density, and the population's wellbeing as indexed by such variables as condition, birth rate, survival rate, and growth rate. The value of these population characteristics will change with distance from the edge of the range according to which environmental factor determines the position of the range boundary. The intrinsic rate of increase is difficult to measure in the field and so most attention would be focused on trends in density and wellbeing.

The trend is described as a ramp if the population characteristic takes a low value at the range boundary but increases as the core of the distribution is approached. It is what you might expect for the trend in density of a species most suited to a moist environment if the position of the range boundary is determined by a smooth cline in aridity from the core of the distribution to the boundary.

Alternatively the trend is described as a step if the characteristic is already high at the boundary and shows no tendency to increase as the core of the distribution is approached. In special circumstances it might in fact decrease in value. In either circumstance it is called a step. It is what to expect, for example, if the range is determined by substrate. All species of wild sheep, goats, and goat-antelopes (i.e., *Ovis; Capra, Hemitragus, Pseudois, Rupicapra, Nemorhedus, Oreamnos, Ammotragus;* and *Capricornis*) live only in mountain country, the edge of their range usually coinciding with the break in slope. This is a special case of control of distribution by "substrate." Similar boundaries to distribution can be seen at the sharp transition between habitats, the break between trees and alpine meadow (the *tree line*) on mountains, and the edges of swamps. Another case of substrate or habitat determining distribution is provided by the rock warbler (*Origma solitarea*) which is restricted to the Hawksbury sandstone around Sydney, Australia.

Where the range boundary is determined in this way conditions are either favorable or unfavorable, all or none. There is no trend in favorability and so the population characteristics show no tendency to change in value as the boundary is approached from within the distribution.

Table 5.1 gives the trends to be expected for the three population

Table 5.1 The likely trend in population characteristics from the periphery to the core of a species' distribution, where the range boundary is controlled by a single factor. (After Caughley *et al.* 1988)

Factor affecting range boundary		Population characteristic		
		"Wellbeing" profile	Density profile	r_m profile
Climate	1a	Ramp	Ramp	Ramp
Habitat	1b	Step	Step	Step
Unmodifiable resource	1c	Ramp	Ramp	Ramp
Resource used preemptively	2a	Step	Ramp	Step
Resource used consumptively	2b	Step	Ramp	Step
Parasite or pathogen				
Facultative	3a	Ramp	Ramp	Ramp
Obligate	3b	–	–	–
Predator				
Facultative	3c	Step	Ramp	Ramp
Obligate	3d	–	–	–

characteristics according to which environmental factor controls the position of the range boundary. It is taken from Caughley *et al.* (1988) but modified after Jeff Short (pers. comm.) pointed out to us that the pattern resulting from predation differs from that resulting from the influence of a parasite or pathogen. Figure 5.4 provides a diagrammatic summary.

Table 5.2 gives the geographic trend in several population characteristics for two sympatric species, the eastern grey kangaroo (*Macropus giganteus*) and the southern (= western) grey kangaroo (*M. fuliginosus*). From these trends it can be deduced that, in those sampled portions of the range boundary, the factors controlling the boundary differed between the two species. *M. fuliginosus* exhibited a ramp of density and of most indices of wellbeing which, taken together with additional knowledge of the ecology of that species, suggests that the edge of the range was positioned by a component of climate, perhaps interacting with an unmodifiable resource. In contrast *M. giganteus* exhibited a ramp of density but a step in most indices of wellbeing, implicating a renewable resource as the environmental factor determining the position of the range boundary in that area.

Remember, however, that, as shown in Fig. 5.2, the position of the range boundary may be determined by different environmental factors in different parts of the range.

5.4.3 *Range limited by temperature*

Temperature can limit the distribution of animals through direct effects on their physiology and indirectly by affecting resources. Some distributions can be described empirically by temperature contours (isotherms). Thus the southern limit to northern hemisphere seals is set by sea surface temperatures never $> 20°C$ (Lavigne *et al.* 1989). The

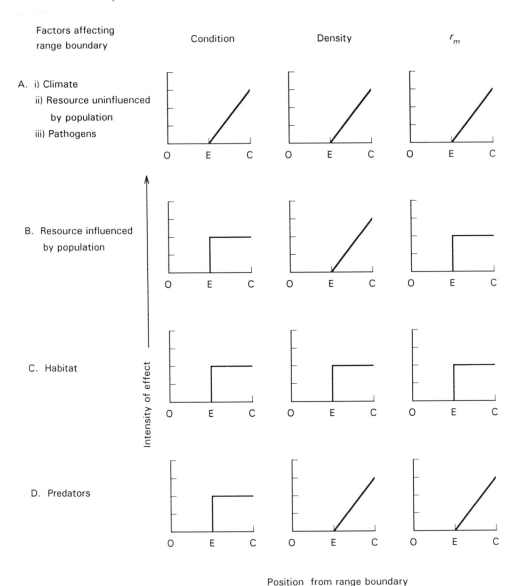

Fig. 5.4 Population condition, density, and intrinsic rate of increase (r_m) can change from the edge of a species range towards the center in either a gradual (ramp) or sudden (step) fashion, depending on which factors determine the distribution. O, Outside range; E, edge of range; C, center. (After Caughley *et al.* 1988.)

reason is unclear but most seals breed in regions of high marine productivity and these are largely restricted to high latitudes. Similarly, the penguins of the southern hemisphere inhabit seas with temperatures $< 23°C$. Most penguin species inhabit latitudes between 45°S and 58°S where marine productivity is high (Stonehouse 1967). They reach the equator at the Galapagos Islands off the Pacific coast of South

Table 5.2 The deduced trend in population characteristics of kangaroos from the periphery of the range to its core. (After Caughley *et al.* 1988)

Parameter	*Macropus giganteus*	*Macropus fuliginosus*
Density	Ramp	Ramp
Wellbeing		
Growth rate	Step	Ramp
Adult weight	Step	Ramp
Recruitment rate	Step	Ramp
Kidney fat	Ramp	Step

America, but only because those shores are bathed by the cold Humboldt current.

The northern limit for rabbits (*Oryctolagus cuniculus*) in Australia is marked by the 27°C isotherm. These temperatures coincide with high humidity, and the combination of the two causes resorption of embryos so that the animals cannot breed.

Cold is clearly an important factor limiting species in the Arctic and subarctic. Although the Arctic is an important breeding ground for birds, most leave during winter. Only four North American species can withstand the cold to reside year-round in the Arctic: the raven (*Corvus corax*), the rock ptarmigan (*Lagopus mutus*), the snowy owl (*Nyctea scandiaca*), and the hoary redpoll (*Acanthis hornimanni*) (Lavigne *et al.* 1989). Amphibians and reptiles are particularly affected by temperature. The American alligator (*Alligator mississipiensis*) cannot tolerate temperatures < 5°C. Although several species of amphibians and reptiles tolerate freezing temperatures, in general there is a negative relationship between the number of species and the latitude. The direct effect of cold in limiting the distribution of these groups is probably less important than the availability of hibernation sites remaining above lethal temperatures (Lavigne *et al.* 1989).

Movements of animals can be affected by temperature. In the Rocky Mountains several ungulates such as moose (*Alces alces*), elk, and deer move down hill for the winter. Sometimes a temperature inversion in winter positions a warmer air layer above a colder one, and in these conditions Dall sheep (*Ovis dalli*) in the Yukon climb higher rather than lower.

The limiting effects of temperature are demonstrated by changes in the range of several species during historical times. Temperatures increased in the northern hemisphere from 1880 to 1950. The breeding ranges of herring and black-headed gulls (*Larus argentatus*, *L. ridibundus*) moved north into Iceland, and that of green woodpeckers (*Picus viridis*) extended into Scotland. Temperatures have declined since 1950 and the breeding ranges of snowy owls and ospreys (*Pandion haliaetus*) have moved south (Davis 1986). On the American prairies the warming period was associated with severe droughts in the

1930s. As a result the cotton rat (*Sigmodon hispidus*) has spread north (Davis 1986).

Cold temperatures themselves may be less important than the consequent changes in snow pack. Caribou must expend greater amounts of energy in exposing ground lichens when snow develops a crust (Fancy and White 1985). Even further north on Canada's High Arctic Islands the warming temperatures of spring melt the surface snow. As the water trickles through the snow pack it freezes as it hits the frozen ground and forms an impenetrable layer. The caribou abandon feeding in those areas and may migrate across the sea-ice to areas where the wind has blown the shallow snow away (Miller *et al.* 1982).

Deep snow limits other species also. North American mountain sheep (*Ovis canadensis, O. dalli*) are usually found in winter on cold windswept ledges where there is little snow. Deer (*Odocoileus* species) are limited by snow cover of moderate depths (< 60 cm) whereas moose can walk through meter-deep snow (Kelsall and Prescott 1971). Both move to coniferous forest in late winter because the snow is less deep there (Telfer 1970; Rolley and Keith 1980).

The stress of cold temperature has resulted in various adaptations to conserve energy, the most notable being the hibernation of ground squirrels during winter and the dormancy and lowering of body temperature of bears. Hummingbirds also lower body temperature overnight to about 15°C or, when resting in cold conditions, a state called *torpor*. The limiting effect of temperature on ground squirrels operates indirectly through soil type, slope, and aspect. Squirrels need to dig burrows deep enough to avoid the cold and this requires sandy, friable soil. They also need to avoid being swamped by melt water in spring, so burrows are situated on slopes where water can drain away. Similarly, in Australia, the distribution of rabbits within the 27°C isotherm is influenced by soil type, soil fertility, vegetation cover, and distribution of water (Parer 1987).

5.4.4 *Range limited by water loss and heat stress*

High temperatures are often combined with high solar radiation and restricted water supplies. In high rainfall areas the last factor is important for restricting distribution; in arid regions all three have interrelated effects on animals. These effects are expressed as heat loads build up in the body, and there are various adaptations to overcome them.

Adaptations to high temperatures include behavioral responses such as using shade in the middle of the day and restricting feeding to the hours of darkness. Both eland (*Taurotragus oryx*) and impala (*Aepyceros melampus*) reduce heat stress by feeding at night in East Africa (Taylor 1968a). At the driest times of year both species boost water intake by switching from grazing grasses and forbs to browsing on succulent shrubs (Taylor 1969; Jarman 1973).

Solar radiation restricts the movements of animals that are large and have dark coats. Elephant and buffalo are examples where they seek shade in the heat of the day to cool off (Sinclair 1977). Coat color and structure can reduce heat loads. The lighter tan-colored coat of hartebeest (*Alcelaphus buselaphus*) reflects 42% of short-wave solar radiation as against only 22% for the darker coat of eland. In both species reradiation of long-wave thermal radiation is greater than that absorbed, and this represents 75% of total heat loss (Finch 1972).

High heat loads can be avoided by sweating when water is abundant. African buffalo, eland, and waterbuck use sweating for evaporative cooling (Taylor 1968a; Taylor et al. 1969b). Buffalo keep body temperature in the range 37.4–39.3°C and allow body temperature to rise to 40°C only when water is restricted. They cannot reduce water loss from sweating when water is restricted (Taylor 1970a; 1970b). Waterbuck show similar physiologic adaptations. When water is restricted for 12 h at 40°C ambient (environmental) temperature they lose 12% of their bodyweight compared with the 2% for beisa oryx (*Oryx beisa*) which is a desert-adapted species (Taylor et al. 1969b). As a consequence both buffalo and waterbuck must remain within a day's walk of surface water.

Large animals can afford to lose water by sweating but smaller animals such as the gazelles cannot. They employ panting instead, as do species in arid areas (e.g., the beisa oryx) or those on open plains with high solar radiation, such as wildebeest (Robertshaw and Taylor 1969; Taylor et al. 1969a; Maloiy 1973).

Some species can adapt to extreme arid conditions by allowing their body temperature to rise before they start panting: up to 43°C for Thomson's gazelle (*Gazella thomsonii*) and 46°C for Grant's gazelle (*G. granti*) (Taylor 1972). Other adaptations for water conservation include restriction of urine output, concentrating the urine, and reabsorbing water from the feces. Dikdik, a very small antelope that lives in semiarid scrub away from water, had the lowest fecal water content and the highest urine concentration of all antelopes, followed by hartebeest, impala, and eland (Maloiy 1973).

Grazing ungulates in Africa are restricted to areas within reach of surface water and all show behavioral adaptations such as night feeding or migration (Sinclair 1983). Those that can do without water are all browsers (Western 1975). Beisa oryx and Grant's gazelle select hygroscopic shrubs (*Disperma* species). They eat them at night because these shrubs contain only 1% free water in the day but absorb water from the air at night to boost the water content of the leaves to 43% (Taylor 1968b).

Perhaps not apparent at first sight is the restricted availability of water for wildlife in cold regions. Not only are many of those regions deserts, as their rainfall is low, but during winter the moisture is available only as snow and valuable energy is needed to melt it. Arctic mammals go to some lengths to conserve water. Caribou recycle nitrogen to reduce the formation of urine, thereby conserving water.

5.4.5 *Range limited by day length and seasonality*

The distribution of many North American birds is limited at northern latitudes by season length, the number of days available for breeding above a certain temperature. This is another aspect of temperature limitation. However, the southern boundary is limited by day length, the number of hours available for feeding themselves and the young (Emlen *et al.* 1986; Root 1988).

Seasons are highly predictable in the northern temperate latitudes of North America and Eurasia, and many birds and mammals have evolved a response to *proximate factors* (i.e., the immediate factors affecting an animal), particularly day length (photoperiod), which trigger conception and result in the production of young during optimum conditions. Such conditions are the *ultimate factors* (i.e., the underlying selection pressure) to which an animal is adapted by breeding seasonally (Baker 1938).

Increasing photoperiod determines the start of the breeding season in many bird species (Perrins 1970), while declining photoperiod triggers the rut in caribou; the rut is so synchronized that most conceptions occur in a mere 10-day period starting around the first day of November (Leader-Williams 1988). Moose and elk also have highly synchronized birth seasons (Houston 1982), which suggests photoperiodic control of reproduction.

Only the wildebeest among tropical ungulates is known to respond to photoperiod. In southern Africa it uses solar photoperiod to synchronize conceptions, but near the equator where solar photoperiod varies by only 20 minutes in the year it is cued by a combination of lunar and solar photoperiod (Spinage 1973; Sinclair 1977).

In variable environments with less predictable seasons, as in the tropics and arid regions, animals tend not to use photoperiod to anticipate conditions but rather adjust their reproductive behavior to the current conditions. Thus, tropical birds begin breeding when the rainy season starts, responding to the increase in insect food supply and the growth spurt of the vegetation (Sinclair 1978). In some arid areas such as Western Australia the seasonality of rain is relatively predictable but its location is not. Emus (*Dromaius novaehollandiae*) there travel long distances searching for areas that have received rain (Davies 1976).

Most ungulates produce their young during the wet season in Africa and South America but put on fat prior to giving birth. This fat is then used during lactation, the period when the energy demands on the female are highest (Ojasti 1983; Sinclair 1983). Therefore nutrition in the seasonal tropics becomes both the proximate and ultimate factor determining the timing of births. An example is provided by the lechwe (*Kobus lechee*), an African antelope that lives on seasonally flooded riverine grasslands (Fig. 5.5). During the peak of the floods animals are confined to the less preferred surrounding woodlands. The greatest area of flood plain is exposed at the low point in the flood cycle and it is at this time, corresponding with greatest availability of

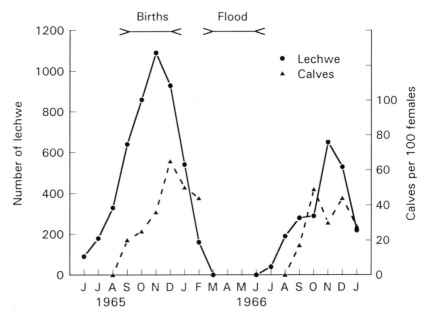

Fig. 5.5 The numbers of lechwe, a flood plains antelope of southern Africa (●), increase on the flood plains as water recedes in the Chobe river exposing the greatest area of high-quality food. The recruitment of newborn per 100 females (▲) shows that births occur at this time. (After Child and von Richter 1969.)

food, that births take place. In Zambia the peak of births occurs in the dry season 3 months after the rains; in the Okavango swamp of Botswana it occurs in the middle of the wet season 9 months after the previous rains; but, both occur when the swamp grasslands are most available.

5.5 **Summary**

The distribution is the area occupied by a population or species, the dispersion is the pattern of spacing of the animals within it; and dispersal, migration, and local movement are the actions that modify dispersion and distribution. Dispersion and distribution are states; dispersal, migration, and local movement are processes. The edge of the distribution is that point at which, on average, an individual just fails to replace itself in the next generation. Its position may be set by climate, substrata, food supply, habitat, predators, or pathogens. The limiting factor can often be identified by the trend in density from the range boundary inward.

6 **Resources and herbivory**

6.1 **Introduction**

We first discuss those resources an animal needs to survive and reproduce, classifying them according to whether they are consumable or nonconsumable and whether they are living or inert. This leads on to a description of the structure and dynamics of plant–herbivore systems where both the animals and their resources, the plants, interact in complex ways. We show how to analyze such systems by breaking them down into their dynamic components.

6.2 **Two views of ecology**

In this chapter we consider the three major influences on the dynamics of an animal population: the dynamics of its limiting resource, the functional response of an animal to the level of the resource, and the numerical response, in terms of rate of increase, to the level of the resource.

An alternative approach is to consider rate of increase as a function of density. Such "density-dependent" models with their strong emphasis on competition are used in Chapters 8–10. It may be noted that the underlying philosophy of that approach is quite different from the one underpinning this chapter. The contrast is best seen in a comparison of this treatment of plant–herbivore systems with the treatment of prey–predator systems in Chapter 10. In fact, the two systems are almost interchangeable as far as their ecological relationships are concerned, but our treatment of them is very different. It reflects a difference of approach between the two authors of this book. Both of us recognize the importance of negative-feedback loops in ecological processes. One of us chooses to express these in terms of density-dependent factors and equilibria, whereas the other stresses the importance of resources in determining the dynamics and centripetality of ecological systems. That is our choice. It is not to be taken as an indication that we necessarily view nature differently. More likely, we just use different words to describe it.

This particular dichotomy in ecological thinking has been with us for 40 years and is unlikely to go away soon. It need not get in the way of good research and sound management. We think it imparts a richness and diversity to the field and so wish to share it with our readers.

6.3 **Definitions**

A *resource* is defined as something that an animal needs. The most obvious example is food, and to that may be added shelter, water, nesting sites, and a particular range of temperature. By definition a resource is beneficial. As the availability of resources rises, the fecundity and probability of survival of an individual is enhanced.

Some components of the environment are resources at one level but become harmful when they are in superabundance, at which level they are no longer resources but *malentities*. Heat is a clear example. It is possible to have too much of a good thing.

A *habitat* comprises all those physical attributes of the environment that make an area habitable for a species. It is not in itself a specific resource but the sum of all physical resources for that species. "Habitat" is what wildlife managers see when they declare an area suitable for a particular species. They might not be able to explain logically what it is that they are seeing because they are summing up an amalgam of all the species' physical resources. One of us is quite good at picking red kangaroo habitat from the air. He is not certain how he judges whether that species will be present or absent but apparently it involves density and spacing of trees, color of exposed soil, color of the grass, and a general impression of the lie of the land. All naturalists understand this and share the frustration of not being able to list precisely what clues the eyes have picked up.

Since habitat is the sum of a species' resources it encapsulates a general idea and, being general, is not useful when we come down to specifics. Then we must consider individual resources, and sometimes their interaction, to determine the effect on the population of changing the level of one or more resources.

6.4 Quality and quantity of a resource

Often a resource such as food is described by two attributes: the amount of food available to an animal and the suitability of that food to the animal's requirements. For example, quality may be described as the percentage of digestible protein in the food whereas quantity may be measured as dry weight of standing food per hectare. There is often then a discussion on whether quality or quantity of the food is the most important to the animal.

In most cases that question is meaningless. It indicates that the resource is being measured in the wrong units. If the resource is in fact digestible protein, then that is what should be measured. The availability of the resource should be expressed as dry weight of digestible protein per hectare. Its measurement may entail measuring dry weight of herbage as an intermediate step, but that does not make herbage the resource.

6.5 Kinds of resources

It is necessary at this stage to give a classification of resources because the interaction between the resource and the animals that depend upon them can take several forms. These in turn influence the dynamics of the population in different ways. The relationship between the use of resources and the growth of the population is outlined in Section 4.6.

A resource may be used by an animal in a way that does not subsequently reduce the level of that resource available to itself or other individuals. This is a *nonconsumable resource*. A reptile needs a certain level of ambient temperature before it is active enough to seek food and to digest that food. Its use of that heat has no effect on the heat available to other reptiles, neither then nor subsequently. The alternative is a *consumable resource*, whose level is reduced by the individual utilizing it. Food provides the most obvious example.

The use of a resource may be *preemptive*. An example is the use of nesting holes by parrots. Individuals are either winners or losers. On the other hand the use of a resource may be *consumptive*: all individuals have access to the resource and each individual's use of it reduces the level of the resource available to others. An example is the use of plants by herbivores. Note that both preemptive and consumptive use of a resource removes a component of the resource from use by other individuals. Consumptive use removes the component permanently whereas preemptive use removes it temporarily. Preemptive use is equivalent to accepting a loan.

To complete the classification, there may be an *interactive relationship* between the population and the resource in that the level of the resource influences the rate of increase of the population, and reciprocally the level of the population's density influences the rate of increase of the resource. The dynamics of the animals interact with the dynamics of the resource, i.e., the relationship between a herbivore and its food supply and between a predator and its prey resource. In a *reactive relationship*, however, the rate of increase of the animal population reacts to the level of the resource (as before) but the density of the animals has no reciprocal influence on the rate of renewal of the resource. The relationship between a scavenger and its food supply or between a herbivore and salt licks are examples of reactive relationships.

6.6 Dynamics of the resource

A resource may be inert, as with suitable soil for rodents to burrow into, or it may be growing. If the latter, the intrinsic pattern of that growth will profoundly affect the dynamics and the density of the population utilizing it. In general, the higher the rate of renewal of the resource the higher will be the average density of the animal population using it.

The dynamics of a renewable resource may be quite complicated, containing elements of seasonality, intrinsic growth pattern, and the modification of those two by the animals using the resource. We provide a simple example: the growth of the herbage layer utilized by kangaroos in the arid zone.

Figure 6.1 shows Robertson's (1987) estimate of the *plant growth response*, the growth of the ungrazed plants of the herb layer in response to rainfall. He sampled growth rates on a kilometer grid over 440 km^2 of the arid zone of Australia. The measurements were

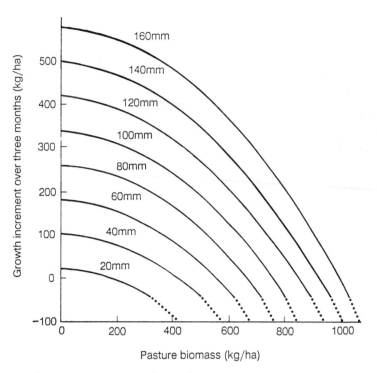

Fig. 6.1 Plant growth over 3 months as a function of plant biomass at the beginning of the interval and rainfall during the interval, for pastures on Kinchega National Park. (After Robertson 1987.)

repeated every 3 months for 3.5 years and rainfall was recorded for each 3-month interval. Look at the curve labeled 100 mm. It indicates that the higher the biomass at the start of the 3-month period the lower is the increment of further biomass added over the next 3 months. That is to be expected because plants compete for space, water, light, and nutrients. The 100-mm curve is just one of a whole family of curves each representing that trend for a given rainfall over 3 months. We can summarize the figure by saying that the higher the rainfall the higher the growth increment but, for a given rainfall, the higher the starting biomass the lower the growth increment. Hence growth over 3 months is the resultant of the rainfall over the 3 months and the biomass at the beginning of the 3-month period.

Figure 6.1 is a graphic representation of a regression analysis which estimated the relationship between growth increment in kilograms per hectare over 3 months (Y) on the one hand and starting biomass (X) and rainfall in millimeters (R) on the other:

$$Y = -55.12 - 0.01535X - 0.00056X^2 + 3.946R$$

Suppose we wished to know what might be the threshold combinations of rainfall and starting biomass above which the vegetation would grow and below which it would die back. That can be determined easily by setting Y to zero in the growth equation. It yields:

$$R = 14 + 0.004X + 0.00014X^2$$

which is called the isocline of zero growth (Fig. 6.2).

6.7 The functional response of consumer to resource

Having established how fast the resource grows in the absence of grazing and browsing we would then wish to know what happens to it when a herbivore is present. The amount a herbivore eats per day is a constant only when it is faced by an *ad libitum* supply. Herbivores are seldom so lucky. The trend of daily intake against food availability is therefore curved, being zero when the level of food is zero and rising with increasing food to a plateau of intake. From there on no increase in food supply has any effect on daily intake because the animal is already taking in a satiating diet. Such a curve is called a *functional response* or *feeding response*, the trend of intake per individual against the level of the resource. It can be represented symbolically by an equation such as:

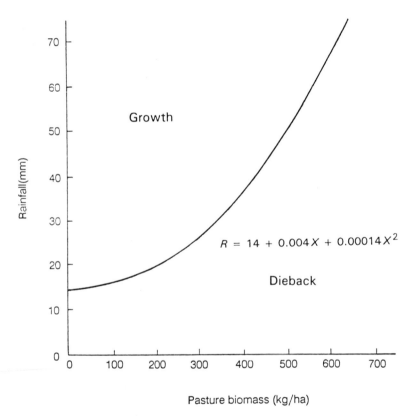

Fig. 6.2 The isocline of zero growth for pasture on Kinchega National Park in terms of rainfall (R) over 3 months and pasture biomass (X) at the beginning of those 3 months. (After Robertson 1987.)

$$I = c(1 - \exp(-bV))$$

where:

I = consumption per day,

c = the maximum (satiating) daily intake,

V = the level of the resource,

b = the slope of the curve, a measure of grazing efficiency.

The last has another meaning. Its reciprocal $1/b$ is the level of the resource V at which 0.63 (i.e., $1 - e^{-1}$) of the satiating intake is consumed.

Figure 6.3 shows the daily dry-weight intake in g/day (I) of a red kangaroo and a rabbit at various levels of pasture biomass when both species are grazing annual grasses and forbs interspersed with scattered shrubs (Short 1987). The equation for a 35-kg kangaroo is:

$$I = 892(1 - \exp(-0.012V))$$

and for a 2-kg rabbit:

$$I = 102(1 - \exp(-0.007V))$$

The grazing efficiency of kangaroos (0.012) is higher than that for rabbits (0.007), but not greatly so. As would be expected from the size disparity between the two species, satiating intake is quite different: 892 g/day for kangaroos as against 102 g/day for rabbits. To compare them in any meaningful sense we estimate what the satiating intake

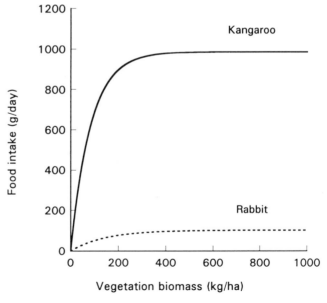

Fig. 6.3 Food intake per individual per day for kangaroos and rabbits at varying levels of food availability. (After Short 1987.)

would be if both animals weighed 1 kg. That cannot be achieved simply by dividing the satiating intake by bodyweight (W) because intake scales as metabolic rate, which varies as the 0.75 power of weight. Hence the appropriate metric is not W but $W^{0.75}$ (which is called *metabolic weight*, see Sections 7.5.2 and 7.8), where W is here expressed as bodyweight in kilograms. On the metabolic weight basis ($g/W^{0.75}$ per day) the satiating intake is 62 g/day for kangaroos and 68 g/day for rabbits, i.e., remarkably similar. Figure 6.4 shows the two functional responses of Fig. 6.3 when expressed in terms of metabolic weight. Both species achieve their satiating intake when pasture biomass exceeds about 300 kg/ha. This value is important. It tells us that the two species will compete for food only when pasture biomass falls below that level.

Short (1987) estimated these two functional responses by allowing high densities of kangaroos and rabbits to graze down pasture in enclosures, the offtake per day being estimated as the difference between successive daily estimates of vegetation biomass corrected for trampling. Daily intake could be estimated for progressively lower levels of standing biomass because the vegetation was progressively defoliated during the experiment.

Although the functional response has been discussed here in the context of a plant–herbivore system, all of that discussion carries over to prey–predator systems. They are exactly analogous. The only difference lies in the difficulty of measuring a predator's food intake. An

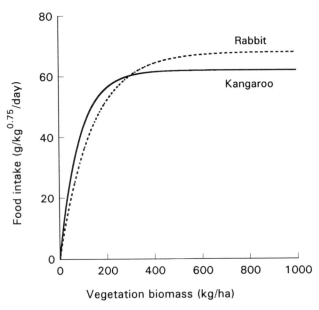

Fig. 6.4 Food intake per individual per day, scaled for body size, at varying levels of food availability. (After Short 1987.)

ability to measure intake by way of radioactive tracers has greatly simplified that problem. A good example is Green's (1978) use of radio-sodium to estimate how much meat a dingo eats in a day.

6.8 The numerical response of consumer to resource

The functional response gives the effect of the animal upon a consumable resource. In contrast, the *numerical response* gives the effect of the resource on the animals. If the resource is used in a preemptive rather than a consumptive way (e.g., remember the example of nesting holes for parrots), the numerical response is best portrayed as the trend of density of the animals against the level of the resource (e.g., nesting holes per hectare). However, if the animals' use of the resource is consumptive there is no necessarily stable relationship between the level of a resource and the density of the animals because the animals themselves change the level of the resource. In this case the relationship between the animals and the resource is best portrayed as the trend of the population's instantaneous rate of increase against the level of the resource. That relationship will be stable except in special circumstances.

Figure 6.5 shows the numerical response relationship between rate of increase of red kangaroos and the biomass of pasture. Bayliss (1987) estimated rates of increase from successive aerial surveys, and pasture biomass from ground surveys. As with the functional response the numerical response has an asymptote: there is an upper limit to how fast a population can increase and no extra ration of a resource will

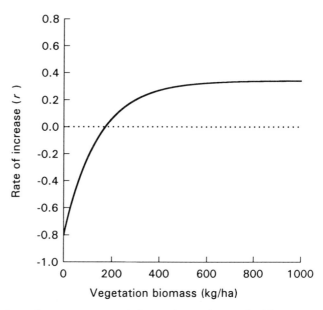

Fig. 6.5 Rate of increase on a yearly basis of a population of red kangaroos at varying levels of food availability. (After Bayliss 1987.)

force that rate higher. The numerical response differs from the functional response in that negative values are both possible and logically necessary. If not, the population would increase to infinity.

The numerical response can usually be described by an equation of the form:

$$r = -a + c(1 - \exp(-dV))$$

where:

r = the exponential rate of increase of the animals,
a = the maximum rate of decrease,
c = the maximum extent to which that rate of decrease can be alleviated.

Hence, $c - a = r_m$ is the maximum rate of increase. Demographic efficiency, the ability of the population to increase when resources are in short supply, is indexed by d. For the present example the constants were solved (Bayliss 1987, modified by Caughley 1987) as:

$$r = -1.6 + 2.0(1 - \exp(-0.007V)).$$

The intrinsic rate of increase on a yearly basis is therefore estimated as $r_m = 2.0 - 1.6 = 0.4$, and so the population's maximum finite rate of increase over a year is $\exp(0.4) = 1.49$, a 49% increase.

6.9 A plant–herbivore system

So far we have taken a plant–herbivore system and dissected it into its component processes: the growth of the plants that comprise the limiting resource of the herbivore; the functional response (i.e., the feeding response) of the herbivore to the level of that resource; and the numerical response of the herbivore, in terms of its rate of increase, to the biomass of the plants.

The evaluation of these component influences upon a population's dynamics provides two bonuses. First, they furnish a tight summary of the dynamic ecology of the system. Second, they furnish that summary in terms of causal relationships rather than correlations. What follows is a short summary of the statistics of an arid-zone plant–herbivore system described in detail by Caughley (1987).

Rainfall

	Mean (mm)	SD (mm)
December–February	62	59
March–May	57	47
June–August	59	34
September–November	61	44
Annual	239	107

These figures summarize 100 years of weather. There was no significant correlation of rainfall from one quarter to the next, nor between consecutive years.

Plant growth response

$$\Delta V = -55.12 - 0.01535V - 0.00056V^2 + 2.5R$$

where:

ΔV = the growth increment to ungrazed standing biomass in kilograms per hectare over 3 months,

V = the standing biomass in kilograms per hectare at the beginning of those 3 months,

R = rainfall in millimeters over those 3 months.

The 2.5 coefficient of R used here differs from Robertson's (1987) 3.946 for reasons given by Caughley (1987).

Feeding response

$$I = 86(1 - \exp(-0.029V))$$

where I = intake of food in kilograms dry-weight over 3 months, per red kangaroo, assuming a mean bodyweight of 35 kg and no shrubs in the pasture layer (Short 1987).

Numerical response

$$r = -0.4 + 0.5(1 - \exp(-0.007V))$$

where r = exponential rate of increase of red kangaroos on a 3-monthly basis. It is the same equation as that in Section 6.8 except that there r is on a yearly basis and here that equation is divided through by 4.

The numerical response of the herbivore allows us to calculate the equilibrium level to which plant biomass will converge in a constant environment under the influence of an unrestrained population of herbivores. It is the x-intercept of the regression of rate of increase of the herbivores against plant biomass or, put another way, the plant biomass at which rate of increase of the herbivore is zero (see Fig. 6.5). In the absence of seasonality, and of year-to-year variation in rainfall and temperature, this will be the equilibrium plant biomass imposed by grazing. The numerical response curve of this example was fitted as:

$$r = -a + c(1 - \exp(-dV))$$

and the level of plant biomass V at which $r = 0$ is solved simply by setting r to zero and solving for V. Thus:

$$V = (1/d) \log_e(c/(c - a))$$

which, when loaded with the values of the constants given here on a 3-monthly basis, yields:

$$V = (1/0.007) \log_e(0.5/(0.5 - 0.4))$$
$$= 230 \text{ kg/ha dry-weight plant biomass.}$$

That value is immensely important ecologically. It is the equilibrium level of plant biomass imposed by grazing in a constant environment. This is of some theoretical interest but of little practical importance because environments are not constant. But it is also the level of plant biomass above which the herbivore population will increase and below which it will decrease (the *critical threshold*), and that is true whether the environment is constant or variable and whether the density of herbivores is high or low.

We will now reassemble the response functions of the system in their proper relationships, quantify the system's driving variable – the rainfall – and see what these in combination reveal about the system's dynamic behavior.

Population trends can be built with these equations. Rainfall is simulated as a sequence of random draws from the mean and standard deviation of the seasonal rainfall given above, and the consequent changes in plant biomass and kangaroo numbers can be calculated accordingly. Initial conditions are not highly influential: the system remembers previous plant biomass for only 3 years but the memory of kangaroo density can linger for 10 years.

Figure 6.6 shows a typical time trend for plants and kangaroos as generated by the equations describing the unpredictable rainfall and the responses to it of the plants and herbivores. The only external input other than starting conditions are the random draws from the 3-monthly rainfall distributions whose observed means and standard deviations are given above. The trajectories of pasture biomass, rate of increase of kangaroos, and kangaroo numbers are a mathematical consequence of that rainfall as its effect feeds through to plant growth, herbivore population growth, and grazing pressure.

The rainfall of this region takes the form of high-amplitude, high-frequency fluctuations. The herb layer, whether grazed or ungrazed, generates a similar trace of high-amplitude, high-frequency fluctuations as it reacts speedily to rainfall or the lack thereof. The fluctuations are paralleled by similar but more constrained fluctuations in the kangaroos' rate of increase as the population reacts dynamically to variations in food supply. Those rates are truncated above, reflecting the strict physiologic constraints on the maximum rate at which a population can increase, contrasted with the lack of any equivalent constraint on precipitous decline. The trend of kangaroo density differs from the other three, comprising fluctuations of high amplitude but low frequency. Perhaps that might have been predictable from first principles: present density is an integration of past rates of increase, not of present conditions.

Figure 6.7 shows the rates of increase given in Fig. 6.6, depicted first against kangaroo density and then against the level of their limiting resource. It makes the point that, in this fluctuating and unpredictable environment, the rate of increase of the population is seldom predicted by its density. On the other hand there is a close relationship

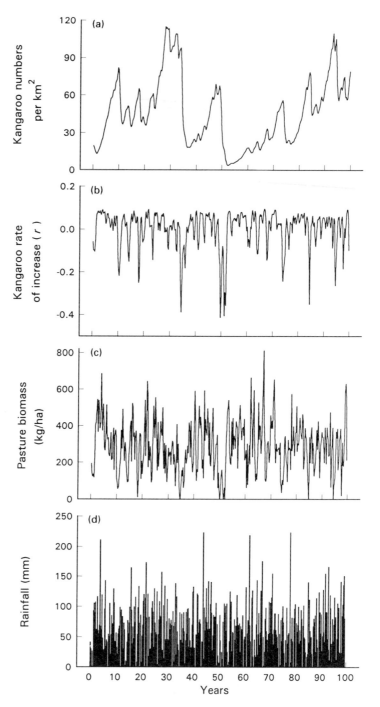

Fig. 6.6 (a) The trend of kangaroo numbers generated by the run of rates of increase in (b). (b) Rate of increase of red kangaroos, on a 3-monthly basis, as it reacts to plant biomass. (c) Biomass of grasses and forbs that would be generated by that rainfall and reduced by the grazing of kangaroos. (d) Three-monthly random draws from an Australian arid-zone climate. (After Caughley and Gunn 1993.)

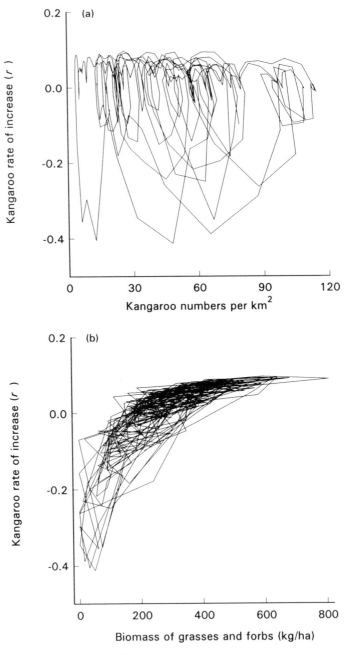

Fig. 6.7 Rate of increase of red kangaroos given every 3 months for the 100-year simulation presented in Fig. 6.6. The trends are graphed (a) against kangaroo density and (b) against the biomass density of grasses and forbs, showing that in this modeled system the latter is a tighter predictor of herbivore rate of increase. (After Caughley and Gunn 1993.)

between the rate of increase and the level of the limiting resource, the graph tracing out a fuzzy version of the kangaroos' numerical response.

6.10 **Summary**

A resource is something an animal needs. Its effect on the animal population depends on whether it is consumed by the animals or simply utilized. Herbivores and their plant resources form a system in which the rate of increase of the plants is determined by the density of the animals eating them, and the rate of increase of the animals is determined by the density of the plants. Such a complex system can be studied only by breaking it down to its dynamic components, of which two dominate. First, there is the functional response of the animal, the rate of intake of plant material by a single herbivore as a function of the standing crop of vegetation. Second, there is the numerical response of the animal, the rate at which its population increases as a function again of the standing crop of vegetation. With these two functions solved, and with supplementary information on growth rate of the plants, the full dynamic behavior of the system can be described and its behavior managed.

7 Food and feeding

The three main areas of wildlife management (conservation, sustained yield, and control) require a knowledge of the food and nutrition of animal populations. Some of the important questions are as follows.

1 Is there enough food to support and conserve an endangered species?

2 What food supply is needed to support a particular sustained yield?

3 Can we alter the food supply to control pest populations more effectively?

The field of animal nutrition covers subjects such as anatomy, physiology, and ecology, and there are several good reviews of these areas, e.g., Hofmann (1973) deals with the anatomy of ruminants, while Robbins (1983) addresses the physiology of wildlife nutrition. From the point of view of wildlife management, however, we are interested in two main types of information to answer the above questions: we need to know the availability of the food and the requirements of the animals. By matching the two sets of information we can answer the questions. Sections 7.2–7.4 deal with availability, and Sections 7.5–7.9 address animal requirements.

Energy is measured in units of calories or joules (1 cal = 4.184 J). The energy content of foods can be found by oxidizing a sample in a bomb calorimeter. Differences in the energy content of different plant and animal materials are due to the differences in their constituents. The energy content of some of the common components of food is given in Table 7.1. We can see that fats and oils have the highest content (>9 kcal/g), with proteins coming next (around 5 kcal/g), and sugars and starches (carbohydrates) close to 4 kcal/g. The gross energy of tissues depends on the combination of these basic constituents, particularly in animals. In plant tissues, energy content remains relatively uniform, in the region of 4.0–4.2 kcal/g. Plant parts with a high oil content such as seeds (>5 kcal/g), and evergreen plants with waxes and resins such as conifers and alpine plants (4.7 kcal/g), are the exceptions (Golley 1961; Robbins 1983).

Energy flow through animals can be measured with isotopes of hydrogen (^{3}H) and oxygen (^{18}O) by the *doubly labeled water method* (Nagy 1983; Bryant 1989). First, water labeled with ^{3}H and ^{18}O is injected and allowed to equilibrate in the animal, taking 2–8 hours depending on body size. A blood sample is then collected to establish

Table 7.1 Approximate energy content of food components. (After Robbins 1983)

Food component	Energy (kcal/g)
Fat	9.45
Protein	5.65
Starch	4.23
Cellulose	4.18
Sucrose	3.96
Urea	2.53
Leaves	4.23
Stems	4.27
Seeds	5.07

the starting concentrations of the two isotopes. Analysis of ^{3}H is done by liquid scintillation spectrophotometry and of ^{18}O by proton activation of ^{18}O to ^{18}F (the isotope of fluoride) with subsequent counting of γ-emitting ^{18}F in a γ-counter. A second blood sample is collected several days later. The timing of the second collection does not need to be exact but should be when approximately half of the isotope has been flushed from the body. Thus, timing depends on body size and the flow rates of the isotopes. Oxygen leaves the body via carbon dioxide and water, and this rate is measured by dilution of the ^{18}O. The rate of water loss is measured from the dilution of ^{3}H. Thus the difference between the total oxygen loss and the oxygen loss in water gives the rate of carbon dioxide production, which is a measure of energy expenditure. The method and its validation have been described by Nagy (1980; 1989).

7.2.2 Protein

Protein is a term covering a varied group of high-molecular-weight compounds: these are major components in cell walls, enzymes, hormones, and lipoproteins. They are made up of about 25 amino acids that are linked together through nitrogen–carbon peptide bonds. Most animal species have a relatively similar gross composition of amino acids.

Animals with simple stomachs require 10 *essential amino acids*, these being the forms that cannot be synthesized by the animal and must be obtained in the diet: arginine, histidine, isoleucine, leucine, threonine, lysine, methionine, phenylalanine, tryptophan, and valine. *Nonessential amino acids*, therefore, are those that can be synthesized in the body. Ruminants, and other species that rely on fermentation through the use of microorganisms, synthesize many of the amino acids themselves and so have a shorter list of essential amino acids.

Although there is some variability in the nitrogen content of amino acids (ranging from 8% to 19%), the average is 16%. Thus in analyzing tissues for *crude protein*, the proportion composed of nitrogen is multiplied by the constant 6.25 (i.e., 100/16). The crude protein content of plant material tends to vary inversely with the proportion of fiber. Since one of the major constituents of fiber is the indigestible compound lignin, fiber content can be used as an index of the nutritive value of the plant food. In many plant tissues such as leaves and stems, protein and digestible energy content (i.e., the nonfiber component) tend to vary together. However, some plant parts such as seeds are high in energy but quite low in protein.

7.2.3 Water

The water content of birds and mammals is a function of bodyweight (W) to the power of 0.98 when comparing across species, but more restricted groups vary in the exponent. Robbins (1983) found that the water content of white-tailed deer and several rodents varied as a function of $W^{0.9}$.

Water is obtained from three sources:

1 *free water* from external sources such as streams and ponds;
2 *preformed water* found in the food;
3 *metabolic water* produced in the body from the oxidation of organic compounds.

Preformed water is high in animal tissues such as muscle (72%) and in succulent plants, roots, and tubers. Because of this carnivores may not have to drink often; and herbivores such as the desert-adapted antelope, and the beisa oryx, which eat fleshy leaves and dig up roots, can also live without free water (Taylor 1969; Root 1972).

The highest rate of production of metabolic water in animals is from the oxidation (catabolism) of proteins because of the initially high water content of these tissues. Catabolism of fats produces 107% of the original fat weight as water, but the low preformed water content (3–7%) means that the absolute amount produced is less than that from protein (Robbins 1983).

Measures of free water intake from drinking underestimate total water turnover. More accurate methods use the ^3H or deuterium oxide isotopes of water. A known sample of isotopic water is injected into an animal and, after a period of 2–8 hours for equilibration (depending on size of animal), a blood sample is collected. The concentration of isotope in the blood is then measured using a liquid scintillation spectrometer. A second blood sample is collected a few days to a few weeks later (again depending on body size) to obtain a new value of isotope concentration. Because water is lost through feces, urine, and evaporation, the isotope is diluted by incoming water. Therefore, the rate of dilution is a measure of water turnover. These techniques have been described by Nagy and Peterson (1988) and used on a wide range of animals including eutherian mammals, marsupials, birds, reptiles, and fishes.

7.2.4 Minerals

Minerals make up only 5% of body composition but are essential to body function. Some minerals (roughly in order of abundance: calcium, phosphorus, potassium, sodium, magnesium, chlorine, sulfur) are present or required in relatively large amounts (mg/g) and are called *macroelements*. Those that are required in small amounts (μg/g) are called *trace elements* (iron, zinc, manganese, copper, molybdenum, iodine, selenium, cobalt, fluoride, chromium). So far very little is known about the mineral requirements for wildlife species, but Robbins (1983) has provided a summary of available information.

Calcium and phosphorus are essential for bones and eggshells. Cervids have a very high demand for these minerals during antler growth. Calcium is also needed during lactation, for blood clotting, and for muscle contraction. Phosphorus is present in most organic compounds. Deficiencies of calcium result in osteoporosis, rickets, hemorrhaging, thin eggshells, and reduced feather growth. Carnivores that normally eat flesh of large mammals need to chew bone to obtain their calcium. Mundy and Ledger (1976) found that the chicks of Cape vultures (*Gyps coprotheres*) in South Africa developed rickets when they were unable to eat small bone fragments. This has an important management consequence: bone fragments from large carcases are made available to vultures by large carnivores, in this case lions and hyenas. Where carnivores have been exterminated on ranch land, carcases were not dismembered and bones were too large for the chicks to swallow. This is a good example of how the interaction of species should be considered in the management and conservation of habitats.

Sodium is required for the regulation of body fluids, muscle contraction, and nerve impulse transmission. Sodium is usually present in low concentrations in plants, so herbivores face a potential sodium deficiency. In areas of low sodium availability, herbivores consume soil or water from mineral licks (Weir 1972; Fraser and Reardon 1980). Sodium is easily obtained by carnivores from their food. They are unlikely to experience sodium deficiency. Isotopic sodium has been used as a measure of food intake rates of carnivores such as lions (Green *et al.* 1984), seals (Tedman and Green 1987), crocodiles (Grigg *et al.* 1986), and birds (Green and Brothers 1989). This approach is possible because sodium remains at a relatively constant concentration in the food supply. The technique is similar to that for isotopic water described in Section 7.2.3.

Both potassium and magnesium are abundant in plants, and deficiencies in free-living wildlife are therefore unlikely. The same is true for chloride ions and sulfur. Trace element deficiencies are unusual under normal free-ranging conditions but they occur locally from low concentrations in the soil: there are some reports of iodine and copper deficiencies and of toxicity from too much copper and selenium (Robbins 1983).

7.2.5 *Vitamins*

Vitamins are essential organic compounds that occur in food in minute amounts and cannot normally be synthesized by animals. There are two types of vitamins: fat soluble (vitamins A, D, E, K) and water soluble (vitamin B complex, C, and several others). Fat-soluble vitamins can be stored in the body. Water-soluble vitamins cannot be stored and hence must be constantly available. Overdose toxicities can arise only from the fat-soluble vitamins.

Vitamin A, a major constituent of visual pigments, can be obtained from β-carotene in plants. Vitamin D is needed for calcium transport and the prevention of rickets. Vitamin E is an antioxidant needed in many metabolic pathways. It is high in green plants and seeds, but decreases as the plants mature. Vitamin K is needed to make proteins for blood clotting. Deficiencies are unlikely to occur because it is common in all foods. The vitamin K antagonist, warfarin, causes hemorrhaging. It is used as a rodenticide.

Little is known about the B-complex vitamins and whether deficiencies occur in free-living wildlife species, although cases of thiamin (B$_1$) deficiency have been reported for captive animals which can be observed more closely (Robbins 1983). Vitamin C differs from the others in that most species can synthesize it in either the kidneys or liver. Exceptions include primates, bats, guinea pigs, and possibly whales. Vitamin C is not as commonly available as the B-complex vitamins but is found in green plants and fruit. It is absent in seeds, bacteria, and protozoa.

7.3 **Variation in food supply**

7.3.1 *Seasonality*

Food supply varies with season. To some degree all environments are seasonal, including those of the tropics. Food supply is greatest for herbivores when plants are growing, during the summer at higher latitudes (temperate and polar regions), and during the rainy season in lower latitudes (tropics and subtropics). The protein content in grass and leaves declines from high levels of 15–20% in young growth to as little as 3% in mature flowering grass, or even 2% in dry senescent grass. Leaves from mature dicots maintain a higher protein content of about 10%. Thus, herbivores such as elk in North America and eland and elephant in Africa will switch from grazing in the growing season to browsing in the nongrowing season.

Animals adjust their breeding patterns so that their highest physiologic demands for energy and protein occur during the growing season. Thus northern ungulates give birth in spring so that lactation can occur during the growing period of plants, whereas tropical ungulates produce their young during or following the rains allowing the mother to build up fat supplies to support lactation (Spinage 1973). Although most birds complete their entire breeding cycle during one season, the timing of breeding is closely associated with food supply (Perrins 1970). Very large birds such as ostrich behave like ungulates and start their reproductive cycle in the previous wet season so that the precocial chicks hatch at the start of the next wet season (Sinclair 1978).

Some carnivores also adapt their breeding to coincide with maximum food supply. Schaller (1972) records that lions have their cubs on the Serengeti plains of Tanzania when the migrant wildebeest are giving birth. In the same area birds of prey have their young coinciding with the appearance of other juvenile birds and small mammals that form their prey (Sinclair 1978).

7.3.2 Year-to-year variation in food supply

A particular kind of variability in food supply occurs with the production of prolific seed crops by some tree species. This seed is termed *mast*. It occurs when the majority of trees in a region synchronize their seed production. Beech trees (*Fagus, Nothofagus*) and many northern hemisphere conifers (e.g., white spruce, *Picea glauca*) produce their seeds at the same time, these mast years occurring every 5–10 years. Birds that depend on these conifer seeds, e.g., the crossbill (*Loxia curvirostra*), breed throughout the winter when a mast cone crop occurs. In the following year, when few cones are produced, the crossbills disperse to find regions with a new mast crop, sometimes traveling many hundreds of kilometers (Newton 1972).

Red squirrels (*Tamiasciurus hudsonicus*) also respond to cone masts in white spruce. This species caches unopened cones in food tunnels in the ground and uses them throughout the next winter. The survival rate of squirrels is high during these mast winters.

An unusual form of variability in food supply occurs in the bamboo species that form the main food of the giant panda (*Ailuropoda melanoleucus*). The bamboo synchronized flowering in much of southern China during the early 1980s (Schaller *et al.* 1985). The plants died after flowering and there was little food available for a few years. With the giant panda now confined to a few protected areas, the population suffered from this sudden drop in food supply. Knowledge of such events is important for conservation. It tells us that reserves must be sufficiently diverse in environment, habitat, and food species to avoid the type of restriction in food supply produced by the synchronous flowering of bamboo. Presumably in prehistoric times giant pandas were able to range over a much wider area and so take refuge in regions where bamboo was not flowering. They cannot now move in this way and most of their former range in the lowlands is no longer available.

In the Canadian boreal forest, lynx and great horned owls breed prolifically during the peak of the 10-year snowshoe hare cycle, and cease breeding during the low phase.

7.3.3 Plant secondary compounds

Many plants produce chemicals that deter herbivores from feeding on them. These chemicals are called *secondary compounds*. Their production is associated with growth stage, but this association differs between plant species. Although secondary compounds are found in some grasses (monocots), most are found in dicots. Tannins are low in young oak leaves but abundant in mature leaves (Feeny and Bostock 1968). Conversely, various secondary compounds are abundant in the juvenile twigs of willows, birches, and white spruce in Alaska and Canada, but sparse in mature twigs 3 years and older (Bryant and

Kuropat 1980). Thus, the palatability and availability of food for herbivores differs between seasons and between years because of changes in the concentration of secondary compounds.

There are three major classes of secondary compounds: terpenes; soluble phenol compounds; alkaloids, cardenolides, and other compounds.

Terpenes

These are cyclic compounds of low molecular weight and usually with one to three rings. They inhibit activity of rumen bacteria (Schwartz *et al.* 1980) and are bitter tasting or volatile. Examples are essential oils from citrus fruits, carotene, eucalyptol from eucalyptus, papyriferic acid in paper birch, and camphor from white spruce. Camphor and papyriferic acid act as antifeedants to snowshoe hares (Bryant 1981; Sinclair *et al.* 1988), and α-pinene from ponderosa pine deters tassle-eared squirrels (*Sciurus alberti*) (Farentinos *et al.* 1981).

Soluble phenol compounds

The main groups of chemicals are the hydrolyzable and condensed tannins (McLeod 1974). They act by binding to proteins and thus making them indigestible. The name "tannin" comes from the action of polyphenols on animal skins, turning them into leather that is not subject to attack by other organisms, a process called tanning.

Tannins are widespread amongst plant species, occurring in 87% of evergreen woody plants, 79% of deciduous woody species, 17% of annual herbs, and 14% of perennial herbs. Tannins have negative physiologic effects on elk (Mould and Robbins 1982), and may determine food selection by browsing ungulates in southern Africa (Owen-Smith and Cooper 1987; Cooper *et al.* 1988) and by snowshoe hares in North America (Sinclair and Smith 1984). Domestic goats (*Capra hircus*) learn to avoid young twigs of blackbrush (*Coleogyne ramosissima*) because of condensed tannins (Provenza *et al.* 1990).

Alkaloids, cardenolides, and other compounds

These are cyclic compounds with nitrogen atoms in the ring. They occur in 7% of flowering plants, and some 4000 compounds are known (Robbins 1983). Some alkaloids are nicotine, morphine, and atropine. They have several physiologic effects but they act more as toxicants or poisons than as digestion inhibitors. Some alkaloids, such as cardenolides in milkweed (Asclepiadaceae), are sequestered by insects like the monarch butterfly (*Danaus plexippus*) whose larvae feed on milkweed. These noxious cardenolides act as emetics to birds. Young, inexperienced blue jays (*Cyanocitta cristata*) at first eat these insects, then regurgitate them, and thereafter avoid them (Brower 1984). Cyanogenic glycosides, which release hydrocyanic acid on hydrolysis in the stomach, are sequestered by *Heliconius* butterflies from their passionflower (*Passiflora* species) food plants. These insects are avoided by lizards, tanagers, and flycatchers (Brower 1984).

7.4 **Measurement of food supply**

7.4.1 *Direct measures*

The amount of food available to animals may be measured directly. For carnivores, some form of sampling of their food may be used: insect traps for insectivores; counts of ungulates available to large carnivores. For grazing ungulates one could obtain clippings of grass from exclosure plots to measure the available production as was done for Thomson's gazelle on the Serengeti plains (McNaughton 1976). Winter food supply for snowshoe hares was estimated from the abundance of twigs with a diameter of 5 mm on the two most common food plants, grey willow (*Salix glauca*) and bog birch (*Betula glandulosa*) (Smith *et al.* 1988). Pease *et al.* (1979) used a different approach by feeding a known quantity of large branches to hares in pens and measuring the amount eaten from these branches. Using this measure as the edible fraction from large branches, they then estimated the total available biomass of edible twigs from the density of large branches in the habitat of hares.

The most serious problem with direct measures is that they all depend on the assumption that we can measure food in the same way that the animal comes across it. It is rare that this assumption is valid: insects that enter pitfall traps or are collected by sweepnets are not the same as those that a shrew or bird sees; ungulate censuses do not indicate which animals are actually available to carnivores, for we can be sure that not all are catchable.

If the food supply is relatively uncomplicated, such as the short green sward which is grazed uniformly by African plains antelopes, then we can clip grass in a way resembling the feeding of animals. However, with woody plants we cannot measure food in the same way as an animal feeds. Thus, in most cases our estimates are simply crude indices of food abundance. Our errors can both over- and underestimate the true availability of food: we may include material that an animal would not eat, so producing an overestimate; or we may overlook food items because animals are better at searching for their own food than we are, so producing an underestimate. We can never be sure on what side of the true value our index lies unless we calibrate it with another method.

7.4.2 *Fecal protein and diet protein*

A second method, which has been applied so far only to herbivores, allows the animal to choose its own food and so avoids the problems discussed above. Diet protein, energy, or other nutrients can be estimated by observing what animals eat and then determining the chemical composition of that diet. These indirect estimates of intake are compared with an estimate of requirements either from direct physiologic experiments or inferred from the literature. Examples are from reindeer on South Georgia Island (Leader-Williams 1988) and greater kudu (*Tragelaphus strepsiceros*) in South Africa (Owen-Smith and Cooper 1989). Energy intake for the jerboa (*Allactaga elater*) in north cis-Caspian, former Russia (Fig. 7.1), dropped below requirements in midsummer and so bodyweight declined (Abaturov and Magomedov 1988). For greater kudu (Fig. 7.2), energy intake during

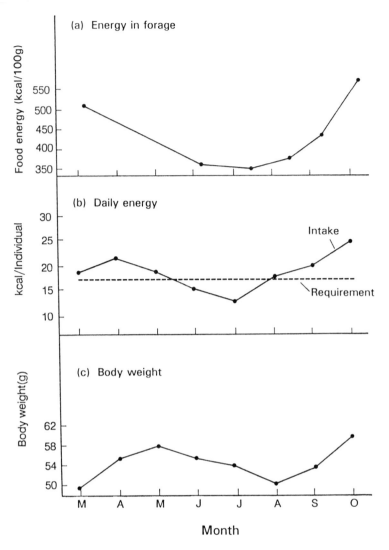

Fig. 7.1 Seasonal changes in energy and bodyweight of the jerboa (*Allactaga elater*) in the cis-Caspian, former Russia, during 1985. (a) Percentage energy of forage in the stomach. (b) Daily energy intake, and daily energy requirement. (c) Bodyweight. (After Abaturov and Magomedov 1988.)

winter was below the minimum requirement, but protein intake was in adequate supply. In contrast, protein intake of African buffalo in tropical dry seasons was below the minimum requirement (Fig. 7.3).

These indirect measures of food intake can often be inaccurate because they are an amalgam of several different measurements. One way around this is to use a physiologic index from the animal to indicate the quality of the food it has eaten. Nitrogen in the feces predicts nitrogen in the diet down to the minimum level of nitrogen balance. If nitrogen intake falls below this level it is not reflected in

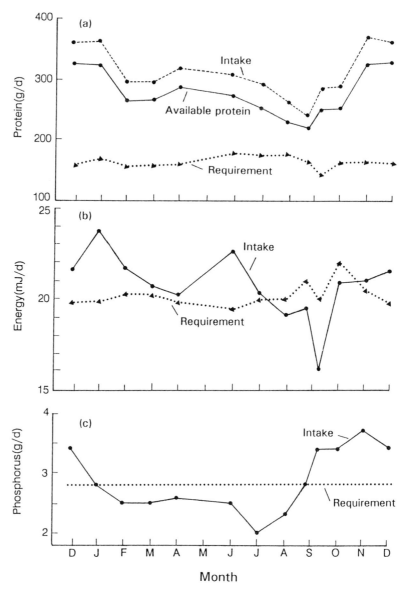

Fig. 7.2 Monthly changes in the estimated daily nutrient intakes of greater kudu relative to estimated maintenance requirements. (a) Crude protein intake (------); available protein (———); requirement for metabolic turnover, fecal loss, and growth (······). (b) Metabolizable energy intake (———); metabolizable energy requirement for resting, activity, and growth (······). (c) Phosphorus intake (———); phosphorus requirement (······). (After Owen-Smith and Cooper 1989.)

the feces because metabolic nitrogen (from microorganisms and gut cells) continues to be passed out irrespective of intake.

In tropical regions this relationship has been found for cattle (Bredon *et al.* 1963), buffalo, and wildebeest (Sinclair 1977), and in North America for cattle, big-horn sheep (*Ovis canadensis*), elk, and

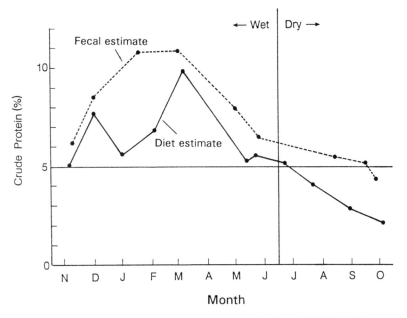

Fig. 7.3 The proportion of crude protein in the diet of African buffalo declines below the estimated 5% minimum requirement in the dry season. Estimates from diet selection with 95% confidence limits (———); estimates from fecal protein (------). (After Sinclair 1977.)

deer (Fig. 7.4) (Leslie and Starkey 1985; Howery and Pfister 1990). These relationships apply to ruminants eating natural food. Similar relationships have been found for experimental diets on rabbits in Australia (Myers and Bults 1977), snowshoe hares (Sinclair *et al.* 1982), elk, and sheep (Mould and Robbins 1981; Leslie and Starkey 1985), although the slopes of the regression lines differ from the natural diets.

A potential problem with this approach is that plant secondary compounds such as tannin may obscure the relationship by causing higher amounts of metabolic nitrogen to be passed out (Robbins *et al.* 1987). This has been observed in experimental diets with high amounts of these compounds (Mould and Robbins 1981; Sinclair *et al.* 1982). But these are abnormal situations. When animals are allowed to choose their own diet the relationship holds up. The regression has been determined for only a few species on natural diets, so more work is needed in this area. A second potential problem could arise if fecal samples are exposed to the weather and the nitrogen leached out. This is not a problem for white-tailed deer feces in fall if samples are collected < 24 days after defecation (Jenks *et al.* 1990).

The relationship between fecal and dietary nitrogen can be used to estimate whether animals are obtaining enough food for maintenance. In African buffalo the estimate of dietary nitrogen using fecal regression was compared with estimates of dietary nitrogen from rumen contents (Fig. 7.3). The two are similar.

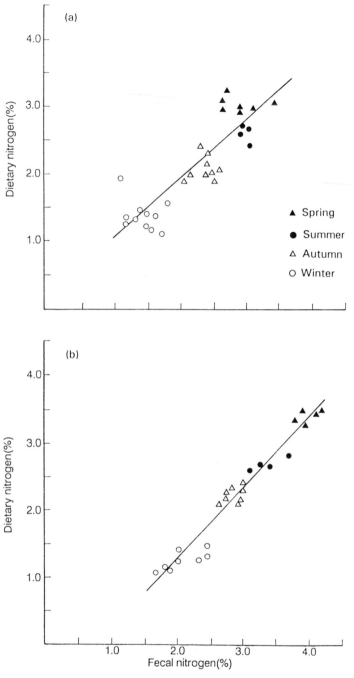

Fig. 7.4 Correlation of dietary nitrogen with fecal nitrogen in (a) elk and (b) black-tailed deer. Nitrogen increases with season. Spring (▲); summer (●); fall (△); winter (○). (After Leslie and Starkey 1985.)

A similar approach has related fecal nitrogen directly to weight loss. Thus Gates and Hudson (1981) found that elk lost weight below about 1.6% fecal nitrogen (Fig. 7.5a) during late winter when there was deep snow (Fig. 7.5b).

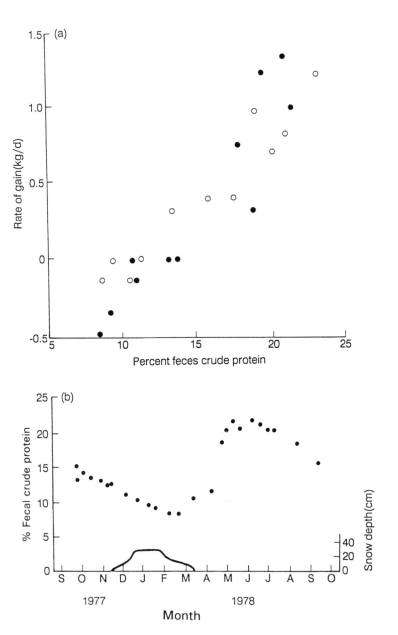

Fig. 7.5 (a) Bodyweight gain of male elk (●) and calves (○) in Alberta can be predicted from the percentage of fecal crude protein. (b) Seasonal changes in the percentage of fecal crude protein are related to snow depth. (After Gates and Hudson 1981.)

7.5 **Basal metabolic rate and food requirement**

7.5.1 Energy flow

The flow of energy through the body is illustrated in Fig. 7.6. Energy starts as consumption energy or intake energy. Part of this is digested in the gut and passes through the gut wall as digestible energy, the rest being passed out in the feces as fecal energy. Part of the digestible energy is lost in the urine and the remainder, called metabolic or assimilated energy, can then be used for work. The work energy can be divided into two: respiration energy which is used for the basic maintenance of the body (resting energy) and for activity, and production energy for growth and reproduction.

The flow chart for protein is similar except that protein is normally used only for production. Protein is not used in respiration except under special conditions of food shortage when protein is broken down (catabolized) to provide energy.

Metabolic energy (M) can be measured in two ways:

1 in the laboratory by measuring resting energy and activity to obtain the respiration component (R), and from growth and population studies to obtain production (P), so that:

$$M = R + P;$$

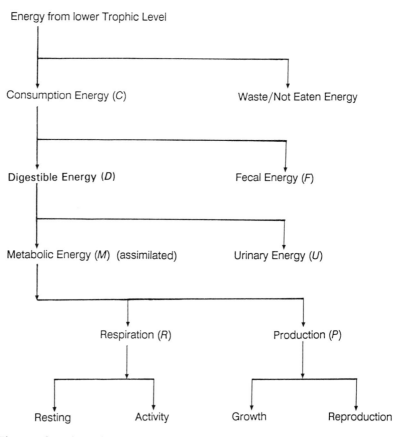

Fig. 7.6 Flow chart of energy through the body.

2 in the field by measuring consumption (C), fecal (F), and urinary (U) outputs, so that:

$$M = C - F - U$$

Basal metabolism is the energy needed for basic body functions. The energy comes from oxidation of fats, proteins, and carbohydrates to produce water and carbon dioxide. Thus, maintenance energy can be measured from expired air volume and composition because intake air has a stable, composition of 20.94% oxygen, 0.03% carbon dioxide, and 79.03% nitrogen. Since 6 mol carbon dioxide and water are produced with 673 kcal of heat, the carbon dioxide level in expired air can be used to calculate the rate of energy used for maintenance. Measurements can be obtained either in chambers or from gas masks, and the animal must be in its thermoneutral zone (not shivering, panting, or sweating), resting, and not digesting food. Such conditions give the basal metabolic rate (BMR).

The BMRs of different eutherian mammalian groups, such as those in Fig. 7.7, when plotted against the log of bodyweight, fall on a line whose slope is approximately 0.75. Thus, Kleiber (1947) produced the general equation:

$$BMR = 70 \ W^{0.75}$$

where BMR is in kilocalories per day and W is bodyweight in kilograms. This is an average over all mammals. Specific groups may

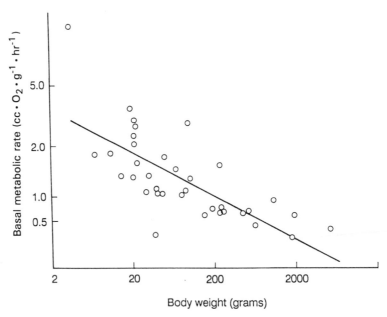

Fig. 7.7 Relationship of basal metabolic rate and bodyweight in different groups of small mammals. (After Mace 1979.)

differ: desert-adapted mammals have lower rates; marine mammals higher rates. Large nonpasserine birds are similar to eutherians, but rates in the smaller passerines are 30–70% higher. The constant, 70, also differs: in marsupials it is 48.6 and in the echidna (a monotreme) it is 19.3 (Robbins 1983).

Hibernating mammals, such as ground squirrels, can lower their body temperatures to a few degrees above ambient temperature, but no lower than about 0°C. Hummingbirds can lower their body temperature to about 15°C, a process called torpor. Both hibernation and torpor save energy (Kenagy 1989; Kenagy *et al.* 1989).

To this point we have discussed resting or maintenance requirements. Activity adds a further energy cost to maintenance. Standing is on average 9% more costly than lying for mammals and 13.6% more costly for birds (Robbins 1983). The cost of locomotion is similar for bipedal and quadrupedal animals (Fedak and Seeherman 1979). The cost of locomotion (LC), expressed as kilocalories per kilogram per kilometer, declines linearly with increasing log body size. Thus:

$$LC = 31.10\,W^{-0.34} \quad (W \text{ in grams})$$

Hence the cost of moving is higher per unit body mass for smaller species and for juveniles.

The average daily metabolic rate (ADMR; the sum of resting and activity rates) is approximately two times the BMR in captive mammals, but is difficult to measure for free-living animals. For captive passerines the ADMR is 1.31 times the BMR and for captive nonpasserines it is 1.26 times the BMR. As a rough approximation, free-living birds and small mammals have a metabolic rate two–four times the BMR.

7.5.3 *Variation in food requirements*

The ADMR or other average measures of metabolic rate hides seasonal fluctuations in food and energy demands. The costs of reproduction add considerably to those for normal daily activity. In the red deer or wildebeest, the rut imposes a considerable energetic cost upon males, which spend several weeks fighting, defending territories, and herding females while eating very little (Sinclair 1977; Clutton-Brock *et al.* 1982). Males put on large amounts of body fat before the rut and use it to cover the extra energy requirements of the rut. Mule deer males (Fig. 7.8a) deposit kidney fat in the fall and use it in November during mating (Anderson *et al.* 1972).

Female mammals use additional energy for lactation and for growing a fetus. Like males they accumulate body fat, especially in the mesentery and around the kidneys, before birth and lactation. During the last third of gestation metabolic costs are two times the ADMR, and during lactation they are three times the ADMR. In female mule deer (Fig. 7.8b) fat is built up in fall and early winter, and used between late winter and summer during gestation, birth, and lactation. Similar

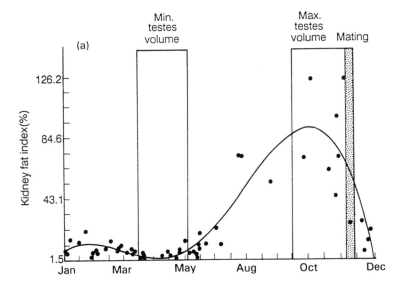

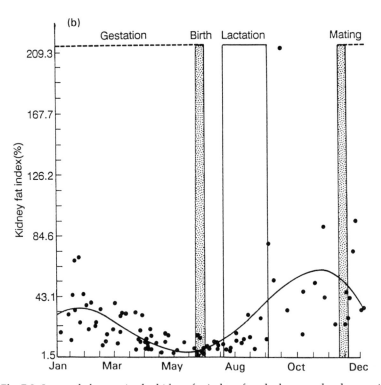

Fig. 7.8 Seasonal changes in the kidney fat index of mule deer are closely associated with reproduction and season. (a) Males; (b) females. (After Anderson *et al.* 1972.)

results have been found for many tropical ungulates such as African buffalo and wildebeest (Sinclair 1977, 1983). Thus the timing of reproduction in ungulates is influenced in part by the need to obtain good food supplies and to build up fat reserves. Other examples for birds are given in Section 7.9.

7.6 Morphology of digestion

7.6.1 Strategies of digestion

Carnivores and omnivores digest their food in the stomach and small intestine. The small intestine is relatively short in these species. Herbivores, which make up most (about 90%) of the mammals (Björnhag 1987), need to digest large amounts of fairly indigestible cellulose and hemicellulose, and to do so they have adapted the gut to increase retention time. One strategy is to evolve a much longer small intestine. An exception is the giant panda which evolved from bears and has retained the short intestine. In this species organic matter digestibility is only 18%, one of the lowest recorded (Schaller et al. 1985). Another adaptation is to use microorganisms (bacteria, fungi, protozoa) which digest cellulose through fermentation. Plant material must be retained in a fermentation chamber long enough for the microorganisms to cause fermentation. Squirrels eat high-energy foods such as seeds, fruits, and insects, and so do not need such mechanisms. Some species have unusually low metabolic rates and hence longer retention times. Most are arboreal folivores: koalas (Phascolarctos cinereus) (Dawson and Hulbert 1970), sloths, and hyraxes (Rubsamen et al. 1979; Björnhag 1987). Reviews of digestive adaptations can be found in Hornicke and Björnhag (1980), Hume and Warner (1980), Robbins (1983), and Björnhag (1987).

7.6.2 Ruminants

True ruminants, which include the bovids (cattle, sheep, antelopes), cervids (deer), tylopodids (camels), and giraffes, have an extension of the stomach divided into three chambers. One of these is the rumen, which acts as the fermentation chamber. Plant food is gathered without chewing and stored in the chamber during a feeding period. This is followed by a rumination period during which portions of compacted food (bolus) are returned to the mouth for intensive chewing. In this way coarse plant material is broken down mechanically and made available to microorganisms for fermentation. The amount of fiber in the food determines how coarse it is, and the coarser the food the longer the process of grinding and fermentation. There is a limit to how coarse the food can be before fermentation takes so long that the animal uses more energy than it gains. On average a ruminant retains food in the gut for about 100 hours.

Microorganisms break down cellulose into short-chain fatty acids, and proteins into amino acids and ammonia, using these to produce more microorganisms. The host animal obtains its nutrients by digesting the dead microorganisms in the stomach and short intestine. The system is efficient, and digestibilities of organic matter and protein of around 65–75% are achieved for medium to good quality food (i.e.,

relatively low in fiber). Another advantage is that nitrogen can be re-
cycled as urea. A disadvantage is that microorganisms digest nutrients
that could be used directly by the host, and this leads to a loss of
energy through production of methane. Another is that ruminants
cannot digest very high fiber diets.

7.6.3 *Hindgut fermenters*

In contrast to the foregut fermenters, or ruminants, a number of animal
groups have developed an enlarged colon or cecum or both to allow
fermentation. Large animals (> 50 kg) are in general colon fermenters,
while small ones (< 5 kg) which feed on fibrous food are cecum
fermenters.

Colon fermenters
In most cases both the colon and cecum are enlarged to hold fiber for
microbial digestion. There is little separation of material into small
particles and microbes on the one hand and fiber on the other, and
there is little evidence that microbial proteins are digested and absorbed, although fatty acids can be absorbed.

Animals in this group are perissodactyls (horses, rhinos, tapirs),
macropods (kangaroos), and perhaps elephants, wombats (*Vombatus
ursinus*), and dugongs (*Dugong dugon*). These are all large animals and
so do not need to ingest high energy and much protein per unit of
bodyweight (see Section 7.5.2). Since food material can be retained in
the gut for longer periods in large animals, the rate of passage may
be slow enough for fermentation and absorption of fatty acids to
take place. None of these animals eats its feces, a practice called
coprophagy.

Cecum fermenters
Small animals (< 5 kg) have relatively high metabolic rates. Those
species that feed on high-fiber diets such as grass and leaves need to
use the microbial protein produced by hindgut fermentation. They
do this by coprophagy. In conjunction with this process there is a
sorting mechanism in the colon that separates fluids, small particles
of food, and microbes from the fiber. The fluids and microbes are returned by antiperistaltic movements to an enlarged cecum for further
fermentation and digestion. This mechanism, therefore, retains the
nutrients long enough for fermentation. It is necessary because small
animals cannot hold food material long enough for fermentation under
normal passage rates.

Dead microbial material is passed out in the form of special soft
pellets, *cecotrophs*, and these high-nutrient feces are eaten directly
from the anus, a behavior called *cecotrophy*. The sorted high fiber is
passed out as hard pellets which are not reingested.

Animals that both ferment food in the cecum and practise cecotrophy include myomorph rodents (voles, lemmings, brown rat), lagomorphs (hares, rabbits), some South American rodents (coypu, guinea

pig, chinchilla), and some Australian marsupials such as the ringtail possum (*Pseudocheirus peregrinus*) (Chilcott and Hume 1985).

Two marsupials, the koala and greater glider (*Petauroides volans*), feed on arboreal leaves and have cecal fermentation and a colonic sorting mechanism (Cork and Warner 1983; Foley and Hume 1987). Neither practises cecotrophy. At least in the koala, both the metabolic rate and the passage time are slow enough that cecotrophy is not necessary.

Björnhag (1987) has identified four strategies employed by small mammals that feed on plants.

1 They eat only highly nutritious plant parts such as seeds, berries, buds, and young leaves. Squirrels fall in this group.

2 They have a low metabolic rate for their size so that fermentation is prolonged. Koala and tree sloths are examples of this group.

3 Digesta are separated in the colon, and easily digestible food particles plus microorganisms are retained to allow fermentation and fibrous material to be sorted and passed out.

4 Only the microorganisms are separated and retained to allow rapid fermentation.

Both (3) and (4) involve the recirculation of protein-rich fecal material by reingestion through cecotrophy. Examples are voles, lemmings, and lagomorphs.

7.7 Food passage rate and food requirement

The passage rate of food through an animal depends on the *retention time*, which is the mean time an indigestible marker takes to pass through. Various markers can be used, e.g., dyes, glass beads, radioisotopes, and polyethylene glycol. Certain rare earth elements (samarium, cerium, lanthanum) bind to plant fiber and provide useful markers to measure passage times of fiber (Robbins 1983).

The rate of food intake by herbivores depends on the nutritive quality of the food. For example, in domestic sheep (Fig. 7.9), the intake rate of white-tailed and mule deer first increases then decreases as the energy quality of the food declines (Sibly 1981; Robbins 1983). This relationship occurs because both energy and protein are inversely related to fiber content.

Estimation of fecal protein can be used as a means of determining whether a population is obtaining enough food because protein intake is related to the amount of protein in the feces (see Fig. 7.4). This method has been used to predict the change in bodyweight of elk (see Fig. 7.5b) (Gates and Hudson 1981) and to monitor food requirements in snowshoe hares (Sinclair *et al.* 1988).

7.8 Body size and diet selection

The gut, e.g., the rumen, large intestine and cecum, crops of hummingbirds, and cheek pouches of heteromyid rodents, has a capacity that is a linear function of bodyweight ($W^{1.0}$) (Clutton-Brock and Harvey 1983; Robbins 1983). Energy requirements, however, are a function of metabolic bodyweight ($W^{0.75}$). Thus the difference between

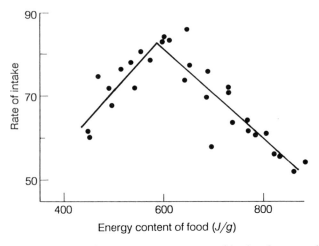

Fig. 7.9 Relationship of intake rate to energy content of food in domestic sheep. Below an energy content of 590 J/g, intake rate falls because of a finite gut capacity and declining fermentation rates. Rate of intake is dry matter/bodyweight$^{0.75}$ per day. (After Dinius and Baumgardt 1970.)

the exponents ($W^{1.0}/W^{0.75} = W^{0.25}$) means that a larger animal can obtain more food relative to requirement than a smaller one. This can be expressed in two ways: (i) on the same quality of diet a larger animal needs to eat less food per unit of bodyweight than the smaller; and (ii) a larger animal on a lower nutrient diet can extract the same amount of nutrient per unit of bodyweight as a smaller animal on a higher nutrient diet. Thus, larger animals can eat higher-fiber diets, a feature that allows resource partitioning in African ungulates (Bell 1971; Jarman and Sinclair 1979).

Jarman (1974) extended the relationship between body size and diet of African ungulates to explain interspecific patterns of social and antipredator behavior. We can identify five categories, from selective browsers to unselective grazers.

1 Small species (3–20 kg), solitary or in pairs, that are highly selective feeders on flowers, buds, fruits, seed pods, and young shoots. Their habitats are thickets and forest which provide cover from predators. There is little sexual dimorphism and both species help in defending a territory. This group includes duikers (*Cephalophus* species), suni antelope (*Nesotragus moschatus*), steinbok (*Raphicerus campestris*), dikdik (*Madoqua* species), and klipspringer (*Oreotragus oreotragus*).

2 Small to medium species (20–100 kg) that can be both grazers and browsers, but are very selective of plant parts as in (1). Habitat is riverine forest, thicket, or dense woodland. Group size is larger, from two to six, one male and several females. They are usually territorial and include lesser kudu (*Tragelaphus imberbis*), bushbuck (*T. scriptus*), gerenuk (*Litocranius walleri*), reedbuck (*Redunca* species), and oribi (*Ourebia ourebia*). There is some sexual dimorphism. Predators are avoided by hiding and freezing.

3 Medium size species (50–150 kg) that are mixed feeders, changing from grazing in the rains to browsing in the dry season. Habitats are varied and range from dense woodland and savanna to open flood plains. There is one male per territory. Female group size is variable (6–200) and these groups do not defend a territory but wander through the male territories. Nonterritorial male groups are excluded from territories and behave like female groups. Females have a large home range that is smaller in the dry season than in the wet season. Species typical of this group include impala, greater kudu, sable, kob (*Kobus* species), lechwe, and gazelles (*Gazella* species). They are sexually dimorphic in the extreme. Predators are avoided by group vigilance and by running.

4 Medium to large species (100–250 kg) that are grazers selecting high-quality grass leaves. Males are single and territorial or form large bachelor groups. Female groups range from six to many hundred. They have a large home range, often divided into wet- and dry-season ranges separated by a considerable distance. Habitats are generally open savanna and treeless plains. Predators are avoided by group vigilance and by running. Sexual dimorphism is present but less extreme than in (3). Wildebeest, hartebeest, topi (*Damaliscus korrigum*), and Grevy's zebra (*Equus grevyi*) are in this group.

5 Large species (>200 kg) are unselective grazers and browsers of low-quality food. Habitats are closed woodland and open savanna. Movements are seasonal. Males are nonterritorial and form a dominance hierarchy. Females form groups of ten to several hundred and have a large home range. Active group defense against predators is shown by African buffalo and African elephant, while other species use group vigilance and running to avoid predation. Burchell's zebra (*Equus burchelli*), giraffe (*Giraffa camelopardalis*), eland (*Taurotragus oryx*), gemsbok (*Oryx gazella*), and Roan antelope (*Hippotragus niger*) are included in this group.

Jarman's (1974) categories relate body size inversely to food supply because low-quality food is more abundant. In turn this allows species to form larger groups to avoid predators, and the size of group then determines how a male obtains his mate. In small species, males keep females in their territories year-round and this may be the only way of finding females in estrus. When female groups are larger (group 3), females cannot remain within one territory. Hence males compete for territories within the females' home range to provide an opportunity for mating when females move through the territory. These territories are for mating and not to provide year-round food.

Finally, interspecific competition for male mating territories may have led to larger males with elaborate weapons. Since these selection pressures have not operated on the females, which have remained at a smaller size, sexual dimorphism develops. Thus we see a connection between body size, food quality, group size predator defense, and mating system.

7.9 Indices of body condition

7.9.1 Bodyweight and total body fat

Bodyweight and fat reserves affect survival and reproduction in mammals (Hanks 1981; Dark *et al.* 1986) and birds (Johnson *et al.* 1985). Bodyweight can be measured directly for small mammals such as the jerboa (see Fig. 7.1) and for birds. Weight changes seasonally in response to changes in food supply and hence intake.

Bodyweight is a function of genetic determinants, age, and the amount of fat and protein stored in the body. To monitor fat and protein changes it is better to take out the effects of body size (the genetic and age components). This can be done by using a ratio of weight to some body measure that is a function of size. Thus, for cottontail rabbits (*Sylvilagus floridanus*), Bailey (1968) found a relationship between predicted bodyweight (PBW in grams) and total length (*L* in centimeters) such that:

$$PBW = 16 + 5.48 \ (L^3).$$

The condition index for the rabbits would then be the ratio of observed to predicted weight. A similar relationship has been found between foot length and weight for snowshoe hares (O'Donoghue 1991). At the other extreme of size, blubber volumes of fin whales (*Balaenoptera physalus*) and sei whales (*B. borealis*) have been calculated from body length, girth, and blubber thickness measured at six points along the carcase (Lockyer 1987).

Although bodyweight alone is a satisfactory measure of condition for such birds as sandhill cranes (*Grus canadensis*) and white-fronted geese (*Anser albifrons*) (Johnson *et al.* 1985), it is usually better to account for body size using some measure such as wing length, tarsus length, bill length, or keel length.

In female mallards (*Anas platyrhynchos*) fat weight (F), an index of condition, is related to bodyweight (BW) and wing length (WL) by:

$$F = (0.571BW) - (1.695WL) + 59.0$$

and a similar relationship holds for males (Ringelman and Szymczak 1985).

In maned ducks (*Chenonetta jubata*) of Australia, the bodyweight and total fat of females increase before laying. Some 70% of the fat is used during laying and incubation. Protein levels, however, do not change (Briggs 1991). Among northern hemisphere ducks there are four general strategies for storing nutrients prior to laying: (i) fat is deposited before migration and supplemented by local foods on the breeding grounds, as demonstrated by mallards; (ii) reserves are formed entirely before migrating to the breeding area, as in lesser snowgeese (*Chen caerulescens*); (iii) reserves are built up entirely on the breeding grounds with no further supplementation, as illustrated by the common eider (*Somateria mollissima*); and (iv) body reserves are both formed in the breeding area and supplemented by local food, as seen in the wood duck (*Aix sponsa*) (Owen and Reinecke 1979; Thomas 1988).

Both ducks and game birds can alter the length of their digestive system in response to changes in food supply. Under conditions of more fibrous diets during winter, gut lengths increased in three species of ptarmigan (*Lagopus* species) (Moss 1974), gadwall (*Anas strepera*), and mallard (Paulus 1982; Whyte and Bolen 1985).

In passerine birds energy is stored in various subcutaneous and mesenteric fat deposits, and protein is stored in flight muscles. The latter varies with total body fat, as in the yellow-vented bulbul (*Pycnonotus goiavier*) in Singapore (Ward 1969), and the house sparrow (*Passer domesticus*) in Britain (Jones 1980). In the gray-backed camaroptera (*Camaroptera brevicaudata*), a tropical African warbler, both total body fat and flight muscle protein vary in relation to laying date (Fig. 7.10) (Fogden and Fogden 1979).

As in bodyweight measures, flight muscle weights are corrected for body size by dividing by a standard muscle volume (SMV). For example, Davidson and Evans (1988) used the formula:

$$SMV = H(L \times W + 0.433C^2)$$

for shorebirds of the genus *Calidris* where:

H = height of keel of sternum,
L = length of keel of sternum,
W = width of raft of sternum (one side only),
C = distance from keel to end of coracoid.

Direct measures of bodyweight are feasible with birds and small mammals, but impractical for large mammals where some other index of body condition and food reserves must be used. These have been reviewed by Hanks (1981). Large mammals store fat subcutaneously,

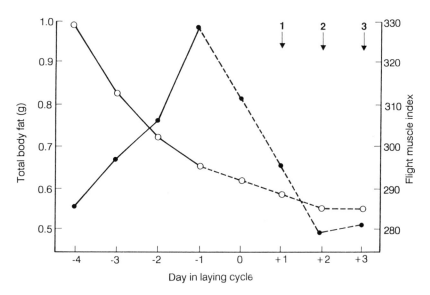

Fig. 7.10 Total body fat (●) and flight muscle index (○) are related to laying day in the female gray-backed camaroptera (a tropical African warbler). Eggs are laid on days 1–3. Flight muscle index is the ratio of (lean dry flight muscle weight)/(flight muscle cord[3]). The broken line indicates estimates. (After Fogden and Fogden 1979.)

in the gut mesentery, around the kidneys and heart, and in the marrow of long bones. The fat stores are used up in that order (Mech and DelGiudice 1985). Because of this sequential use no single fat deposit is a perfect indicator of total body fat. However, some of these fat stores are of interest for specific purposes such as reproduction (kidney fat) or starvation (bone marrow fat). For these purposes they provide a reasonable guide for managers, total body fat being less useful.

7.9.2 Kidney fat index   Ungulates accumulate fat around the kidney and in other places in the body cavity in anticipation of the demands of reproduction. We saw in Section 7.5.3 how the fat reserves of mule deer change according to the stage of the reproductive cycle (see Fig. 7.8), the timing of these changes differing between the sexes.

Although there is little relationship between kidney fat and total body fat in some species (Robbins 1983), others, such as most African ruminants, show a close relationship (Smith 1970; Hanks 1981). For white-tailed deer the percentage of fat in the body is related to the kidney fat index (KFI) by:

$$\text{Percentage of fat} = 6.24 + 0.30\text{KFI}$$

(Finger *et al.* 1981). In mule deer both the weight of kidney fat and the KFI are correlated with total body fat (Anderson *et al.* 1969; 1972; 1990).

Although a similar relationship is evident for the brush-tailed possum (*Trichosurus vulpecula*) in New Zealand (Bamford 1970), a better correlation was found between total body fat and mesenteric fat.

Kidney fat index has been measured in two ways.

1 The kidney is pulled away from the body wall by hand and the surrounding connective tissue with embedded fat tears away along a natural line posterior to the kidney. A cut along the midline of the kidney allows the connective tissue to be peeled away cleanly. The KFI is the ratio of connective tissue plus fat weight to kidney weight summed for both kidneys.

2 The connective tissue is cut immediately anterior and posterior to the kidney, so that only the fat immediately surrounding the kidney is used. This has a small advantage of being more objective than (1), but has the great disadvantage of discarding most of the tissue where fat is deposited, so much of the relevant variability in fat deposition is lost. Hence, the first method may be the more useful index.

7.9.3 Bone marrow fat   Bone marrow fat does not decline until after most of the kidney fat has been used (Fig. 7.11) in temperate and tropical ruminants, and in some marsupials (Ransom 1965; Bamford 1970; Hanks 1981). Consequently, a decline in bone marrow fat reflects a relatively severe depletion of energy reserves and therefore provides an index of severe nutritional stress.

Mobilized marrow fat is replaced by water. Hence, the ratio of dry

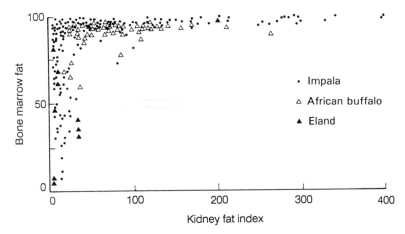

Fig. 7.11 Kidney fat index (KFI) is depleted almost entirely before bone marrow fat declines in ruminants. KFI is the ratio of kidney-plus-fat weight/kidney weight. (After Hanks 1981.)

to wet weight of marrow is a good measure of fat content. A number of studies on both temperate and tropical ruminants (Hanks 1981) have indicated that, as a close approximation:

Percentage of marrow fat = percentage of dry weight − 7

The dry weight of marrow is measured from the middle length of marrow in one of the long bones, avoiding the hemopoitic ends. The method has been used on wildebeest (Fig. 7.12) and deer (Klein and Olson 1960) to establish whether they died from lack of food.

Broad categories of marrow fat content in ruminants are provided by the color and texture of the marrow (Cheatum 1949). This method is quick (it avoids collection of marrow) and it is sufficient to determine whether an animal has been suffering from undernutrition at death (Sinclair and Duncan 1972; Verme and Holland 1973; Kirkpatrick 1980). The following are categories with approximate fat values.

1 Solid, white, and waxy: the marrow can stand on its own and contains 85–98% fat. Such animals are not suffering from undernutrition.
2 White or pink, opaque, gelatinous: the marrow cannot stand on its own and covers a broad range of fat values (15–85%). It indicates that such animals have depleted fat reserves.
3 Yellow, translucent, gelatinous: the clear, gelatinous appearance is distinctive and indicates there is <15% fat and often only 1% fat. Such animals are starving.

7.9.4 Blood indices of condition

Blood parameters as indices of condition and food intake are potentially useful for living animals that are too large to be weighed easily. However, blood characteristics are not well known for most species. More work is needed. Different parameters have been examined in different studies. Plasma nonesterified fatty acid, protein-bound iodine,

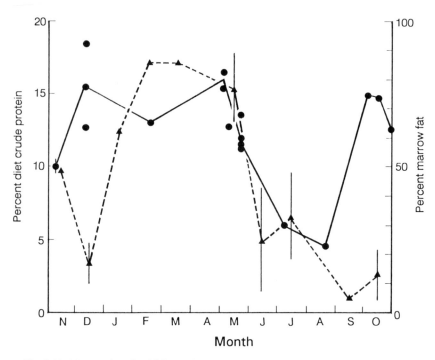

Fig. 7.12 Marrow fat of wildebeest (------) dying from natural causes in Serengeti is related to season and the percentage of crude protein in their diet (———, with 95% confidence limits). (After Sinclair 1977.)

and serum total protein were all related to nutrition in Australian tropical cattle (O'Kelly 1973). The body condition of moose has been related to various sets of blood parameters ranked according to their sensitivity (Franzmann and LeResche 1978). Starting with the single best parameter, sensitivity increased by adding other measures: (i) packed cell volume (PCV); (ii) (PCV plus hemoglobin content (Hb); (iii) PCV, Hb, calcium (Ca), phosphorus (P), and total blood protein (TP); and (iv) PCV, Hb, Ca, P, TP, glucose, albumin, and β-globulin.

Protein loss from the body was strongly correlated with body-weight loss in whitetailed deer on experimentally restricted diets. Serum urea nitrogen levels and the ratio of urinary urea nitrogen to creatinine were the best blood and urine indicators of undernutrition and protein loss (DelGiudice and Seal 1988; DelGiudice *et al.* 1990).

7.9.5 *Problems with condition indices*

We have already noted that it is generally impractical to obtain measures of total body fat in large mammals. Various single indices such as kidney fat and bone marrow fat have been used but these are useful for specific purposes and cannot be used over the whole range of total body fat values. Kidney fat is more appropriate for estimating the upper range of body fat values and bone marrow fat represents the lower values. A combination of six indices of body fat deposits in carcasses has been proposed (Kistner *et al.* 1980). This method is useful for complete carcases but it cannot be used for animals dying naturally, because the soft parts are usually eaten by predators and scavengers, or they decompose. Under these conditions the only index that remains uniformly useful is that of bone marrow fat.

Bone marrow fat as an index is biased towards the low-body-fat values. It cannot reflect changes in the higher levels of body fat, so that very fatty bone marrow does not necessarily mean the animal is in good condition (Mech and DelGiudice 1985).

Although blood indices may be useful as a means of assessing condition and nutrition in living animals, they require careful calibration. Many of the blood characteristics are influenced by season, reproductive state, age, sex, and hormone levels. More importantly, they can be altered rapidly by the stress of capture and handling. All of these could act to obscure and confound changes in nutrition.

All estimates of body condition taken from a sample of the live population are poor indicators of the nutritional state of the population for two reasons. First, such samples are biased towards healthy animals because those in very poor condition are either dead or dying and not available for sampling. Second, the age groups that are most sensitive to density-dependent restriction in food supply – the very young and very old – form a small proportion of the live population. Thus, even a strictly random sample of the population will include a majority of healthy animals and consequently the mean value of condition will be very insensitive to changes in food supply. Therefore, it is unlikely that one can assess whether a population is regulated by food supply or by predators based solely on body condition samples of the live population. To make this assessment one should look at the condition of animals that have died.

7.10 Summary

Before we can make decisions about whether a wildlife population has enough food or whether the food supplies can be changed to manipulate a population we need to know about food availability and requirements. The availability of food is subject to seasonal variation, either directly or through changes in plant defenses. Food requirements depend much on the gastric anatomy and body size of an animal. During the annual cycle, breeding, moulting, and lactation are other costs that change food requirements. The balance between food requirements and availability is reflected in an animal's body condition, which can be indexed by weight, fat reserves, and blood parameters.

8 Competition within species

8.1 Introduction

Next we examine one of the major causes of regulation, namely competition between individuals for resources, or *intraspecific competition*. Other causes of regulation such as predation will be dealt with in Chapter 10. Chapter 6 has already outlined an alternative approach.

8.2 Stability of populations

If we look at long-term records of animal populations we see that some populations remain quite constant in size for long periods of time. Records of gray herons (*Ardea cinerea*) in England and Wales (Fig. 8.1) illustrate that although the population fluctuates it remains within certain limits (2100–4800) and the average population is relatively constant. Other populations, such as those of insects or house mice (*Mus domesticus*) in Australia (Fig. 8.2) fluctuate to a much greater extent and furnish no suggestion of an equilibrium population size. Nevertheless, such populations do not always become extinct and they remain in the community for long periods. Occasionally one finds unusual situations where populations show regular cycles. The snow-shoe hare (*Lepus americanus*) in northern Canada shows the clearest (Fig. 8.3), as indicated by the furs collected by trappers for the Hudson Bay Company over the past two centuries (MacLulich 1937).

This relative constancy of population size or fluctuations within limits is in contrast to the intrinsic ability of populations to increase rapidly. The fact that population increase is limited suggests that there is a mechanism in the population that slows down the rate of increase and so regulates the population. We discuss first the theory for how populations might be limited and regulated.

8.3 The theory of population limitation and regulation

Populations have inputs of births and immigrants and outputs of deaths and emigrants. For simplicity we will confine discussion to a self-contained population having only births (B) and deaths (D) per unit time.

8.3.1 Density dependence

If either the proportion of the population dying increases or the proportion entering as births decreases as population density increases, then we define these changes in proportions as being *density dependent*. The underlying causes for the changes in these rates are called *density-dependent factors*.

Births and deaths as a proportion of the population (B/N_t, D/N_t) can

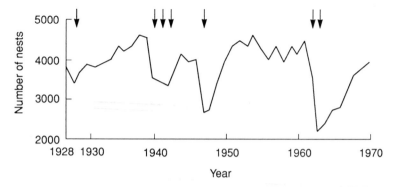

Fig. 8.1 Some populations remain within relatively close bounds over long time periods. The gray heron population of England and Wales (estimated by the number of occupied nests) shows a steady level despite some perturbations due to severe winters. The occurrence of severe winters is shown by arrows. (After Stafford 1971.)

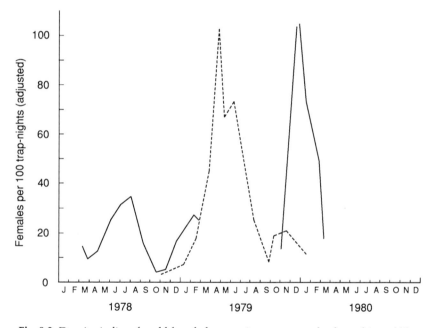

Fig. 8.2 Density indices for old female house mice on contour banks and in stubble fields of rice crops in southeastern Australia. Dashed lines distinguish the crop cycle cohort of 1978–79 from that of 1977–78 and 1979–80. The extent of the peak in January 1980 is unknown, owing to a poisoning campaign. (After Redhead 1982.)

be related to the instantaneous birth (b) and death (d) rates in the following way.

The change in population per unit time is:

$$N_{t+1} - N_t = B - D$$

the instantaneous rate of increase (r) is given by:

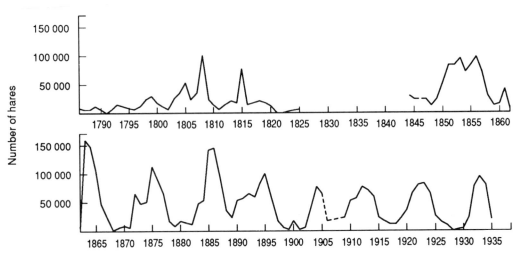

Fig. 8.3 Snowshoe hares in the boreal forest of Canada show regular fluctuations in numbers with a 10-year periodicity. Data are from the Hudson Bay Company's fur records up to 1903, and from questionnaires thereafter. (After MacLulich 1937.)

$$r = b - d$$

and the finite rate of increase (R) (sometimes called λ, see Chapter 4) is given by:

$$R = N_{t+1}/N_t = e^r$$

Therefore:

$$e^{b-d} = (N_{t+1}/N_t) = (B - D + N_t)/N_t$$

If $d = 0$, $D = 0$, then:

$$e^b = (B + N_t)/N_t = (1 + (B/N_t))$$

and by taking the natural log:

$$b = \ln (1 + (B/N_t)).$$

Similarly, if $b = 0$, $B = 0$, and D/N_t is much less than one, then:

$$d = \ln (1 + (D/N_t))$$

If B and D fall in the range of 0–20% of the population then b and d are linear on N, and they remain approximately linear even if B and D are 20–40% of N. This range covers most of the examples we see in nature, so for our purposes we can say that D/N_t and B/N_t change with density in the same way as do b and d, and both go through the origin.

In Fig. 8.4a we plot b against density (or population size) N as a constant so that it is a horizontal line. If we now plot d as an increasing function of density, we see that where the two lines cross, $b = d$, and the population is stationary, at the equilibrium point K. The difference between the b and d lines represents r, and this declines linearly as density increases, in the same way as it does for the logistic curve (see

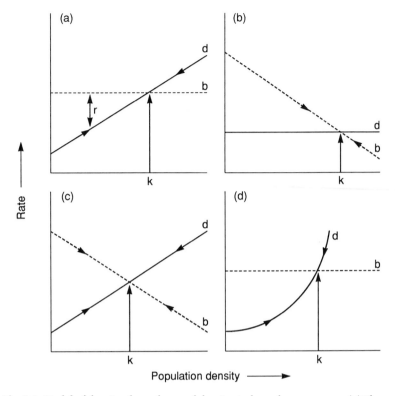

Fig. 8.4 Model of density-dependent and density-independent processes. (a) The birth rate, *b*, is held constant over all densities while the mortality rate, *d*, is density dependent. The population returns to the equilibrium point, *K*, if disturbed. The instantaneous rate of increase, *r*, is the difference between *b* and *d*. (b) As in (a) but *b* is density dependent and *d* is density independent. (c) Both *b* and *d* are density dependent. (d) *d* is curvilinear so that the density dependence is stronger at higher population densities.

Section 4.6.3). In Fig. 8.4a the decline in *r* is due solely to *d* being density dependent. Since *b* (or B/N_t) is constant in this case, we describe it as *density independent* (i.e., it is unrelated to density). In real populations density-independent factors such as the weather may affect birth and death rates randomly. Rainfall acted in this way on greater kudu (*Tragelaphus strepsiceros*) in Kruger National Park, South Africa, causing a mortality of juveniles which was unrelated to population size (Owen-Smith 1990).

We can apply the same arguments if we assume that *b* is density dependent and *d* is density independent (Fig. 8.4b), or if both are density dependent (Fig. 8.4c). So far we have assumed that the density-dependent factor has a linear effect on the rate of increase, as in the logistic curve. However, density-dependent mortality is more likely to be curvilinear, as in Fig. 8.4d.

8.3.2 *Limitation and limiting factors*

In Fig. 8.5 we take the explanation a little further. Let us assume a constant (density-independent) birth rate *b*. Shortly after birth a

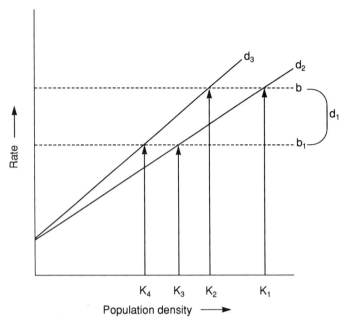

Fig. 8.5 Model showing that the equilibrium point, K, can vary with both density-dependent and density-independent processes. The birth rate, b, is held constant over all densities. In sequence, a density-independent mortality d_1 reduces the input to the population to b_1. There follows a density-dependent mortality d_2 or d_3. The intercept of b or b_1 with d_2 or d_3 determines the equilibrium (K_1-K_4).

density-independent mortality d_1 (depicted here as a constant) kills some of the babies so that inputs are reduced to b_1. There follows a density-dependent mortality d_2, and the population reaches an equilibrium at K_3. If mortality d_1 had not occurred (or was smaller), the equilibrium population would be at K_1. Therefore, the presence or absence of the density-independent factor causing d_1 alters the size of the equilibrium population.

The strength or severity of the density-dependent factor is indicated by the slope of d_2. If the density-dependent factor becomes stronger such as to produce d_3 instead of d_2, the slope becomes steeper and the equilibrium population drops from K_3 to K_4 (or from K_1 to K_2 if d_1 is absent). Thus, altering the strength of density-dependent factors also alters the size of the equilibrium population. For example, increasing the strength of density-dependent factors (i.e., increasing the slope) results in a lower K.

We define the process that determines the size of the equilibrium population as *limitation*, and the factors producing this are *limiting factors*. We can see, therefore, that both density-dependent and density-independent factors affect the equilibrium population size and so they are all limiting factors. Any factor that causes mortality or affects birth rates is a limiting factor.

8.3.3 *Regulation*

If the population is disturbed from its equilibrium, K, by temporary changes in limiting factors (a severe winter or drought or influx of predators might reduce the population; a mild winter or good rains might increase it), the return to K is due entirely to the effect of the density-dependent factor, and this process is called *regulation*. Therefore, regulation is the process whereby a density-dependent factor tends to return a population to its equilibrium. We say "tends to return" because the population may be continually disturbed so that it rarely reaches the equilibrium. Nevertheless, this tendency to return to equilibrium results in the population remaining within a certain range of population sizes. Superficially it appears as if the population has a boundary to its size and fluctuates randomly within this boundary. However, it is more constructive to picture random fluctuations in both the density-independent (d_1) and density-dependent (d_2) mortalities as the shaded range in Fig. 8.6a. This results in a fluctuation of the equilibrium population indicated by the range of K. Figure 8.6a shows that this range of K is relatively small when the density-dependent mortality is strong (steep part of the curve). Figure 8.6b shows the range of K when the density-dependent mortality is weak. We can see that the range of K (which we see in nature as fluctuations in numbers), is very much greater when the density-dependent mortality is weak than when it is strong. Note in Fig. 8.6a and 8.6b that differences in the amplitude of fluctuations are due to changes in the strength of the density-dependent mortality because we have held density-independent (random) mortality constant in this case.

8.3.4 *Delayed and inverse density dependence*

Some mortality factors do not respond immediately to a change in density but act after a delay. Such *delayed density-dependent factors* can be predators whose populations lag behind those of their prey, and food supply where the lag is caused by the delayed action of starvation. Both causes can have a density-dependent effect on the population but the effect is related to density at some previous time period rather than the current one. For example, in the feral Soay sheep (*Ovis aries*) on the Atlantic island of Hirta (St Kilda), winter starvation mortality is dependent on population size in the previous summer (Clutton-Brock *et al.* 1991). A similar relationship was found with winter mortality of red grouse (*Lagopus lagopus*) in Scotland (see Fig. 8.12). Delayed density dependence is indicated when the mortality rate is plotted against current density and the points show an anticlockwise spiral if they are joined in temporal sequence (see Fig. 8.12). These delayed mortalities usually cause fluctuations in population size, as seen with the Soay sheep.

Predators can also have the opposite effect to density dependence, called an *inverse density-dependent* or *depensatory* effect. In this case predators have a destabilizing effect because they take a decreasing proportion of the prey population as it increases, thus allowing the prey population to increase faster as it becomes larger. Conversely, if a

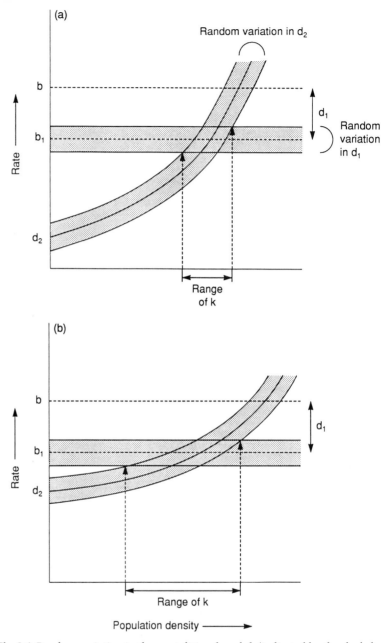

Fig. 8.6 Random variation in the mortalities d_1 and d_2 (indicated by the shaded area) are the same in (a) and (b). In (a) there is stronger density dependence at the intercept of b_1 and d_2 than in (b), and this difference results in a smaller range of equilibria, K, in (a) than in (b).

prey population is declining for some reason, predators would take an increasing proportion and so drive the prey population down even faster towards extinction. In either case we do not see a predator–prey equilibrium. We explore this further in Chapter 10.

8.3.5 *Carrying capacity*

The term *carrying capacity* is one of the most common phrases in wildlife management. It does, however, cover a variety of meanings and unless we are careful and define the term, we may merely cause confusion (Caughley 1976a; 1981). Some of the more common uses of the term carrying capacity are discussed below.

Ecological carrying capacity

This can be thought of abstractly as the K of the logistic equation. In reality it is the natural limit of a population set by resources in a particular environment. It is one of the equilibrium points that a population tends towards through density-dependent effects from lack of food, space (e.g., territoriality), cover, or other resources. As we discussed above, if the environment changes briefly it deflects the population from achieving its equilibrium and so produces random fluctuations about that equilibrium. A longer-term environmental change can affect the resources, which in turn alters K. Again, the population changes by following or tracking the environmental trend.

There are other possible equilibria that a population might experience through regulation by predators, parasites, or disease. Superficially they appear similar to that equilibrium produced through lack of resources because if the population is disturbed through culling or weather events it may return to the same population size. To distinguish the equilibria produced by predation, by resource limitation, and by a combination of the two, we need to know whether predators or resources or both are affecting b and d.

Economic carrying capacity

This is the population level that produces the maximum offtake (or maximum sustained yield) for culling or cropping purposes. It is this meaning that is implied when animal production scientists and range managers refer to livestock carrying capacity. We should note that this population level is well below the ecological carrying capacity. For a population growing logistically its level is $\frac{1}{2}K$ (Caughley 1976a).

Other senses of carrying capacity

We can define carrying capacity according to our particular land use requirements. At one extreme we can rate the carrying capacity for lions on a Kenya farm or wolves on a Wyoming ranch as zero (i.e., farmers cannot tolerate large predators killing their livestock).

A less extreme example is seen where the esthetic requirements of tourism require reducing the impact of animals on the vegetation. Large umbrella-shaped *Acacia tortilis* trees make a picturesque backdrop to the tourist hotels in the Serengeti National Park, Tanzania. In the early 1970s elephants began to knock over these trees. Whereas elephants could be tolerated at ecological carrying capacity in the rest of the park, in the immediate vicinity of the hotels the carrying capacity for elephants was much lower and determined by human requirements for scenery.

8.3.6 *Measurements of birth and death rates*

Birth rates are inputs to the population. Ideally we would like to measure conception rates (*fecundity*), pregnancy rates in mammals (*fertility*), and births or egg production. In some cases it is possible to take these measurements, as in the Soay sheep of Hirta (Clutton-Brock *et al.* 1991). Pregnancies can be monitored by a variety of ways including ultrasonography, radiography, blood protein levels, urine hormone levels, and rectal palpation of the uterus (in large ungulates). In many cases, however, these are not practical for large samples from wild populations.

Births can be measured reasonably accurately for seal species where the babies remain on the breeding grounds throughout the birth season. Egg production, egg hatching success, and fledgling success can also be measured accurately in many bird populations. However, in the majority of mammal species birth rates cannot be measured accurately either because newborn animals are rarely seen (as in many rodents, rabbits, and carnivores), or because many newborn animals die shortly after birth and are not recorded in censuses (as in most ungulates). In these cases we are obliged to use an approximation to the real birth rate, such as the proportion of the population consisting of juveniles first entering live traps for rodents and rabbits, or juveniles entering their first winter for carnivores and ungulates. These are valid measures of recruitment.

Death rates are losses to the population. Ideally they should be measured at different stages of the life cycle to produce a life table (see Section 4.4). Once sexual maturity is reached, age-classes often cannot be identified and all mortality after that age is therefore lumped as "adult" mortality. Mortality can be measured directly by using mortality ratios which indicate when an animal has died, as was done by Boutin *et al.* (1986) and Trostel *et al.* (1987) for snowshoe hares in northern Canada. Survivorship can be calculated over varying time periods by the method of Pollock *et al.* (1989).

Mortality caused by predators can also be measured directly if the number of predators (numerical response) and the amount eaten per predator (functional response) are known (see also Chapters 6 and 10). Such measurements are possible for those birds of prey that regurgitate each day a single pellet containing the bones of their prey. With appropriate sampling, the number of pellets indicates the number of predators, and prey per pellet shows the amount they eat. This method was used for raptors (in particular the black-shouldered kite, *Elanus notatus*) eating house mice during mouse outbreaks in Australia (Sinclair *et al.* 1990).

8.3.7 *Implications*

We should be aware of a number of problems associated with the subject of population limitation and regulation.

1   Much of the literature uses the terms limitation and regulation in different ways. In many cases the terms are used synonymously but the meanings differ between authors. Since any factor, whether density dependent or density independent, can determine the equilibrium point for a population, any factor affecting b or d is a limiting one. It is, therefore, a trivial question to ask whether a certain cause of mortality limits a population – it has to. The more profound question is in what way do mortality or fecundity factors affect the equilibrium.

2   Regulation requires, by our definition, the action of density-dependent factors. Density dependence is necessary for regulation but may not be sufficient. First, the particular density-dependent factor that we have measured, such as predation, may be too weak, and other regulating factors may be operating. Second, some density-dependent factors have too strong an effect and cause fluctuations rather than a tendency towards equilibrium.

3   The demonstration of density dependence at some stage in the life cycle does not indicate the cause of the regulation. For example, if we find that a deer population is regulated through density-dependent juvenile mortality, we do not have any indication from this information alone as to the cause of the mortality. Correlation with population size is merely a convenient abbreviation that hides underlying causes. Density itself is not causing the regulation; the possible underlying factors related to density are competition for resources, competition for space through territoriality, or an effect of predators, parasites, and diseases.

8.4 Evidence for regulation

There are three ways of detecting whether populations are regulated. First, as we have seen in Section 8.3.3, regulation causes a population to return to its equilibrium after a perturbation. Perturbation experiments should therefore detect the return towards equilibrium. Second, if we plot separate and independent populations at their natural carrying capacity against some index of resource (often a weather factor), there should be a relationship. Third, and more commonly, we can try to detect density dependence in the life cycle.

8.4.1 Perturbation experiments

If a population is moved experimentally to either below or above its original density and then returns to that same level, we may conclude that regulation is occurring. An example of downward perturbation is provided by the northern elk herd of Yellowstone National Park (Houston 1982). Prior to 1930 the population estimates ranged between 15 000 and 25 000. Between 1933 and 1968, culling reduced the population to 4000 animals (Fig. 8.7). Culling then ceased and the population rebounded to around 20 000 in 1988. This result is consistent with regulation through intraspecific competition for winter food (Houston 1982) since there are currently no natural predators of elk in Yellowstone.

Density is usually recorded as numbers per unit area. If space is the

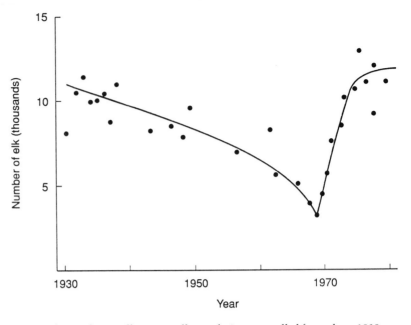

Fig. 8.7 The northern Yellowstone elk population was culled from about 1930 to 1968. Thereafter the population returned to approximately the level it started from, demonstrating regulation. (After Houston 1982, and D. Houston, pers. comm.)

limiting resource (as it might be in territorial animals), or if space is a good indicator of some other resource such as food supply, numbers per unit area will suffice in an investigation of regulation. However, space may not be a suitable measure if density-independent environmental effects (e.g., temperature, rainfall) cause fluctuations in food supply. It may be better to record density as animals per unit of available food or per unit of some other resource.

The Serengeti migratory wildebeest experienced a perturbation (Fig. 8.8) when an exotic virus, rinderpest, was removed. The population increased fivefold from 250 000 in 1963 to 1.3 million in 1977 and then levelled out (Dublin *et al.* 1990a). This example is less persuasive than that of the Yellowstone elk because the prerinderpest density (before 1890) was unknown, but evidence on reproduction and body condition suggests that rinderpest held the population below the level allowed by food supply, a necessary condition for a perturbation experiment implicating a disease.

A case of a population perturbed above equilibrium is provided by elephants in the Tsavo National Park, Kenya (Laws 1969; Corfield 1973). From 1949 until 1970 the population had been increasing, due in part to immigration from surrounding areas where human cultivation had displaced the animals. A consequence of this artificial increase in density was depletion of the food supply within reach of water. In 1971 the food supply ran out and there was starvation of

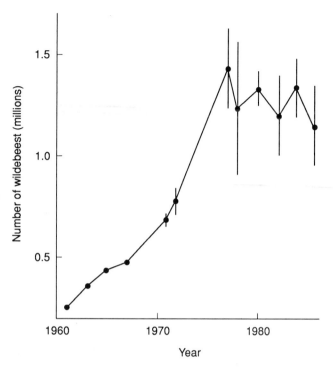

Fig. 8.8 The wildebeest population in Serengeti increased to a new level determined by intraspecific competition for food, after the disease rinderpest was removed in 1963. (After Sinclair 1989.)

females and young around the water holes. After this readjustment of density, the vegetation regenerated and starvation mortality ceased.

8.4.2 *Mean density and environmental factors*

A population uninfluenced by dispersal and unregulated (i.e., no density-dependent factors affecting it) will fluctuate randomly under the influence of weather and will eventually drift to extinction (DeAngelis and Waterhouse 1987).

Just by chance there may be for a time a correlation between density and environmental factors. However, if we take many separate populations, the probability that all of them are simultaneously correlated with an environmental factor by chance alone is very small. Therefore, if we find a correlation between mean densities from independent populations with an environmental factor, there is a strong inference that weather is influencing some resource for which animals are competing, and which results in regulation about some equilibrium point.

An example of this approach is shown in Schluter's (1988) study of seed-eating finches in Kenya (Fig. 8.9): finch abundance from various populations is correlated with seed abundance. Other examples of density correlated with weather factors are given in Sinclair (1989).

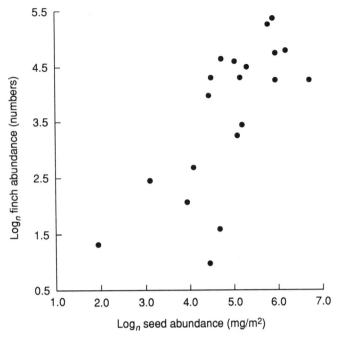

Fig. 8.9 The total abundance of seed-eating finches in savanna habitats of Kenya is related to the abundance of the food supply. Such a positive relationship in unconnected populations may demonstrate regulation. (After Schluter 1988.)

8.4.3 *Examples of density dependence*

As we discussed in Section 8.3.7, density dependence is a necessary but not sufficient requirement to demonstrate regulation. There are an increasing number of studies in the bird and mammal literature demonstrating density-dependent stages in the life cycle. For birds (Fig. 8.10a), the long-term study on great tits (*Parus major*) at Oxford, England has shown that winter mortality of juveniles is related to the number of juveniles entering the winter (McCleery and Perrins 1985). In contrast (Fig. 8.10b), early chick mortality in summer is density dependent for the English partridge (*Perdix perdix*) (Blank *et al.* 1967).

For mammals, density-dependent juvenile mortality has been recorded for red deer on Rhum, Scotland (Clutton-Brock *et al.* 1985) (Fig. 8.11a), for reindeer in Norway (Skogland 1985) (Fig. 8.11b), for feral donkeys (*Equus asinus*) in Australia (Choquenot 1991), and for greater kudu in South Africa (Owen-Smith 1990). Adult mortality was density dependent for African buffalo in Serengeti (Sinclair 1977). In each case the cause was lack of food at critical times of year. Reproduction is known to be density dependent in both birds (Arcese and Smith 1988) and mammals (Clutton-Brock *et al.* 1991). Figure 8.11c shows that the proportion of Soay sheep that give birth at 12 months of age declines with density. Fowler (1987) has reported over 100

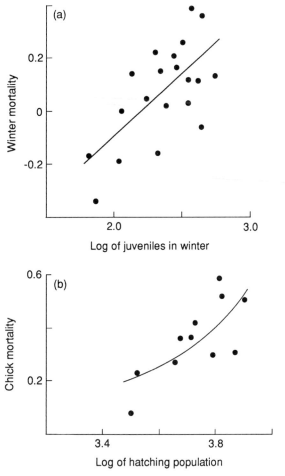

Fig. 8.10 Examples of density-dependent mortality in birds. (a) The great tit (*Parus major*) over-winter mortality (log of [juveniles in winter/first year breeding population]) plotted against log juvenile density in winter. (After McCleery and Perrins 1985.) (b) Chick mortality of the European partridge (*Perdix perdix*) (measured as log hatching population/log population at 6 weeks) plotted against log hatching population in Hampshire, England. (From Blank *et al.* 1967.)

studies of terrestrial and marine mammal populations where density dependence was detected.

Delayed density dependence has been recorded in winter mortality of snowshoe hares in the Yukon and in overwinter mortality of red grouse in Scotland (Watson and Moss 1971) (Fig. 8.12). For the hares the delay appears to have been due to a lag of 1–2 years in the response of predator populations to changing hare numbers (Trostel *et al.* 1987), while for the grouse the delay came from density responding to food conditions in the previous year (see Section 8.6.3).

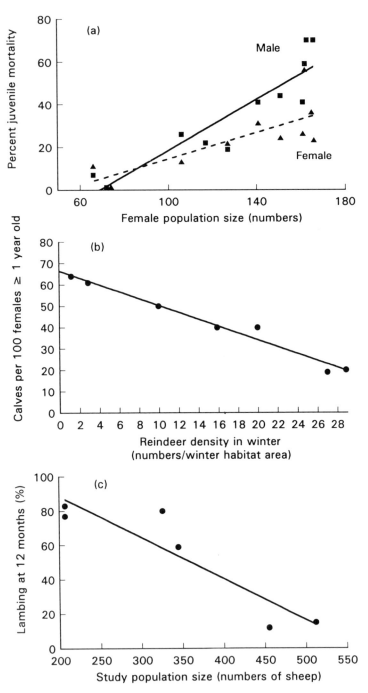

Fig. 8.11 Density dependence in large mammals. (a) The juvenile mortality rate of male and female red deer on the islands of Rhum, Scotland. (From Clutton-Brock *et al.* 1985.) (b) Juvenile recruitment per 100 female reindeer older than 1 year in Norway. (From Skogland 1985.) (c) The fertility rate of 1-year-old Soay sheep on St Kilda island. (After Clutton-Brock *et al.* 1991.)

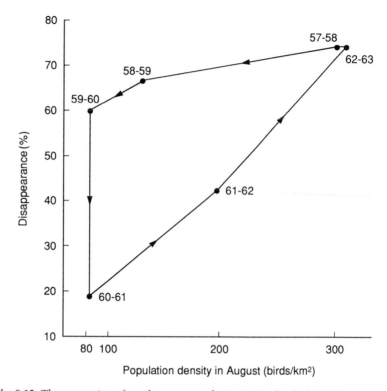

Fig. 8.12 The proportion of a red grouse population in Scotland which disappears over winter (August–April) is related to population density in the previous August in a complex way. Mortality varied according to whether the population was increasing or decreasing. By joining the points sequentially an anticlockwise cycle is produced, indicating a delayed density-dependent effect in the cause of the mortality. By plotting the percentage disappearance against density 1 year earlier, a closer fit can be obtained for a regression line. Thus the delay is 1 year. The numbers at the points are years. (After Watson and Moss 1971.)

8.5 Applications of regulation

Causes of population change can be divided into: (i) those that disrupt the population and often result in "outbreaks," and which can be either density dependent or density independent; and (ii) those that regulate and therefore return the population to its original density after a disturbance. These are always density dependent.

Knowledge of regulation may be useful for the management of house mice (*Mus domesticus*) plagues in Australia. In one experimental study (Barker *et al.* 1991), mice in open-air enclosures were contained by special mouse-proof fences. The objective was to create high densities, thus mimicking plague populations, in order to test the regulatory effect of a nematode parasite (*Capillaria hepatica*). It turned out that they could not test the effect of the parasite because other factors regulated the population and thus obscured any parasite effect. The replicated populations declined simultaneously. Why did this happen? By dividing up the life cycle into stages they found that late juvenile and adult mortality was strongly density dependent but that other stages, including fertility and newborn mortality, were not. This allowed them to discount causes that would affect reproduction and to focus more closely on what was happening among adults, in particular the social interactions of mice.

Other studies suggest that mouse populations in Australia may be regulated by predators, disease, and juvenile dispersal (Redhead 1982; Sinclair et al. 1990). Under conditions of superabundant food following good rains, the reproductive rate of females increases faster than the predation rate and an outbreak of mice occurs. The implication of these results for management is that if reproduction could be reduced, e.g., through infections of the *Capillaria* parasite, then predation may be able to prevent outbreaks even in the presence of abundant food for the mice.

8.6 **Intraspecific competition**

Regulation can occur by a number of mechanisms such as predation or parasitism, but a commoner cause is competition between individuals for resources. Such resources can be food, shelter from weather or predators, nesting sites, and space to set up territories. We have seen some examples already in Figs 8.10 and 8.11.

8.6.1 *Definition*

Intraspecific competition occurs when individuals of the same species utilize common resources that are in short supply; or, if the resources are not in short supply, competition occurs when the organisms seeking that resource nevertheless harm one or other in the process (Birch 1957).

8.6.2 *Types of competition*

When individuals use a resource so that less of it is available to others, we call this type of competition *exploitation*. This includes both removal of a resource (consumptive use) when food is consumed and occupation of a resource (preemptive use) when resources such as nesting sites are used (see Section 6.5). Individuals competing for food need not be present at the same time: an ungulate can reduce the food supply of another that arrives later.

Another type of competition involves the direct interaction of individuals through various types of behavior. This is called *interference* competition. One example of behavioral interference is the exclusion of some individuals from territories. Another is the displacement of subordinate individuals by dominants in a behavioral hierarchy.

8.6.3 *Intraspecific competition for food*

Experimental alteration of food supply
Food addition experiments provide the best evidence for intraspecific competition. Krebs et al. (1986) supplied extra food to snowshoe hares in winter from 1977 to 1985. This raised the mean winter density fourfold at the peak of the 10-year population cycle. Similarly Taitt and Krebs (1981) increased the density of vole populations (Fig. 8.13) by giving them extra food. The elk population at Jackson Hole,

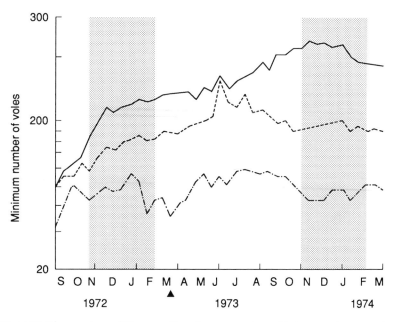

Fig. 8.13 The numbers of Townsend's voles on trapping grids increase in proportion to the amount of food provided, indicating that intraspecific competition regulates the population. Control (–·–·–); low food addition (------); high food addition (———); shaded area indicates winter. (After Taitt and Krebs 1981.)

Wyoming, is kept at a higher level than otherwise would be the case by supplementary feeding in winter (Boyce 1989). These examples show that food is one of the factors limiting density.

The dense shrubland (chaparral) of northern California contains two shrubs, chamise (*Adenostema taxiculatum*) and oak (*Quercus wislizenii*), that are preferred food for black-tailed deer (*Odocoileus hemionus*). These shrubs resprout from root stocks after burning to provide the new shoots, which are the preferred food. Taber (1956) showed that on plots thinned by experimental burning, the herbaceous food supply increased to 78 kg/ha from the 4.5 kg/ha found on control plots, and the shrub component increased to 460 kg/ha from 165 kg/ha. Deer densities consequently increased from 9.5/km² on the experimental controls to 22.9/km² on the treatment plots, while the fertility of adult females increased from 0.77 to 1.65 young per female.

Red grouse (*Lagopus lagopus*) live year round on heather (*Calluna vulgaris*) moors in Scotland. Their diet consists almost entirely of heather shoots. Watson and Moss (1971) described experiments where some areas were cleared of grouse, fertilized with nitrogen in early summer, and were then left to be recolonized. Fertilizing increased the growth and nutrient content of heather. The size of their territories did not differ between fertilized and control areas when grouse set up their territories in the fall. However, territorial grouse that had been present all winter reared larger broods on the fertilized than on the control areas, indicating that reproduction was affected by overwinter nutrition. Territory sizes did decline in the following fall and densities increased, showing the 1 year lag of density responding to nutrition. On other areas, old heather was burned every 3 years, creating a higher food supply of young regenerating heather. Territory size on these plots decreased (as density increased) in the same year as the treatment, so there was a more immediate response than on the fertilized plots.

Direct measures of food

Snowshoe hare populations in the boreal forests of Canada and Alaska reach high numbers every 10 years or so. Measurements of known food plants, and feeding experiments, suggested that the animals ran short of food at peak numbers (Pease *et al.* 1979). Other measures such as the amount of body fat (Keith *et al.* 1984) and fecal protein levels (Sinclair *et al.* 1988) also identified food shortage at this time (see Section 7.9).

African buffalo graze the tropical montane meadows of Mt Meru in northern Tanzania, keeping the grass short. Grass growth rates and grazing offtake were measured by use of temporary exclosure plots. Growth in the rainy season was more than sufficient for the animals, but in the dry season available food fell below maintenance requirements (Sinclair 1977).

Murton *et al.* (1966) measured the impact of wood-pigeons (*Columba palumbus*) on their clover (*Trifolium repens*) food supply. Food supply was measured directly by counting clover leaves in plots. Pigeons consumed over 50% of the food supply during winter. They fed in flocks, those at the front of the flock obtaining more food than those in the middle or at the back. The proportion of underweight birds (< 450 g) was related directly to the overwinter change in numbers (Fig. 8.14) and inversely related to the midwinter food supply. Thus competition within flocks resulted in some animals starving, and the change in numbers was related to the proportion that starved.

Indirect measures of food shortage

Indirect evidence for competition for food comes from indices of body condition (see Section 7.9). The last stores of body fat that are used by ungulates during food shortages are in the marrow of long bones such as the femur. Bone marrow fat can be measured directly by extraction with solvents. However, since there is an almost linear relationship between fat content and dry weight (Hanks 1981) (see Section 7.9.3), it is easier to collect a sample of marrow from carcases found in the field and oven dry it. A cruder but still effective method is to describe the color and consistency of the marrow, a method introduced by Cheatum (1949).

Other fat stores such as those around the heart, mesentery, and kidney are used up before the bone marrow fat starts to decline (see

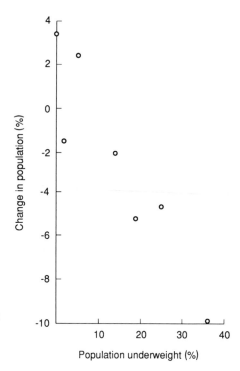

Fig. 8.14 The percentage change in a wood pigeon population in England is related to the proportion of the population that is underweight. (After Murton *et al.* 1966.)

Section 7.9). The relationship between kidney and marrow fat holds for many ungulate species (see Fig. 7.11). If both kidney and marrow fat can be collected, a range of body conditions can be recorded. However, often the marrow fat is all that is found in carcases because scavengers have eaten the internal organs.

Klein and Olson (1960) used bone marrow condition indices to conclude that deer in Alaska died from winter food shortage, as did Dasmann (1956) for deer in California. Similarly, migratory wildebeest in Serengeti that died in the dry season were almost always in poor condition, as judged by the bone marrow, and this was correlated with the protein level in their food (see Fig. 7.12). This dry season mortality was density dependent and was sufficiently strong to allow the population to level out (Sinclair *et al.* 1985).

Problems with measurement of food supply

To determine whether competition for resources such as food is the cause of regulation we need to know what type of food is eaten, how much is needed, and how much is available. What is needed must exceed what is available for competition to occur. The types of food eaten form the basis for many studies on diet selection, sometimes called *food habit studies*. These in themselves do not tell us what is needed in terms of digestible dry matter, protein, and energy. We should note that such requirements are unknown for most wild species and we have to use approximations from other, often domestic, species.

The amount of food available to animals is particularly difficult to assess because we are unlikely to measure potential food in the same way as an animal does. On the one hand animals are likely to be far more selective than our crude sampling and so we are likely to record more "food" than the animal sees. Our measures of food supply are often seriously flawed, and this is one of the reasons why direct evidence for intraspecific competition for food is rare. There is far more indirect evidence for competition provided by indicators such as body condition.

8.7 **Summary**

Regulation is a biotic process which counteracts abiotic disturbances affecting an animal population. Two common biotic feedback processes are predation and intraspecific competition for food. These are called density-dependent factors if they act as negative feedbacks. Negative feedback imparts stability to the population. Abiotic disturbances are usually provided by fluctuating weather. They are called density-independent factors and will cause populations to drift to extinction if there are no counteracting density-dependent processes operating. For wildlife management it is necessary to know: (i) what are the causes of the density-dependent processes that stabilize the population, and what are the causes of fluctuations and instability; and (ii) which age and sex groups are most influenced by these stabilizing or destabilizing processes. A common cause of regulation is intraspecific competition for food. Competition occurs when the needs of the population exceed availability. To measure such competition we need to know how much food is available and how much is needed.

9 Competition and facilitation between species

Species do not exist alone. They live in a community of several other species and some of these will interact. There are various forms of interaction between species; competition, commensalism (facilitation), mutualism (symbiosis), predation, and parasitism are the main ones. These are defined by the way each species affects the other, as shown in Table 9.1. In competition each species suffers from the presence of the other, although the interaction need not be balanced. With commensalism or facilitation one species benefits without affecting the other, while in mutualism both benefit. These can be thought of as the converse of interspecific competition. With predation and parasitism one species benefits to the disadvantage of the other. We shall discuss predation and parasitism in Chapter 11 and will confine ourselves here to interspecific competition and mutualism.

9.1.1 Definition

Interspecific competition is similar to intraspecific competition. It occurs when individuals of different species utilize common resources that are in short supply; or, if the resources are not in short supply, competition occurs when the organisms seeking that resources nevertheless harm one or other in the process (Birch 1957).

9.1.2 Implications

Interspecific competition deals with the cases when there are two or more species present; we should be aware of a number of implications arising from this definition.

1 Competition must have some effect on the fitness of the individuals. In other words resource shortage must affect reproduction, growth, or survival, and hence the ability of individuals to get copies of their genes into the next generation.

2 Although it is necessary for species to require common resources (i.e., to overlap in their requirements), we cannot conclude there is competition unless it is also known that the resource is in short supply, or that they affect each other.

3 The amount of resource such as food that is available to each individual must be affected by what is consumed by other individuals. Thus two species cannot compete if they are unable to influence the amount of resource available to the other species, or to interfere with that species obtaining the resource.

4 Both exploitation and interference competition (see Section 8.6.2)

Table 9.1 Types of interaction

Species 2	Species 1		
	+	0	−
+	Mutualism	Commensalism	Predation/parasitism
0	Commensalism		(Amensalism) Competition
−	Predation/parasitism	Competition	Competition

can occur between species, although interference between species is relatively uncommon.

9.2 Theoretical aspects of interspecific competition

To obtain an understanding of what might be the expected outcome from a simple and idealized interspecific competition we return to the logistic equation:

$$dN_1/dt = r_{m1} \times N_1 \times (1 - N_1/K_1) \tag{9.1}$$

9.2.1 Graphic models

The term in parentheses $(1 - N_1/K_1)$ describes the impact of individuals upon other individuals of the same species and on the population growth rate dN_1/dt. We must now add a term representing the impact of the second species N_2 on species 1. The equation for the effect of species 2 on population growth of species 1 is:

$$dN_1/dt = r_{m1} \times N_1 \times (1 - N_1/K_1) \times (a_{12} \times N_2/K_1) \tag{9.2}$$

where r_{m1} is the intrinsic rate of increase for species 1.

The ratio N_2/K_1 represents the abundance of species 2 relative to the carrying capacity (K_1) of species 1. It is a measure of how much of the resource is used by species 2 that would have been used by species 1. The *coefficient of competition* a_{12} measures the competitive effect of species 2 on species 1. We discuss some examples in Section 9.3.1. If we define the competitive effect of one individual of species 1 upon the resource use of an individual of its own population as unity, then the coefficient for the effect of other species is expected to be less than unity. We expect this because individuals will compete more strongly with those similar to themselves than with the dissimilar individuals of other species. This does not always occur: when two species differ greatly in size an individual of the larger species (l) may consume far more of a resource than one of the smaller species (s) and in this case the a_{sl} could be greater than unity. The converse effect of species 1 on species 2 is denoted by the coefficient a_{21} in the equation for the other species:

$$dN_2/dt = r_{m2} \times N_2 \times (1 - N_2/K_2) \times (a_{21} \times N_1/K_2) \tag{9.3}$$

These two equations (9.2, 9.3) are called the Lotka–Volterra equations, named after the two authors who produced them (Lotka 1925; Volterra 1926). We can examine the implications of the equations graphically

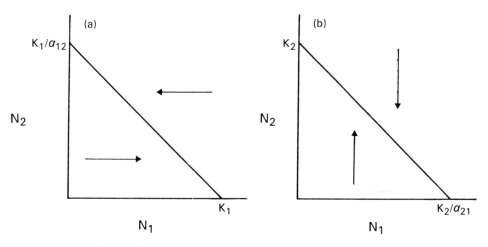

Fig. 9.1 Isoclines for the Lotka–Volterra equations. (a) At any point on the isocline $dN_1/dt = 0$. This indicates where the number of species 1 is held constant for different population sizes of species 2. Species 1 increases to the left of the isocline, but decreases right of it. (b) The isocline where $dN_2/dt = 0$. This shows where the population of species 2 is held constant at different values of species 1. Species 2 increases below the line, but decreases above it.

by plotting the numbers of species 2 against those of species 1, as in Fig. 9.1a. First we plot the conditions for species 1 when dN_1/dt is zero. There are the two extreme points when N_1 is at K_1 so that N_2 is zero, and when N_1 is zero because species 2 has taken all the resource. This latter point can be found from Equation 9.2 by setting dN_1/dt to zero and rearranging so that it simplifies to:

$$N_1 = K_1 - a_{12} \times N_2$$

If the resource is taken entirely by species 2, then:

$$N_1 = 0, \text{ and } N_2 = K_1/a_{12}$$

Of course there can be any combination of N_1 and N_2 so that dN_1/dt is zero; this is seen from the diagonal line joining these two extreme points. To the left of this line dN_1/dt is positive so that N_1 increases, and to the right it is negative and N_1 decreases as indicated by the arrows. At all points on the line (called an *isocline*) the population is stationary. Exactly similar reasoning produces the equivalent diagram for species 2 (Fig. 9.1b). Below the line (isocline) N_1 increases, and above N_2 decreases.

With these two diagrams describing the competitive abilities of the two species independently we can now predict the outcome of competition between them. If we put the two diagrams in Fig. 9.1 together, as in Fig. 9.2a, we see that K_1 is larger than K_2/a_{21}. The latter term is the number of species 1 required to drive species 2 to extinction, and since it is possible for species 1 to exist at higher numbers than this level (i.e., at K_1), species 1 will drive species 2 down. On the other axis

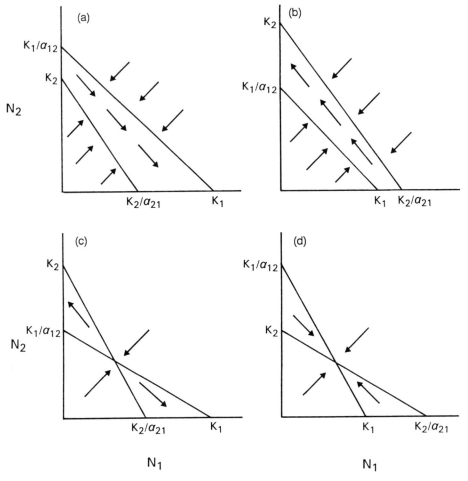

Fig. 9.2 The relationship of the two species' isoclines determines the outcome of competition. (a) Species 1 increases at all values of species 2 so that species 1 wins. (b) The converse of (a) such that species 2 wins. (c) In the region where the isocline of species 1 is outside that of species 2, species 1 wins and vice versa, so that either can win. (d) A stable equilibrium occurs because all combinations tend towards the intersection of the isoclines.

we see that K_2, which is the maximum number of species 2 that the environment can hold, is less than that necessary to drive down species 1. Therefore, species 2 always loses when the two species occur together as can be seen by the resultant arrows, and by the fact that the species 1 isocline is always outside that of species 2.

The above outcome is not the only possible solution, for this depends on the relative positions of the two isoclines which are shown in Figs 9.2b–d. Figure 9.2b is the converse to that of Fig. 9.2a so that species 2 always wins. In Fig. 9.2c we see that $K_2 > K_1/a_{12}$, and $K_1 > K_2/a_{21}$ so that, depending on the exact combination of the two population sizes, either can win. Where the two isoclines cross there is an equilibrium point but this is unstable in the sense that any slight change in the populations will cause the system to move to either K_1 or K_2 and the extinction of one of the species. In nature we would never see such an equilibrium.

Figure 9.2d also shows the two isoclines crossing, but in this case $K_2 < K_1/a_{12}$ and $K_1 < K_2/a_{21}$ (i.e., individuals of the same species affect each other more than do individuals of the other species, and neither is capable of excluding the other). This also means that intraspecific competition is always greater than interspecific competition. Hence, whatever the combination of the two populations, the arrows show that the system moves to the equilibrium point, which is therefore stable. This situation can only occur if there is some form of separation in the resources that they use, which we call niche partitioning (see Section 9.4).

9.2.2 Implications and assumptions

1 We can see from the figures that the outcome of competition depends on the carrying capacities (K_1 and K_2) and the competition coefficients (a_{12} and a_{21}) according to the Lotka–Volterra model. The intrinsic rate of increase has no influence on which species will be the eventual winner.

2 Coexistence occurs when intraspecific competition within both species is greater than interspecific competition between them.

3 These equations can be expanded to include the effects of several species on species 1 by summing the $a \times N$ terms. This assumes that each species acts independently on species 1.

4 There are several other assumptions underpinning the logistic equation, e.g., constant environmental conditions leading to constant r and K, and no lags in competing species' responses to each other. Furthermore, the competition coefficients are constant: the intensity of competition does not change with size, age, or density of the competing species.

These assumptions mean that the Lotka–Volterra equations, like the logistic one, are simplistic and idealized. It is unlikely that the assumptions hold, although they may be approximated in some cases. The real value of these models is that they show how it is possible for coexistence to occur in the presence of competition, and that exclusion is not necessarily predetermined but may depend on the relative densities of the competing species.

9.3 Experimental demonstrations of competition

9.3.1 Perturbation experiments

Much of the work in ecology has assumed that competition has occurred and is necessary for the coexistence of species, and competition is one of the major assumptions in Darwin's theory of natural selection. Nevertheless, it is necessary to demonstrate that interspecific competition does actually take place. One of the most direct approaches is to carry out a *removal experiment* whereby one of the species is removed, or reduced in number, and the responses of the other species are then recorded. If competition has been operating we would expect that either the population, or reproductive rate, or growth rate of the other species would increase.

One such experiment, illustrated in Fig. 9.3, examined the competitive effect of voles (*Microtus townsendii*) on deermice (*Peromyscus maniculatus*). Deermice normally live in forests but one race on the west coast of Canada can also live in grassland, the normal habitat of voles. Redfield *et al.* (1977) removed voles from three plots and compared the population response of the deermice there with that on two control areas. On one control there were no deermice, on the other 4.7/ha. All the removal areas showed increases in deermouse numbers, one going from 7.8/ha before removal to 62.5/ha 2 years later. At the end of the study, when they stopped removing voles, these animals recolonized, reaching densities of 109/ha, while deermice numbers dropped to 9.4/ha. In another experiment, instead of removing voles, Redfield disrupted the social organization of the voles by altering the sex ratio so that there was a shortage of females, but the density remained similar to that of controls. In this area deermice numbers increased from nearly zero to 34/ha. This result suggests that it was *interference competition* due to aggression from female voles that excluded the deermice because the density and food supply remained the same.

A similar type of experiment was conducted on desert rodents in Arizona by Munger and Brown (1981). They excluded larger species from experimental plots while smaller species were allowed to enter. Plots were surrounded by a fence and access was controlled by holes cut to allow only the smaller species to enter. There were two types of

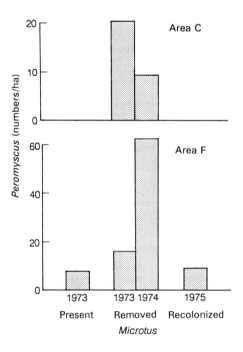

Fig. 9.3 Population densities (numbers per hectare) of deermice (*Peromyscus*) on two areas from which voles (*Microtus*) were removed in the years 1973 and 1974. On area F only deermice were monitored after voles were allowed to recolonize. Deermice were absent or in low numbers before and after the vole removal, but high during the removal. (After Redfield *et al.* 1977.)

small rodents: those that ate seeds (granivores) and those that ate a variety of other foods as well (omnivores). Munger and Brown predicted that if there was exploitation competition for seeds between the large and small granivores then the latter should increase in number in the experimental plots, while the omnivore populations should stay the same; if, however, the increased density of granivores was an artifact of the experiment (e.g., by excluding predators) then the number of small omnivores should also increase. Figure 9.4 shows that after a 1-year delay small granivores reached and maintained densities that

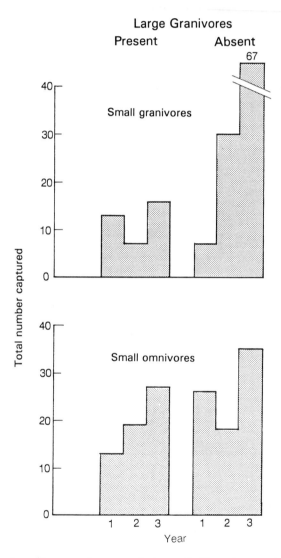

Fig. 9.4 Exclusion of large granivorous rodents resulted in an increase in the small granivorous rodent population relative to control areas, indicating competition. Small omnivorous rodent numbers did not increase significantly, indicating lack of competition. (After Munger and Brown 1981.)

averaged 3.5 times higher on the removal plots than on the controls, but the small omnivores did not show any significant increase. These results are consistent with the interpretation that there was competition between large and small granivorous rodents.

Although the above examples produced results consistent with the predictions of interspecific competition, there was no attempt to measure the competition coefficients. However, Abramsky *et al.* (1979) carried out a similar removal experiment on the shortgrass prairie in Colorado in which a competition coefficient was measured. In this case voles (*M. ochrogaster*) were removed and the response of deermice (*P. maniculatus*) recorded.

Figure 9.5 shows the negative relationship between the number of deermice present in the removal plot and the number of voles present in the previous sampling period 2 weeks earlier, as expected if competition were acting. To measure the competition effect (*a*) of voles on deermice, the Lotka–Volterra equation was used. At equilibrium $dN_1/dt = 0$, and so:

$$K_1 = N_1 + a \times N_2 \times (2^{0.75})$$

where:

K_1 = the carrying capacity of the environment for individuals of deermice when alone,

N_1 = the number of deermice,

N_2 = the number of voles,

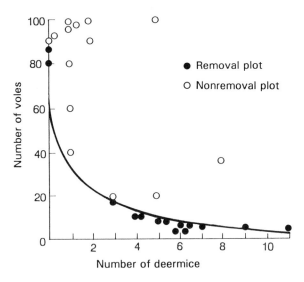

Fig. 9.5 Number of deermice (*Peromyscus maniculatus*) known to be alive at time *t* plotted against number of voles (*Microtus ochrogaster*) known to be alive at time *t* − 1 (*t* − 1 is the sample period 2 weeks earlier than period *t*). (After Abramsky *et al.* 1979.)

$2^{0.75}$ = the conversion factor, and standardizes the species in terms of their metabolic rates.

The bodyweight (W) of voles is about two times that of deermice, and the basal metabolic rate (M) is taken as $M = W^{0.75}$ (see Section 7.5.2). Using various combinations of N_1 and N_2 an average estimate of $a = 0.06$ was obtained.

Properly designed removal experiments are difficult to carry out for practical reasons, so it is not surprising that they have not yet been performed with large mammals.

9.3.2 *Natural experiments*

An easier approach uses natural absences or combinations of species to observe responses that would be predicted from interspecific competition. For example, mallard ducks (*Anas platyrhynchos*) breed on oligotrophic (low-nutrient) lakes in Sweden (Pehrsson 1984). Some of the lakes contained fish, while others did not. In lakes with fish, mallard-pair density was lower, mean invertebrate food size was lower, and emerging insects were significantly smaller. In an experiment where ducklings were released, their intake rate was higher on lakes without fish (Table 9.2). These results imply competition between ducks and fish.

Another type of natural experiment is illustrated by the distributions of two gerbilline rodent species (Abramsky and Sellah 1982). One species, *Gerbillus allenbyi*, lives in coastal sand dunes and is bounded in the north by Mt Carmel, Israel. In the same region the other species, *Meriones tristrami*, is restricted to nonsandy habitats. In the coastal area north of Mt Carmel, *M. tristrami* occurs alone and inhabits several soil types, including the sand dunes. Abramsky and Sellah suggested that *M. tristrami* colonized from the north and was able to bypass Mt Carmel, whereas *G. allenbyi* colonized from the south and could not pass the Mt Carmel barrier. In the region of overlap, south of the barrier, interspecific competition had excluded *M. tristrami* from the sand dunes. They tested this hypothesis by removing *G. allenbyi* from habitats where the two species overlapped, and found that there was no significant increase in *M. tristrami*. They

Table 9.2 Mean dry weight of subaquatic invertebrates available to mallard ducklings and the rate of duckling food intake in lakes with and without fish in Sweden. (After Pehrsson 1984)

	Year	Lakes without fish	Lakes with fish
Mean dry weight (May–June)	1977	119.8 (21.0)	45.3 (13.7)*
	1978	159.0 (9.9)	26.5 (4.8)*
Duckling feeding (food items per minute)	1977	12.4 (0.6)	9.5 (0.5)†
	1978	20.4 (5.1)	7.9 (0.7)*

*$P < 0.01$; †$P < 0.001$.
Values in parentheses are percentages.

concluded that there was no present-day competition occurring. Instead they suggested that competition in the past had resulted in a shift in habitat choice so that there was no longer any detectable competition.

Islands are sometimes used to look at the distributions of overlapping species, because on some islands a species can occur alone while on others it overlaps with related species. The theory of interspecific competition would predict that when alone a species would expand the range of habitats it uses (a process we call *competitive release*), while on islands where there are several species the range of habitats contracts (*competitive exclusion*). A good example of this is seen in ground doves on New Guinea and surrounding islands (Diamond 1975). On the larger island of New Guinea there are three species each with its own habitat (Fig. 9.6): *Chalcophaps indica* in coastal scrub, *C. stephani* in second growth forest further inland, and *Gallicolumba rufigula* in the interior rainforest. On the island of Bagabag *G. rufigula* is absent and *C. stephani* expands into the mature forest. On some islands (New Britain, Karkar, Tolokiwa) only *C. stephani* occurs and it uses all habitats, while on the island of Espiritu Santo only *C. indica* occurs and this species also expands into all habitats. It is assumed that this habitat expansion has been due to competitive release through the absence of the other potential competitors.

9.3.3 *Interpreting perturbation experiments*

Perturbation experiments are designed to measure responses of populations that would be predicted from interspecific competition theory. We should be aware, however, that there are two types of perturbation (Bender *et al.* 1984). One, called a *pulse* experiment, involves a one-

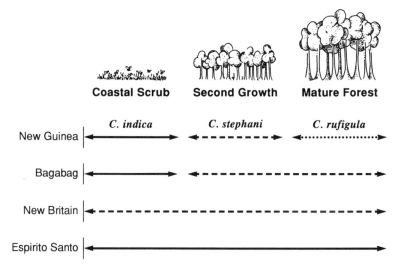

Fig. 9.6 Habitats of ground doves (*Chalcophaps* and *Gallicolumba*) on islands off New Guinea demonstrate "competitive release." (After Diamond 1975.)

time removal of a species. We then measure the rate of return by the various species to the original equilibrium. This requires accurate measurements of rates of population increase which in practice is not easy and in fluctuating environments very difficult. As a consequence few of these experiments are carried out.

The other type of perturbation is the continuous removal, or *press* experiment. Let us assume that species 1 is reduced to a new level and kept there. Other species are allowed to reach a new equilibrium and it is this level that is observed. This type of perturbation avoids having to measure rates, but there are other problems. If there are more than two species in a community, which in most cases there are, an increase in another species' population is neither a necessary nor a sufficient demonstration of competition. First, species 1 and 2 may not overlap and so not compete but they may affect each other through interactions with other competing species: this is *indirect competition*. Second, the two species could be alternate prey for a food-limited predator. Changes in the population of species 1 could affect that of species 2 by influencing the predator population: this has been termed *apparent competition* (Holt 1977) and we will discuss it again below (see Section 9.8).

All of the examples we have discussed above are press experiments and, strictly speaking, to demonstrate competition unequivocally we would need to know that: (i) resources were limiting; (ii) there was overlap in the use of the resources; (iii) other potential competitors were having a negligible effect; and (iv) predator populations were not responding to the experiment. In few cases have all these conditions been met. Because of these difficulties, an entirely different approach to the study of interspecific competition has measured the pattern of overlap in the use of resources. We now consider this approach.

9.4 The concept of the niche

In Chapter 2 we saw that different species on different continents appeared to adopt the same role in the community and often these species have evolved similar morphologic and behavioral adaptations. This place in the community is called the *niche*, defined by Elton (1927) as "the functional role and position of the organism in its community."

For practical reasons the niche has come to be associated with use of resources. Thus, we can plot the range and frequency of seed sizes eaten by different bird species, as a hypothetical example in Fig. 9.7a. Species that exploit the outer parts of the resource spectrum use a broader range of resources because they are less abundant. Some species, e.g., 2, 3, and 4, overlap while others such as 2 and 5 do not. Overlap is necessary (but not sufficient) to demonstrate competition. An example (Fig. 9.7b) is provided by the range of seed sizes eaten by finches in Britain (Newton 1972). In this case we see that, contrary to the theoretical distribution proposed in Fig. 9.7a, there is a broader range of

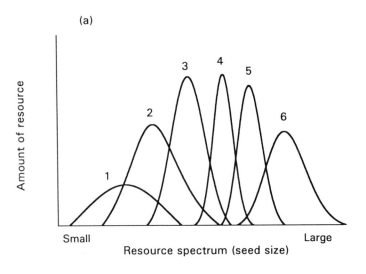

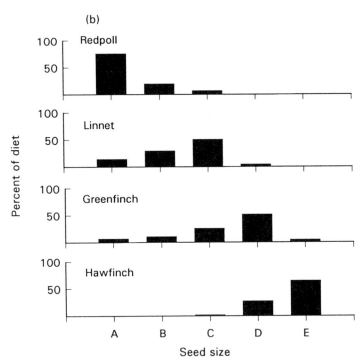

Fig. 9.7 (a) Hypothetical frequency distribution of seeds of different sizes indicating the range and overlap of potential niches for granivores. (After Pianka *et al.* 1979.) (b) Range of seed sizes eaten by British finches feeding on herbaceous plants. Seeds are in five size-categories A–E. The finches are redpoll (*Carduelis flammea*), linnet (*C. cannabina*), greenfinch (*Chloris chloris*), and hawfinch (*Coccothraustes coccothraustes*). (After Newton 1972.)

seed sizes eaten by these finches in the middle range than by birds eating seeds at the extremes.

So far we have considered only one resource axis, i.e., one variable such as seed size. When we consider two or more axes the picture becomes less clearcut in terms of overlap. Take two species, 1 and 2, which overlap along two axes, e.g., moisture and temperature as in Figs 9.8a and 9.8b. If we plot the outline of the two species distributions by considering the two axes simultaneously we see that it is possible for the two distributions to be distinct (Fig. 9.8c) or to overlap (Fig. 9.8d). Which one occurs depends on whether individuals show *complementarity* (i.e., individuals that overlap on one axis do not do so on the other (Fig. 9.8c), or overlap simultaneously on both axes (Fig. 9.8d)).

An example of complementarity is shown in Fig. 9.9. DuBowy (1988) examined the resource overlap patterns in a community of seven North American dabbling ducks all of the genus *Anas*, by plotting habitat overlap against food overlap for pairs of species. In winter, when it is assumed that resources were limiting, points for pairs were below the diagonal line (Fig. 9.9a), indicating complementarity: pairs with high overlap in one dimension had low overlap in the other. In contrast, during summer, species pairs showed high resource

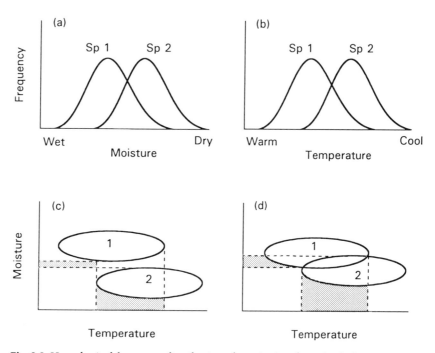

Fig. 9.8 Hypothetical frequency distribution of species 1 and species 2 along two parameter gradients: (a) moisture; (b) temperature. Outline of the species distributions when considering the two parameters simultaneously shows niches that can be either distinct (c), or overlapping (d).

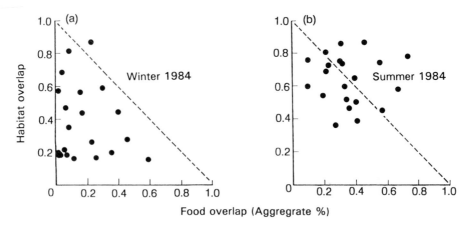

Fig. 9.9 Resource overlaps in seven species of dabbling ducks. Below the dotted line there is complementarity in overlaps. (a) In winter, high habitat overlap between pairs of species tends to be associated with low food overlap, demonstrating complementarity. (b) In summer, there is simultaneous overlap in both dimensions. (After DuBowy 1988.)

overlap in both dimensions (several points are outside the line), indicating that species fed on the same food at the same place. In summary, the change in niche of these duck species from summer to winter results in lower overlap and by implication lower competition at a time when we would expect that resources would be limiting. Note, however, that neither the lack of resources nor interspecific competition was demonstrated, merely that the results conform to what we would predict if competition had been acting.

We have considered only two dimensions of a niche so far, but clearly the niche must include every aspect of the environment that would limit the distribution of the species. We cannot draw all these dimensions on a graph but we could perhaps imagine a sort of sphere or volume with many dimensions which could theoretically describe the complete niche. Hutchinson (1957) described this as the *n-dimensional hypervolume*. This is the *fundamental niche* of the species and is defined by the set of resources and environmental conditions which allow the species to persist.

The fundamental niche is rarely if ever seen in nature because the presence of competing species restricts a given species to a narrower range of conditions. This range is the observed or *realized niche* of the species in the community. It emphasizes that interspecific competition excludes a species from certain areas of its fundamental niche. In terms of the Lotka–Volterra diagrams (see Fig. 9.2) the weaker competitor has no realized niche in Figs 9.2a and 9.2b, and for Fig. 9.2d parts of the fundamental niche are not used.

The difference between the two types of niche can be seen in the study of red-winged blackbirds (*Agelaius phoeniceus*) and yellow-headed blackbirds (*Xanthocephalus xanthocephalus*) by Orians and

Willson (1964). Both species make their nests among reeds in fresh-water marshes of North America, and, if alone, both will use the deep water parts of the marsh (there is greater protection from mammalian predators here). However, when the two species occur together, the yellow-heads exclude the red-winged blackbirds, which are then restricted to nesting in the shallow parts. Thus, the fundamental niche for nesting red-winged blackbirds is the whole marsh, but the realized niche is the shallow-water reed beds. Coexistence occurs from the partitioning of the resource (nesting habitat), and the divergence of realized niches.

9.5 The competitive exclusion principle

In 1934 Gause stated that "as a result of competition two species hardly ever occupy similar niches, but displace each other in such a manner that each takes possession of certain kinds of food and modes of life in which it has an advantage over its competitor" (Gause 1934). In short, two species cannot live in the same niche, and if they try one will be excluded; second, coexisting species live in different niches. This is known as the *principle of competitive exclusion*, or Gause's principle, and has become one of the fundamental tenets of ecology. It is the basis for studies of habitat partitioning and overlap as a way of measuring interspecific competition.

There are, however, two serious problems with Gause's statement. The first is that it is a trivial truism, because we have already identified the two coexisting populations as being different by calling them different species, and, therefore, if we look hard enough we are likely to find differences in their ecology as well. This is called a tautology: having defined the species as being different, it should be no surprise to find they are different.

The second problem is that the principle is untestable. It cannot be disproved because either result (exclusion or coexistence) can be attributed to the principle. To disprove the principle it is necessary to demonstrate that the niches of two species are identical. Yet, as we can see from Fig. 9.8, what appears to be overlap, even complete overlap, may not be so when an additional axis is taken into account. Since we can never be sure that we have measured all relevant axes in describing the niches of two species, we can never be sure that the two niches are the same. Hence, we cannot disprove the principle.

9.6 Resource partitioning and habitat selection

9.6.1 Habitat partitioning

Despite these problems with the competitive exclusion principle, it underlies the numerous studies of habitat partitioning among groups of coexisting species. Lamprey (1963) described the partitioning of habitats by species of savanna antelopes in eastern Africa. A similar study by Ferrar and Walker (1974) showed how various antelopes in Zimbabwe use the three habitat types of grassland, savanna, and woodland (Fig. 9.10). In both cases there is partitioning as well as overlap.

Similar studies by Wydeven and Dahlgren (1985) showed partitioning of both habitat and food in North American ungulates (Fig.

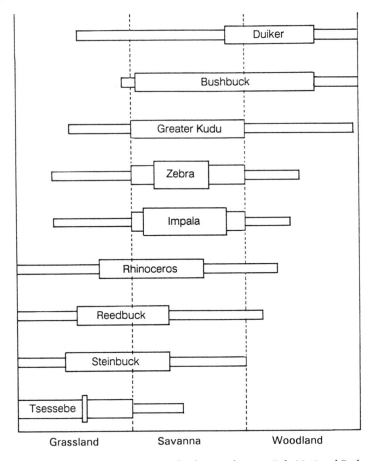

Fig. 9.10 Habitat partitioning and overlap by ungulates in Kyle National Park, Zimbabwe. Width of boxes reflects the degree of preference. (After Ferrar and Walker 1974.)

9.11). In Wind Cave National Park elk and mule deer have similar winter habitat choices, as do pronghorn and bison, but these pairs have very different diets. For example, the diet of bison contains 96% grass as against 4% for pronghorn.

MacArthur (1958), in a now classic paper, described the different feeding positions of five species of warblers within conifer trees in northeastern USA. They varied in both height in the tree and use of inner or outer branches.

9.6.2 Limiting similarity

As we have mentioned above it should not be surprising that species divide up the resource available to them. However, Gause's principle implies that there should be a limit to the similarity of niches allowing coexistence of two species. Earlier studies predicted values of limiting similarity based on theoretical arguments (MacArthur and Levins 1967). If the distance between the midpoints of species distributions along the resource axis is d and the standard deviation of the

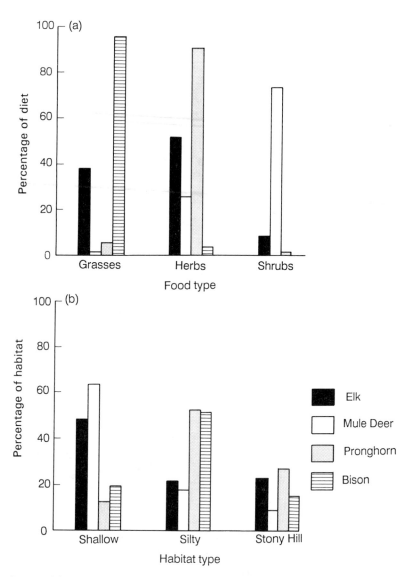

Fig. 9.11 (a) Diet and (b) winter habitat use of elk, mule deer, pronghorn, and bison in Wind Cave National Park, South Dakota. Where habitat choice is similar there are major differences in diet. (After Wydeven and Dahlgren 1985.)

curves (such as those in Fig. 9.7a) is w (the relative width), then limiting similarity can be predicted from the ratio d/w. However, various assumptions, e.g., the curves must be similar, normally distributed, and along only one resource axis, make this approach unrealistic.

Pianka *et al.* (1979) asked: how much would niches overlap if resources were allocated randomly among species in a community? A frequency distribution of niche overlaps generated from randomly constructed communities is shown in Fig. 9.12. This can be compared with distributions of observed overlap in diets of desert lizards from 28

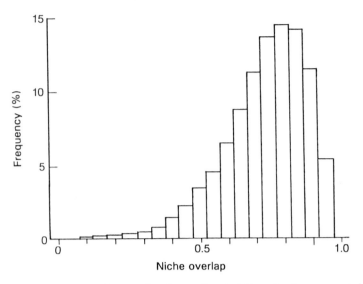

Fig. 9.12 Frequency distribution of niche overlaps in 100 randomly constructed communities with 15 species and five equally abundant resource states. (After Pianka *et al.* 1979.)

sites on three continents (Fig. 9.13). Those in the Kalahari desert of southern Africa showed the greatest degree of overlap (because one food type, termites, comprised a large amount of the diet), and those in Australia the least. In no case were observed distributions similar to the random distributions: there were far more species pairs with small overlaps than would be predicted by chance, implying that interspecific competition was causing niche segregation.

9.6.3 *Habitat selection from field data*

As we have seen, species usually differ from each other by choosing different resources such as food types, habitats, etc. We call this choice *habitat selection*. One approach to measuring the competition coefficients has been to look at the variation of a species density in different habitats. First, the variation in density due to habitats, and other resources, is estimated by statistical procedures such as multiple regression (Crowell and Pimm 1976). Then the remaining variance should be attributable to interspecific competition with another identifiable species. An example of this approach is given by Hallett (1982) in a study of 10 desert rodent species in New Mexico. He measured habitat variables related to the common plants such as number of individuals, plant height, distance to nearest plant from trap, and percentage cover. Regression analysis was used to partition the variance in capture frequency at trap stations due to habitat variables and competitors. Competition was observed within one group of three species, *Perognathus intermedius*, *P. penicillatus*, and *Peromyscus eremicus* (Table 9.3). Although the competitive effects differed from year to year, they were not random. Also the inhibitory effects were

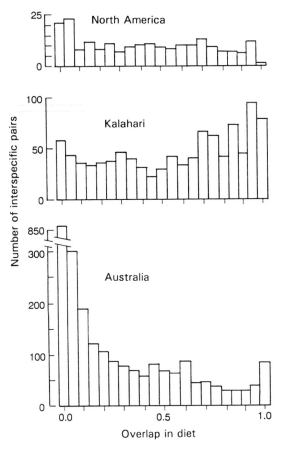

Fig. 9.13 Distributions of observed overlap in the diets of desert lizards. Dietary overlap is higher in Kalahari lizards, where termites comprised 41% of the diet. (After Pianka *et al.* 1979.)

Table 9.3 Matrix of competition coefficients for the *Perognathus–Peromyscus* guild for each year. Entries are the partial regression coefficient after removal of the effects of the habitat variables. The coefficients are the effects of the column species (independent variable) on the row species (dependent variable)

	1971		1972			1973		
	PP	PI	PP	PI	PE	PP	PI	PE
PP	–	−0.43*	–	−0.17	−0.42*	–	−0.12	−0.82*
PI	−0.17	–	−0.09	–	0.05	−0.05	–	−0.39*
PE	NI	NI	−0.09	0.19	–	−0.12*	−0.10	–

* $P < 0.05$.
PP, *Perognathus penicillatus*; PI, *Perognathus intermedius*; PE, *Peromyscus eremicus*; NI, not included in the analysis.

not symmetrical: thus, *P. eremicus* always had a greater effect on the other two species than the reverse, and similarly *P. intermedius* had a greater effect on *P. penicillatus* than vice versa.

Abramsky (1981) used a similar regression method to look at interspecific competition and habitat selection in two sympatric rodents, *Apodemus mystacinus* and *A. sylvaticus*, in Israel. Plotting the densities of the two species in different habitats against each other (Fig. 9.14) indicated a negative relationship and suggested that interspecific competition may be operating. However, he found that species abundances could have been the result of habitat differences alone; the effect of the presence of the other species was negligible in this case, implying no competition.

There are problems with the regression method, some of which are outlined by Abramsky *et al.* (1986). One is that if sympatric populations of different species differ greatly in average abundance, then estimates of their variance and regression coefficients are distorted. In turn, estimates of competition are unreliable. A second problem lies in the assumption of constant competition coefficients; if competition is weak when populations are close to equilibrium (which we assume is when regressions from field populations are estimated), but strong when disturbed from equilibrium (the situation in perturbation experiments), then regression analysis is likely to miss competitive effects while experiments will indicate their presence. A third problem is that we can never be sure that we have accounted for all the variability in density at various sites from environmental factors; there may be some factor that has been overlooked to account for the remaining variability instead of attributing it to interspecific competition.

9.6.4 *The theory of habitat selection*

Since species prefer to use some habitats over others, we ask how does this choice change when resources become limiting? There are two hypotheses that we should consider. We start with the *theory of*

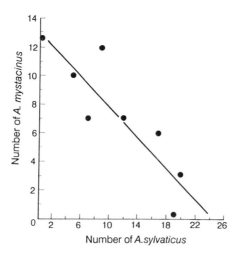

Fig. 9.14 Negative correlation in the numbers of two woodmice species (*Apodemus mystacinus, A. sylvaticus*) in Israel. (After Abramsky 1981.)

optimal foraging which predicts that when resources are not limiting species should concentrate their feeding on the best types of food or the best types of habitat and ignore the others no matter how abundant they are. When resources are limiting then a species should expand its niche to include other types of food, habitat, etc. This is the expected response under intraspecific competition.

When two species are present one might expect both to respond to declining resources by expanding their niches and so increase the overlap. However, Rosenzweig's (1981) *theory of habitat selection* introduces a second hypothesis which predicts that, when resources are limiting, species should contract their niches as a result of inter-specific competition. He considered two different situations: we start by assuming that there are two species, 1 and 2, and two habitats, A and B. In the first case, called the *distinct preference* case, both species use both habitats but each prefers to use a different one (i.e., species 1 treats A as the better and B as the poorer habitat, while species 2 does the reverse). In the second case, the *shared preference* case, again both species use both habitats, but now both treat the same habitat A as the better and B as the poorer one.

In either case we should first consider the habitat choice of a species when no competitors are present. Under conditions of abundant resources, such as food, a species should confine itself to its preferred habitat. As density increases and resources become limiting through the feeding of other individuals, the species will continue to remain in the better habitat (A) so long as the food intake rate is greater than what it would be in the poorer habitat (B). At some point, density in habitat A increases so that intake rate falls until it equals that in habitat B, and at this stage the species should not confine itself to A but should expand its habitat use in such a way that densities keep intake rates similar in the two habitats. The intake rate at which the species changes from one to two habitats is called the *marginal value*.

Now we consider what happens when there is a competitor present and resources are limiting. The outcome depends on which of the two preference cases occurs. First, in the distinct preference case, each species will confine itself to its preferred habitat rather than expand into the other habitat. Therefore, when resources are limiting, species will specialize, contract their niches, and reduce overlap. When resources are abundant they should either use one or both habitats depending on their intake rates in the two habitats. Second, in the shared preference case, we have to assume that one of the species (1) is dominant and can exclude, by behavior or other means such as scent marking, the second species from the preferred habitat A. If species 2, the subordinate, is to coexist with species 1 then it must be more efficient at using the less preferred habitat B than species 1. If the dominant is more efficient in both habitats then it will exclude species 2. Therefore, when resources are limiting, one species – the dominant one – will not change its habitat choice. In contrast the subordinate will change its choice from A to B: the competitive effect is asymmetric, with the dominant having a large effect on the subordinate while the reverse effect is small.

In a test of these hypotheses Pimm and Pimm (1982) recorded the feeding choices of three nectar-feeding bird species (*Himatione, Loxops, Vestiaria*) on the island of Hawaii. There were two main tree species, *Metrosideros* and *Sophora*, which come into flower at different times of the year. The evidence for the distinct preference case is seen in Fig. 9.15. When the number of flowers is high, all three species feed on both trees. When flowers per tree are low (and assuming that this indicated limiting resource) only *Loxops* feed on *Sophora*, and only *Himatione* feed on *Metrosideros*. Thus, both species reduce their niche width and specialize. There was also evidence of shared pref-

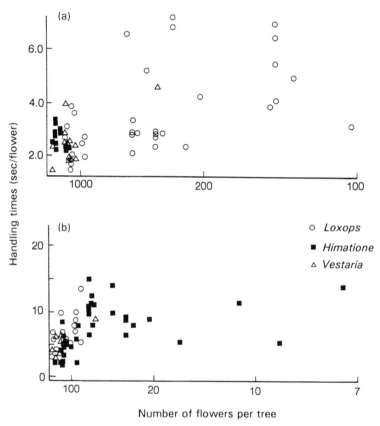

Fig. 9.15 Demonstration of distinct and shared preferences in habitat selection by three species of honeycreepers in Hawaii. The two main flowering trees are (a) *Sophora* and (b) *Metrosideros*. At low flower numbers *Loxops* (○) fed on *Sophora*, and *Himatione* (■) fed on *Metrosideros*, showing "distinct preference." *Vestaria* (△) fed on both trees, excluding the other species, but only at high flower numbers, indicating "shared preference." Note the reverse scale on the x-axis. (After Pimm and Pimm 1982.)

erence. *Vestaria* feed on both tree species but only at high flower numbers, and it physically excludes the other species by visual and vocal displays. In contrast both *Himatione* and *Loxops* spend much of their feeding time on trees with few flowers. Thus these two species are confined to poorer feeding areas during times when resources are low, as predicted by the theory.

Rosenzweig's theory predicts that niches contract when resources are limiting and there is interspecific competition. We have seen that the Hawaiian honeycreepers may conform to the predictions, but what about other species? Information from wildlife both agrees and disagrees with the predictions. The overlap in diet of sympatric mountain goats (*Oreamnos americanus*) and bighorn sheep (*Ovis canadensis*) is high in summer but reduces in winter (Dailey *et al.* 1984), as predicted by the theory. In ducks we have already seen that during winter there is a decrease in overlap (see Fig. 9.9). Burning grasslands increases the nutrient content of regenerating plants and may produce locally abundant food. Under these conditions mountain goats and mule deer (*Odocoileus hemionus*) actually increase dietary overlap (Spowart and Thompson Hobbs 1985). In contrast, elk and deer in natural forests increased dietary overlap in winter when resources are assumed to be least available, contrary to expectation (Leslie *et al.* 1984).

We should note that we do not have actual measures of the food supply in these examples, so we cannot be sure that we are seeing competition. In Serengeti, Tanzania, wildebeest are regulated by lack of food in the dry season, so that overlap with this species at this time should result in competition. However, overlap in both diet and habitat between wildebeest and several other ungulate species increases or does not change between wet and dry seasons (Hansen *et al.* 1985; Sinclair 1985). One interpretation could be that interspecific competition is asymmetric, with the impact of the rarer ungulates on the numerous wildebeest being very slight, while the reverse does not occur because these other ungulates may be regulated by predation (Sinclair 1985).

9.7 Competition in variable environments

So far we have discussed the patterns of occupancy and utilization of habitats as if they were constants for a species, or that they only changed seasonally. However, longer-term studies are now showing that species densities vary in the same habitat and they also change over a longer time scale measured in years. Thus populations may go through periods when there are abundant resources and, although there is overlap with other species even at the supposedly difficult time of year, there is no competition (Fig. 9.16). Occasionally there are periods of resource restriction and it is only at these times that one sees competition and niche separation (Wiens 1977).

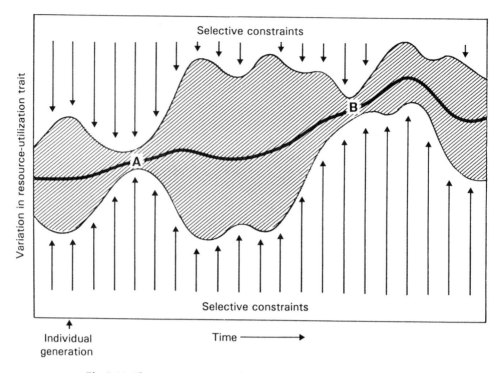

Fig. 9.16 Changes over time in the mean (thick line) and variance (shaded area) of a selective constraint such as resource availability. At times A and B there are "bottlenecks" when competition is more likely. (After Wiens 1977.)

9.8 Apparent competition

9.8.1 Shared predators

Some of the predicted outcomes from interspecific competition include the reduction of populations, the contraction of niches, and exclusion of species from communities. However, these predictions are also to be expected when species have nonoverlapping resource requirements but share predators, especially when predators can increase their numbers fast.

Let us suppose there is a predator that is food limited, and which feeds on two prey species. The prey are both limited by predation and not by their own food supplies. If species 1 increases in number then this should lead to more predators, which in turn will depress species 2 numbers. This result is called *apparent competition* because it produces the same changes in prey populations as would be predicted from interspecific competition (Holt 1977; 1984).

If both prey species live in the same habitat then, at high intensities of predation, coexistence is unlikely. On the other hand coexistence is promoted if the two species select different habitats, i.e., niche partitioning occurs.

9.8.2 Implications

Since the observed responses of prey populations to changes in predator numbers are similar to those from interspecific competition, we cannot infer such competition simply from observations or even experiments that show either changes in species population size or niche shifts. We need to know: (i) whether resources are limiting; and (ii) the predation rates and predator numbers.

9.9 **Facilitation**

9.9.1 *Examples of facilitation*

Facilitation is the process whereby one species benefits from the activities of another. In some cases the relationship is *obligatory* as in the classic example of the Nereid worm (*Nereis fucata*), which lives only in the shell of hermit crabs (*Eupagurus bernhardus*). The crabs are messy feeders and scraps of food float away from the carcase that is being fed upon; these scraps are filtered out of the water by the worm. While the worm benefits, the crab appears not to suffer any disadvantage (Brightwell 1951). In other cases the relationship may be *facultative*, by which we mean that the dependent species does not have to associate with the other in order to survive, but does so if the opportunity arises. Thus cattle egrets (*Ardea ibis*) often follow grazing cattle in order to catch insects disturbed by these large herbivores. Although the birds increase their prey capture rate by feeding with cattle, as they probably do by following water buffalo (*Bubalus bubalus*) in Asia, and elephants and other large ungulates in Africa, they are quite capable of surviving without large mammals (McKilligan 1984). The European starling (*Sturnus vulgaris*) also follows cattle on occasions. In contrast its relative in Africa, the wattled starling (*Creatophora cinerea*) seems always to follow large mammals, and in the Serengeti they migrate with the wildebeest like camp followers.

Vesey-Fitzgerald (1960) suggested that there was grazing facilitation among large African mammals. Lake Rukwa in Tanzania is shallow and has extensive reedbeds around the edges. The grasses, sedges, and rushes can grow to several meters in height, and in this state the vegetation can be fed upon only by elephants. As the elephants feed and trample the tall grass they create openings where there is lush regenerating vegetation. This provides a habitat for African buffalo which in turn provide short grass patches that can be used by the smaller antelopes such as topi. In this case elephants are creating a habitat for buffalo and topi, which would not otherwise be able to live there. Therefore, the presence of elephants increases the number of herbivores that can live in the Lake Rukwa ecosystem. Vesey-Fitzgerald called this sequence of habitat change in the grasslands a *grazing succession*.

Bell (1971) has described a similar grazing succession amongst the large mammals of Serengeti. In certain areas of Serengeti there is a series of low ridges bounded by shallow drainage lines. The ridges have sandy, thin soils and support short, palatable grasses. The drainage lines have fine silt or clay soils that retain water longer than those on the ridges and so support dense but coarse grasses, which remain green long into the dry season. Between these two extremes there are intermediate soil types on the slopes. The whole soil sequence from top to bottom is called a *catena*. In the wet season when all areas are green, all five nonmigratory species (wildebeest, zebra, buffalo, topi, Thomson's gazelle) feed on the ridge tops. Once the dry season starts the different species move down the soil catena into the longer grass in sequence, with the larger species going first (Fig. 9.17). Thus, zebra is one of the first species to move because it can eat the tough tall grass stems. By removing the stems, zebra make the basal leaves in these tussock grasses more available to wildebeest and topi, and these in turn prepare the grass sward for the small Thomson's gazelle. Thus, there is a grazing succession.

Zebra, wildebeest, and Thomson's gazelle also have much larger migratory populations in Serengeti, separate from the smaller resident populations discussed above. It is tempting to think that the movements of these migrants follow the same pattern as those of the resident populations. Indeed, McNaughton (1976) has shown that migrating Thomson's gazelle prefer to feed in areas already grazed by wildebeest because these areas produce young green regrowth not found in ungrazed areas. The gazelle take advantage of this growth which was stimulated by the grazing, and so benefit from the wildebeest.

The relationship between the migrating zebra and wildebeest is more complex. Although zebra usually move first from the short grass plains to the long grass dry season areas, the wildebeest population (1.3 million), which is much larger than that of zebra (200 000), often do not follow the zebra but take their own route and eat the long dry grass. Therefore, most migrant wildebeest obtain no benefit from the zebra. In contrast, zebra may be benefiting from the wildebeest for a completely different reason. In the wet season, when there is abundant food, many zebra graze very close to the wildebeest, and by doing so they can avoid predation because most predators (lions and spotted hyenas (*Crocuta crocuta*)) prefer to eat wildebeest. Only if there are no wildebeest within range will predators turn their attention to zebra.

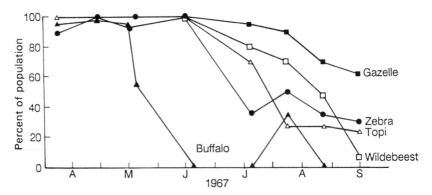

Fig. 9.17 Proportion of the population of different ungulate species using short-grass areas on ridge tops (upper catena) in the Serengeti. The larger species leave before the smaller at the start of the dry season. (After Bell 1970.)

Therefore, it pays zebra to make sure there are wildebeest nearby. In the dry season, however, zebra compete with wildebeest for food. Zebra, therefore, show habitat partitioning and avoid the wildebeest. But by doing so they probably make themselves more vulnerable to predators again (Sinclair 1985). Thus, zebra have to balance the disadvantages of predation if they avoid wildebeest with competition if they stay with the wildebeest. We can see a seasonal change from facilitation in the wet season when there is abundant food, to competition in the dry season when food is regulating the wildebeest.

A clear example of facilitation has been recorded. On the Isle of Rhum, Scotland, cattle were removed in 1957 and reintroduced to a part of the island in 1970 where they graze areas used by red deer. Pasture used by cattle in winter results in a greater biomass of green grass in spring compared with that in ungrazed areas. Gordon (1988) found that deer preferentially grazed areas in spring that had previously been used by cattle (Fig. 9.18), and subsequently there were more calves per female deer.

On the North American prairies both black-tailed prairie dogs (*Cynomys ludovicianus*) and jackrabbits (*Lepus californicus*) benefit from grazing by cattle. If grazing is prevented, the long grass causes prairie dogs to abandon their burrows. At a site in South Dakota where cattle were fenced out, there were half as many burrows as on adjacent areas where grazing was continued (Robinson and Bolen 1984). Snell and Hlavachick (1980) showed that a large prairie dog site of 44 ha could be reduced to a mere 5 ha by the elimination of cattle grazing in summer to allow the grass to grow. Presumably under natural conditions when American bison grazed the prairies there was facilitation by bison allowing prairie dogs to live in the long grass prairies. Facilitation could be mutual because both pronghorn and bison respond to the vegetation changes caused by prairie dogs, both species using prairie dog sites (Coppock *et al.* 1983; Wydeven and Dahlgren 1985).

This example illustrates two management points that follow from the understanding of the interaction (facilitation) between large mammal grazers and prairie dogs: (i) a simple management program (through

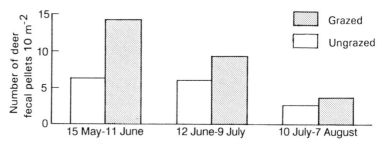

Fig. 9.18 Facilitation of deer grazing by cattle is demonstrated by deer fecal-pellet groups on cattle-grazed plots during each of the 3 months of study. Deer preferred to graze plots used by cattle the previous winter. (After Gordon 1988.)

grazing manipulation) can be devised to control the prairie dogs, without the use of harmful poisons that could affect other species; (ii) in many areas prairie dogs are becoming very scarce and their colonies need to be protected. In addition, the black-footed ferret (*Mustela nigripes*), one of the rarest mammals in the world, depends entirely on prairie dogs and it is thought that their very low population has resulted from the decline in prairie dog populations. The conservation of both species would benefit from the manipulation of grazing practices.

In another example where facilitation improved management for wildlife, Anderson and Scherzinger (1975) showed that ungrazed grassland resulted in tall, low-quality food in winter for elk. Cattle grazing in spring maintained the grass in a growing state for longer. If cattle were removed before the end of the growing season, the grasses could regrow sufficiently to produce a shorter, high-quality stand for elk; their population increased from 320 to 1190 after grazing management was introduced.

In Australia, rabbits prefer very short grass. Rangelands that have been overgrazed by sheep benefit rabbits through facilitation, and rabbit numbers increase. When sheep are removed and long grass returns, rabbit numbers decline.

The saltmarsh pastures of Hudson Bay in northern Canada are grazed by lesser snowgeese (*Chen caerulescens*) during their summer breeding (Bazely and Jefferies 1989; Westoby 1989; Hik and Jefferies 1990). The marshes are dominated by the stoloniferous grass *Puccinellia phryganodes* and the rhizomatous sedge *Carex subspathacea*. At La Perouse Bay some 7000 adults and 15 000 juvenile geese graze the marsh, taking 95% of *Puccinellia* leaves. These are nutritious with high amounts of soluble amino acids. From exclosure plots it was found that natural grazing by geese increased productivity by a factor of 1.3–2.0. Experimental plots with different levels of grazing by captive goslings showed that the above-ground productivity of *Puccinellia* was 30–100% greater than that of ungrazed marsh. In addition, the biomass (standing crop) of the grass was higher if allowed to regrow for more than 35 days following clipping. Immediately after the experimental grazing the biomass was less than that of the ungrazed plots, so that at some point between then and the eventual measurements the biomass on the treated and untreated plots was the same. Even so, the production rate of shoots was higher on the grazed sites. Other experimental plots, where grazing was allowed but from which geese feces were removed, showed that biomass returned to the level of control plots, but no further. Thus, it appears that geese droppings, which are nitrogen rich and easily decomposed by bacteria, stimulate growth of *Puccinellia*. Geese, therefore, benefit each other from their grazing by fertilizing the grass — a form of intraspecific facilitation.

In summary, facilitation occurs when one species alters a habitat or creates a new habitat which allows the same or other species to benefit. We have discussed grazing systems, in particular, but the concept applies in many other cases. For example, many hole-nesting birds and mammals in North America such as wood ducks (*Aix sponsa*) and flying squirrels (*Glaucomys sabrinus*) depend on woodpeckers to excavate the holes, a form of facilitation. Knowledge of such interactions is important for the proper management and conservation of ecosystems.

9.9.2 *Do grasses benefit from grazing?*

If a species such as Thomson's gazelle benefits from the grazing effects of wildebeest due to the increased productivity of the plants, then do the plants themselves benefit? In other words what benefits do the plants receive from being grazed and growing more? In evolutionary terms (see Chapter 3) we have to rephrase this as, "Does herbivory increase the fitness of individual plants?" In ecological terms one may ask, "Does the plant grow more after herbivory?"

The studies of lesser snowgeese on the saltmarshes of Hudson Bay which we have discussed above are now showing that the grass *Puccinellia* comes in different genotypes (Jefferies and Gottlieb 1983; Westoby 1989). Nineteen grazing experiments have shown that under grazing there is selection for those genotypes that are fast growing. These types have the ability to take up the extra nitrogen from the goose feces and seem to outcompete slower growing genotypes. This is, therefore, an evolutionary benefit from grazing. Plots where grazing is prevented show that, after 5 years, change to slower growing genotypes is only just beginning. The more immediate ecological benefit from grazing again comes from the addition of nutrients resulting in a 30–50% increase in biomass.

In general there are few studies that show plants increasing their fitness as a result of herbivory (Belsky 1987). In contrast, we can look at communities of plants and see that if the majority of plants, such as grasses, can tolerate grazing (i.e., survive despite herbivory) a few other intolerant species in that community may not survive due to inadvertent feeding or trampling by large mammals (i.e., apparent competition rather than true competition between plants). This may be simply a consequence of grazing and not necessarily an evolutionary advantage for the grass species. Nevertheless, McNaughton (1986) has argued in opposition to Belsky that grasses and grasslands have evolved in conjunction with their large mammal herbivores, especially in Africa. From an evolutionary point of view a grass individual that by chance evolved an antiherbivore strategy (such as the production of distasteful chemicals) should be able to spread through the grassland. We have to surmise at this stage that antiherbivore adaptations are constrained in some way, e.g., it could be that production of distasteful chemicals results in the plant being less successful, in root competition, or in the uptake of nutrients, as in the example of lesser snowgeese grazing.

On the surface it appears disadvantageous for a grass to grow more as a response to grazing because it would provide more food and invite further grazing. But growth could also be viewed as a damage repair mechanism which is making the best of a bad situation (i.e., the grass may have less loss of fitness by growing than by not doing so).

In summary, we know too little about both the ecological and evolutionary consequences of herbivory on plants. We are left with many questions and opposing views, and more work is needed.

9.9.3 *Complex interactions*

Competition, parasitism, and predation are all processes that have negative effects on a species. However, when they act together they may end up having a beneficial effect. For example, acorns of English oak (*Quercus robur*) are parasitized by weevils and gall wasps, and are eaten by small mammals. Very high mortality rates are imposed on healthy acorns by small mammals, but parasitized acorns are left alone. While most of the parasitized acorns also die, some survive and are avoided by the small mammals. Thus, higher survivorship and hence fitness occurs when the plant is parasitized (Crawley 1987; Semel and Andersen 1988).

9.10 **Applied aspects of competition**

9.10.1 *Applications*

It is important that we should understand the underlying concepts of interspecific competition if we are to comprehend how species might or do actually interact in the field. There are several applications where we need to be aware of potential competition: (i) in conservation where we might have to protect an endangered species from competition with another dominant species; (ii) in managed systems such as rangelands and forests where there could be competition between domestic species and wildlife, e.g., an increase in livestock or the expansion of rangeland might cause the extinction of wildlife species, or wildlife might eat food set aside for the domestic animals; and (iii) if we want to introduce a new species to a system, e.g., a new game-bird for hunting, and there could be competition from other resident species.

9.10.2 *Conservation*

Let us imagine a situation where we want to conserve a rare species but are concerned about possible competition from a common species. For example, roan antelope (*Hippotragus equinus*) (a fairly rare species) were released in Kruger National Park, South Africa, as part of a conservation program (Smuts 1978). There were concerns that the numerous wildebeest would exclude the roan antelope. In this case the management response was to cull the wildebeest. A similar example involved the extremely rare Arabian oryx (*Oryx leucoryx*). A few of the last remaining individuals of this species were captured in Arabia in the early 1960s and taken to the zoo in San Diego with the intention of releasing some to become free ranging on ranchland at a later date.

In both of these examples it would be important to detect whether there was competition with resident species. We have seen that simple measures of overlap, or even changes of overlap with season, may not be good indicators of competition. Similarly observations that an increase in a common species is correlated with a decrease in the rare species does not mean that competition is the cause because of the problem of apparent competition. These measures are necessary but not sufficient: in addition we would need a measure of: (i) resource requirement; (ii) availability of limiting resources and demonstration that one is in short supply; and (iii) the predation rates on both the target species and alternative prey.

A second kind of problem comes from changes in habitat. Assume there is coexistence and habitat partitioning between two species along the lines of Rosenzweig's shared preference hypothesis described above. Since studies of diet and habitat selection would show that both species prefer the same habitat, one may be tempted to manage an area by increasing the preferred habitat at the expense of the other habitats. In this case, however, only one species, the dominant, would benefit and the other would decline. The breeding habitats of yellow-headed and red-winged blackbirds (Orians and Willson 1964) may be a case in point. Both species prefer deeper-water marshes but one may predict that increasing the depth of a marsh where both occur, thereby leaving little shallow water, may well result in the exclusion of the red-winged blackbird.

In Lake Manyara National Park, Tanzania, there is a habitat consisting of open grassland on the lake shore which is used by wildebeest. Adjacent to this is savanna consisting of longer grass with scattered trees and shrubs, and this habitat is preferred by African buffalo. In 1961 heavy rains caused the lake levels to rise and flood the open grasslands, a situation that remained for the rest of the 1960s. The wildebeest were forced to use the savanna habitat which they did not prefer, and after 4 years the population because extinct. On the surface this appeared to be due to competition with buffalo. Closer inspection of the situation showed that the wildebeest had been killed by lions, whose densities were high because of the high buffalo population. Wildebeest normally escape predation by running, which they can do in habitats with short grass and little cover for ambush from predators. Once wildebeest were confined to the savanna they were less able to avoid lions, which killed most of the them. Buffalo, on the other hand, avoid predators by hiding in thickets and defending themselves with their horns, and this they could do in savanna but not on the open grassland. Thus each prey species had their own specialized antipredator habitat which allowed coexistence between the prey species as predicted by Holt's (1977) "apparent competition" hypothesis. Once this habitat partitioning broke down the predator was able to eliminate one of the species. These observations are, therefore, better explained by the process of apparent competition than by true competition.

9.10.3 *Competition between domestic species and wildlife*

There are a number of studies designed to detect whether there is competition between livestock and wildlife. Thill (1984) recorded the seasonal diets of cattle and white-tailed deer in three forested and two clear cut sites in Louisiana pine forests. On the forest sites woody plants made up >85% of the diet for deer throughout the year (Fig. 9.19). For cattle diets these plants made up <16% in summer and fall but rose to 60% in winter and to 48% in spring. The overlap between the two species in overall diet was highest in winter at 46% and lowest in summer at 12%. In contrast, on cleared sites deer continued to eat mainly woody plants but cattle ate >80% grass year-round. Diet overlap was only 17% in summer and fell to 10% in winter. Since the two species were in the same habitat and there did not appear to be predators, there could be real potential for interspecific competition if cattle were confined to forest sites; in fact most of them stayed on the open sites. It is possible that because cattle and deer have not evolved together we do not see the expected decrease in overlap in winter, so that competition is increased rather than avoided at this time.

Thill (1984) points out the advantages for multiple use management derived from the diet partitioning. As forest practices intensify, forest ages decrease and the young stands become impenetrable without artificial clearing. They are also poor areas for deer forage. If cattle were to be used to graze these sites they could be kept open and so benefit deer by improving accessibility and increasing production of the second-growth deer food plants.

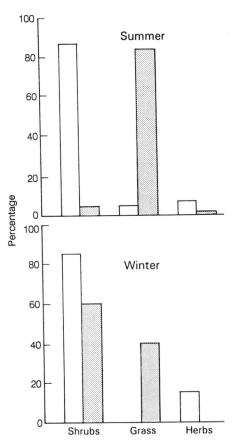

Fig. 9.19 Percentage of the diet of white-tailed deer (open) and cattle (shaded) made up of shrubs, grasses, and herbs. Diet overlap increased in winter when food was limiting. (After Thill 1984.)

Cattle can also have indirect competitive effects by altering habitat structure. In a study of bird communities using the riverine shrub willow habitats in Colorado, Knopf *et al.* (1988) found that cattle grazing altered the structure of the shrubs but not the plant composition. Areas with only summer grazing contained larger bushes widely spaced and with few lower branches when compared with those areas that experienced only winter grazing. The difference in structure affected migratory bird species according to how specific their habitat preferences were. Densities of those with wide habitat preference (e.g., yellow warbler (*Dendroica petechia*), song sparrow (*Melospiza melodia*)), did not change between the sites. Those with moderate niche width (American robin (*Turdus migratorius*), red-winged blackbird (*Agelaius phoeniceus*)), were three times more numerous on the winter-grazed sites; while those with narrow niches (willow flycatcher (*Empidonax traillii*), white-crowned sparrow (*Zonotrichia leucophrys*)), occurred only on the winter-grazed sites.

9.10.4 *Introduction of exotic species*

Exotic species, those that do not normally live in a country, are introduced for a variety of reasons, and very often they become competitors with the native wildlife. Rabbits in Australia are perhaps the most conspicuous example of this for they have been implicated in the decline of native herbivores through either direct competition or apparent competition by supporting exotic predators such as foxes (King *et al.* 1981). Dawson and Ellis (1979) measured the dietary overlap between the rare yellow-footed rock-wallaby (*Petrogale xanthopus*), the euro (*Macropus robustus*, a kangaroo), and two introduced feral species, the domestic goat and European rabbit. During periods of high rainfall the rock-wallaby's diet was mostly of forbs (42–52%) but the proportion of forbs in the ground cover was only 14%. Under drought conditions they still preferred forbs (there was 13% forbs in the diet when forbs were hardly detectable in the vegetation) but trees and shrubs formed the largest dietary component (44% browse). At this season major components for the other species were: euros, 83% grass; goats, 65% browse; rabbits, 25% browse. The rock-wallabies' overall diet overlap was 75% with goats, 53% with rabbits, and 39% with euros. In good conditions dietary overlap was still substantial but lower than when drought prevailed. At that time the overlap was 47% and 45% with goats and euros respectively. Thus, potential competition was greatest with goats and rabbits and least with the indigenous euro.

In North America the introduced starlings (*Sturnus vulgaris*) and house sparrows (*Passer domesticus*) have competed with the native bluebird (*Sialia sialis*) for nesting sites, with the result that bluebird numbers have declined considerably (Zeleny 1976).

Not all introductions result in competition. Chukar partridge (*Alectoris chukar*) have been introduced to North America as a game bird. Their habitat includes semiarid mountainous terrain with a mixture of grasses, forbs, and shrubs. In particular they like the exotic cheatgrass (*Bromus tectorum*). Chukar introductions succeeded only where cheatgrass occurred. These habitat requirements are unlike those of native gamebirds such as sage grouse (*Centrocercus urophasianus*) and thus little competition has taken place (Gullion 1965; Robinson and Bolen 1984).

9.11 **Summary**

Interaction between species can be competitive or beneficial. Competition occurs when two species use a resource that is in short supply, but a perceived shortage in itself should not be used as unsupported evidence of competition. Instead, the relationships must be determined by manipulative experiments reducing the density of one to determine whether this leads to an increase of the other. Care should be taken to eliminate other factors that may cause the response. Facilitation is the process by which one species benefits from the activities of the other. It often takes the form of one species modifying a rank food supply to make it more available to another species, and where one species modifies a habitat making it more favorable to another.

These two effects — competition and facilitation — can often be manipulated by management to increase the density of a favored species.

10 Predation

10.1 Introduction

We start by describing the behavior of predators with respect to prey. With this knowledge we explore some theoretical models for predator–prey interactions. Finally, we examine how the behavior of prey can influence the rate of predation. This chapter presents an alternative to that given in Chapter 6 for analyzing interactions between trophic levels.

10.2 Predation and management

The previous two chapters have dealt with interactions between individuals on the same trophic level. Predation usually involves interactions between trophic levels where one species negatively affects another. With respect to our three issues of management – conservation, control, and harvesting – predators and predation are of great interest. For rare prey species, the presence of a predator can make the difference between survival or extinction of the prey, especially if the predator is an introduced (exotic) species. This type of problem is particularly important on small islands, but also on isolated larger land areas such as New Zealand and Australia. In contrast, where prey are pests, predators may be useful as biological control agents. Ironically, it was for just this purpose that the small Indian mongoose (*Herpestes auropunctatus*) was introduced to Hawaii and the stoat (*Mustela erminea*) to New Zealand. Unfortunately, in these cases the predators found the indigenous birds and small marsupial mammals easier to catch, so that the predators themselves became the problem. Finally, where harvesting of a prey species (e.g., by sport hunters) is the objective, the offtake by natural predators must be taken into account, or one runs the risk of overharvesting and causing a collapse of the prey population.

10.3 Definitions

Predation can be defined as occurring when individuals eat all or part of other live individuals. This excludes detritivores and scavengers that eat dead material.

There are four types of predation.

1 *Herbivory*. This occurs when animals prey on green plants (grazing, defoliation) or their seeds and fruits. It is not necessary that the plants are killed; in most cases they are not. Seed predators (granivores) and fruit eaters (frugivores) often kill the seed, but some seeds require digestion to germinate.

2 *Parasitism.* This is similar to herbivory in that one species, the parasite, feeds on another, the host, and often does not kill the host. It differs from herbivory in that the parasite is usually much smaller than the host and is usually confined to a single individual host. The behavior of nomadic herders in Africa that live entirely on the blood and milk of their cattle would also fit the definition of parasitism. Insect parasites (parasitoids) lay their eggs on or near their host insects which are later killed and eaten by the next generation. We discuss this further in Chapter 11.

3 *Carnivory.* This is the classical concept of predation where the predator kills and eats the animal prey.

4 *Cannibalism.* This is a special case of predation in which the predator and prey are of the same species.

10.4 The effect of predators on prey density

Table 10.1 compares caribou and reindeer populations in areas with different levels of wolf predation (Seip 1991). Densities vary by two orders of magnitude, the highest being in areas with few or no predators. The lowest caribou densities are in areas subject to high and constant predation. Conversely, Fig. 10.1 shows, first, that wolf densities are positively related to moose densities in Alaska and Yukon (i.e., the highest wolf densities are in areas with the highest moose density). This suggests that wolves are regulated by their food supply. Second, when wolves are removed moose densities increase. In general, predator removal experiments show that the prey population increases or that some index such as calf survival increases.

The observations in Table 10.1 and Fig. 10.1 appear to go in opposite directions. No interpretation of cause and effect is possible because they represent correlations. For example, we cannot tell whether the predators are truly regulating the prey at levels well below that allowed by the food supply or whether predators are catching those

Table 10.1 Density of caribou and reindeer populations in relation to the level of predation. (After Seip 1991)

Category	Location	Density per km²
Major predators rare or absent	Slate islands	4–8
	Norway	3–4
	Newfoundland (winter range)	8–9
	South Georgia	2.0
Migratory Arctic herds	George River	1.1
	Porcupine	0.6
	Northwest Territories	0.6
Mountain-dwelling herds	Finlayson	0.15
	Little Rancheria	0.1
	Central Alaska	0.2
Forest-dwelling herds	Quesnel Lake	0.03
	Ontario	0.03
	Saskatchewan	0.03

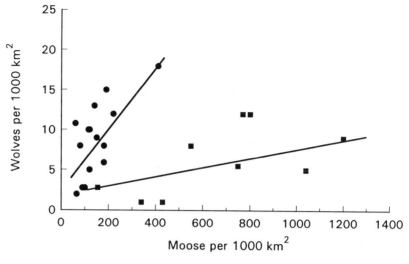

Fig. 10.1 Wolf density is related to moose density in Alaska and Yukon. In areas where wolves are culled (■) moose can reach higher densities than in areas where there is little culling of predators (●). (After Gasaway *et al.* 1992.)

that are suffering from malnutrition (so that predators are not regulating), or whether both processes are occurring. To understand these processes we need to understand the behavior of predators.

10.5 **The behavior of predators**

We must first understand how predators respond to their prey in order to interpret predator–prey interactions. We ask three questions. How do predators respond to: (i) changes in prey density; (ii) changes in predator density; and (iii) differences in the degree of clumping of prey? We look at these in the following three sections.

10.5.1 *The functional response of predators to prey density*

The response of predators to different prey densities depends on: (i) the feeding behavior of individual predators, which is called the *functional response* (see Section 6.7); and (ii) the response of the predator population through reproduction, immigration, and emigration, which is called the *numerical response* (see Section 6.8) (Solomon 1949). We deal with the functional response first.

Understanding of the functional response was developed by Holling (1959). If we imagine a predator that: (i) searches randomly for its prey; (ii) has an unlimited appetite; and (iii) spends a constant amount of time searching for its prey, then the number of prey found will increase directly with prey density as shown in Fig. 10.2a. This is called a type I response. For the lower range of prey densities some predators may show an approximation to a type I response, such as reindeer feeding on lichens (Fig. 10.3), but for the larger range of densities these assumptions are unrealistic. For one thing, no animal has an unlimited appetite. Furthermore, a constant search time is also unlikely. Each time a prey is encountered, time is taken to subdue, kill, eat, and digest it (handling time, h). The more prey that are eaten

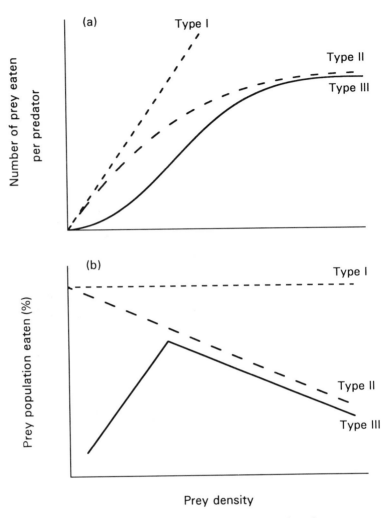

Fig. 10.2 (a) Types of functional response shown as the number of prey eaten per predator per unit time relative to prey density. (b) As for (a) but plotted as the percentage of the prey population eaten.

per unit time (N_a), the more total time (T_t) is taken up with handling time (T_h) and the less time there is available for searching (T_s) (i.e., search time declines with prey density (N)).

Thus, handling time is given by:

$$T_h = hN_a \tag{10.1}$$

and total time is:

$$T_t = T_h + T_s \tag{10.2}$$

The searching efficiency or attack rate of the predator, a, depends on the area searched per unit time, a', and the probability of successful attack, p_c, so that:

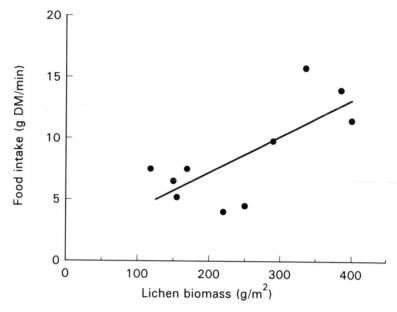

Fig. 10.3 Linear (type I) functional response of reindeer feeding (dry matter intake) on lichen. (After White *et al.* 1981.)

$$a = a'p_c \tag{10.3}$$

The number of prey eaten per predator per unit time (N_a) increases with search time, search efficiency, and prey density, so that:

$$N_a = aT_sN \tag{10.4}$$

Substituting Equations 10.1 and 10.2 into 10.4 we get:

$$N_a = a(T_t - hN_a)N \tag{10.5}$$

or:

$$N_a = (aT_tN)/(1 + ahN) \tag{10.6}$$

This is Holling's (1959) "disc equation" which describes a type II functional response, where N_a increases to an asymptote as prey density increases (Fig. 10.2a).

When there are several prey types (species, sex, or age classes), the multispecies disc equation for prey type i eaten per predator is then:

$$N_{ai} = (a_iT_tN_i)/(1 + \Sigma_j a_j h_j N_j) \tag{10.7}$$

where the sum is across all prey types eaten.

The type II functional response can be constructed from the parameters of the disc equation estimated from observations. Searching efficiency is the product of p_c and a'. The probability of capture is usually low, about 0.1–0.3 in most wildlife cases (Walters 1986). The area of search, a', can be approximated from (distance moved) × (width of reaction field or detection distance). Handling time per prey item, h, can be obtained from direct observation or from maximum feeding rates because the maximum rate = $1/h$. Examples of such calculations are given in Clark *et al.* (1979) and Walters (1986).

The important effect of the type II response is seen when numbers eaten per predator are reexpressed as a proportion of the living prey population alive (Fig. 10.2b). The type II curve shows a decreasing proportion of prey eaten as prey density rises. Figure 10.4a illustrates the type II response of European kestrels (*Falco tinnunculus*) feeding

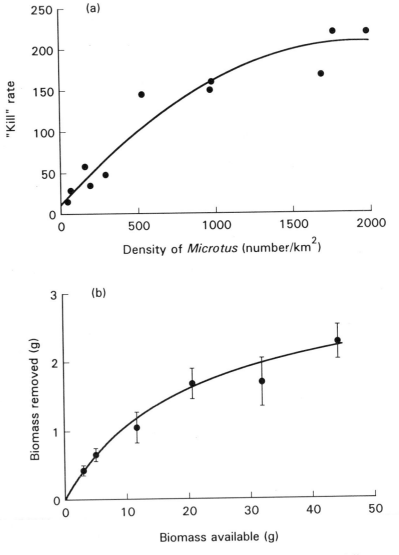

Fig. 10.4 Type II functional response of: (a) European kestrel feeding at different densities of voles (*Microtus* species). "Kill" rate is the number of voles eaten per predator per breeding season. (After Korpimäki and Norrdahl 1991.) (b) Bank voles feeding on willow shoots. (After Lundberg 1988.)

on voles (*Microtus* species) in Finland (Korpimäki and Norrdahl 1991). The functional responses of herbivores are not as well known as those of carnivores but where measured they appear to be type II as in Fig. 10.4b for bank voles (*Clethrionomys glareolus*) feeding on willow shoots (Lundberg 1988). Wickstrom *et al.* 1984 have reported functional responses for ungulates.

Holling found a third type of functional response (type III; Fig. 10.2). The numbers of prey caught per predator per unit time increases slowly at low prey densities, but fast at intermediate densities before leveling off at high densities, producing an S-shaped curve. When those eaten are expressed as a proportion of the live population, the proportion consumed increases first, then declines. Such a curve is shown by the deermouse feeding on cocoons as in Fig. 10.5.

The S shape of this curve is attributed to a behavioral characteristic of predators called *switching*. If there are two prey types, A being rare and B common, the predators will concentrate on B and ignore A. Predators may switch their search from B to A, thus producing an upswing in the number of A killed when A becomes more common. There is often a sudden switch at a characteristic density of A. Birds have a *search image* of a prey species such that they concentrate on one prey type while ignoring another. As the rare prey (A) becomes more common, birds (such as chickadees (*Parus* species) searching for insects in conifers) will accidentally come across A often enough to learn a new search image and switch their searching to this species.

10.5.2 *Predator searching*

The success with which predators catch prey depends on the density of the predator population. Predators usually react to the presence of

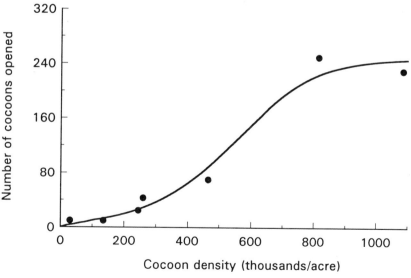

Fig. 10.5 Type III functional response of deermice (*Peromyscus*) feeding on cocoons of European pine sawfly (*Neodiprion sertifer*). (After Holling 1959.)

other individuals of their own kind by dispersing. Mammal and bird predators are usually territorial and evict other individuals once the space has been fully occupied. These examples are forms of "interference," as discussed in Section 8.6.2.

Interference progressively reduces the searching efficiency of the predator as predator density increases. The drop in searching efficiency caused by crowding lowers the asymptote of the functional response curve. Interference also has a stabilizing influence on predator numbers because it causes dispersal once predators become too numerous.

10.5.3 *Predator searching and prey distribution*

Prey usually live in small patches of high density with larger areas of low density in between; in short, prey normally have a clumped distribution. This can be seen in the patchiness of krill preyed upon by whales, of insects in conifers searched for by chickadees, of seeds on the floor of a forest eaten by deermice, of caribou herds preyed upon by wolves, and impala herds hunted by leopards (*Panthera pardus*).

The searching behavior of predators is such that they concentrate on the patches of high density. By concentrating on these patches, predators have a regulating effect on the prey because of the numerical increase of predators by immigration (see Section 10.6).

10.6 Numerical response of predators to prey density

We define the numerical response of predators as the trend of predator numbers against prey density (see Section 6.8 for other ways of looking at this). As prey density increases, more predators survive and reproduce. These two effects, survival and fecundity, result in an increase of the predator population, which in turn eats more prey. An example of this is Buckner and Turnock's (1965) study of birds preying on larch sawfly (*Pristiphora erichsonii*) (Table 10.2). As prey populations increase, the number of birds eating them is also increased by reproduction and immigration. When plotted against prey density, predator numbers increase to an asymptote determined by interference behavior such as territoriality (Fig. 10.6). Territoriality results in dispersal so that resident numbers stabilize. Wolves at high density have high dispersal rates, around 20% for adults and 50% for juveniles (Ballard *et al.* 1987; Fuller 1989).

Table 10.2 The predation rate on larch sawfly in areas of tamarack (*Larix laricina*) (high density) and mixed conifers (low density). Bird predators include new world warblers and sparrows, cedar waxwing (*Bombycilla cedrorum*), and American robin (*Turdus migratorius*). (After Buckner and Turnock 1965)

	High density (N/km²)	Low density (N/km²)
Sawfly larvae	528×10^4	9.88×10^4
Sawfly adults	50.75×10^4	1.16×10^4
Birds	58.1	31.1
Predation of larvae (%)	0.5	5.9
Predation of adults (%)	5.6	64.9

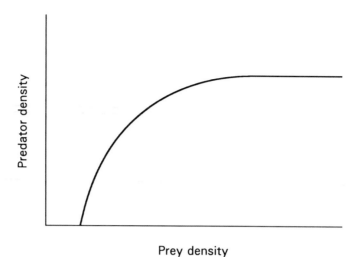

Fig. 10.6 Numerical response may be depicted as the trend of predator numbers against prey density.

The initial increase in numerical response may or may not be density dependent. However, because of the asymptote, the numerical response at higher prey densities can only be depensatory. This means it has a destabilizing effect on the prey population, by either driving the prey to extinction or allowing it to erupt. This is an important characteristic of populations and is illustrated in Buckner and Turnock's (1965) study: the proportion of sawfly eaten by birds in the high prey density area was lower than that in the low density area (i.e., predation was depensatory and, therefore, could not keep the sawfly population down). The conditions when regulation can or cannot occur are discussed in Section 10.7.

10.7 The total response

We can now multiply the number of prey eaten by one predator (N_a, the functional response) with the number of predators (P, numerical response) to give a total mortality, M, where:

$$M = N_a P \tag{10.8}$$

The instantaneous change in prey numbers is:

$$dN/dt = -N_a P \tag{10.9}$$

and an approximation for changes in prey number, over short intervals when prey populations do not change too much ($< 50\%$), is given by

$$N_{t+1} = N_t + N_t e^{-NaP/N_t} \tag{10.10}$$

where $N_t = N$ in Equation 10.6 (Walters 1986).

If we express this total mortality, M, as a proportion of the living prey population, N, we can get a family of curves, as shown in Fig. 10.7, which depend on whether or not there is density dependence in

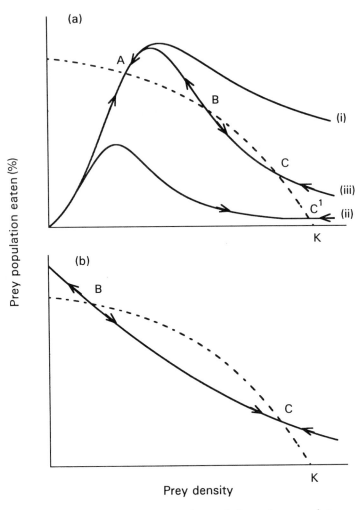

Fig. 10.7 Theoretical total response curves (———) of a predator population, measured as the proportion of the prey population eaten, in relation to prey at different densities, (a) when there is density dependence in either the functional or numerical response and (b) when there is no density dependence. (– – – –) represents the per capita net recruitment of prey $(dN/dt)(1/N)$ assuming logistic growth (i.e., after the effects of competition for food have been accounted for). K is a stable point with no predators; A, C, and C^1 are stable with predators, and B is an unstable boundary point.

the functional and numerical responses. If there is density dependence (e.g., from a type III functional response) then we have a curve with an increasing (regulatory) part followed by a decreasing (depensatory) part. These are called the *total response* curves, and examples are shown for some of Holling's small mammals (Fig. 10.8a) and for wolves eating moose (Fig. 10.8b) (Boutin 1992).

In Fig. 10.7 the total response curves have been superimposed on the per capita net recruitment rate of prey (dN/dt) $(1/N)$. For the case where we have density dependence (Fig. 10.7a) there are several stable

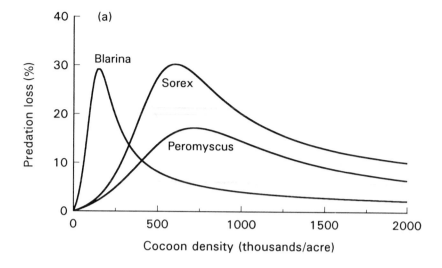

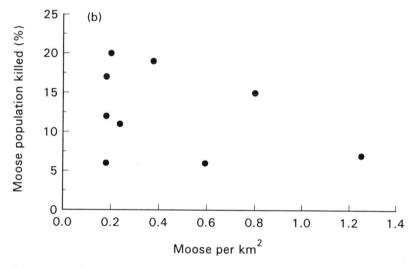

Fig. 10.8 Total response curves of predators at different prey densities: (a) two shrews (*Blarina, Sorex*) and the deermouse (*Peromyscus*) eating European sawfly cocoons. (After Holling 1959.) (b) The proportion of the moose populations killed by wolves in different areas of North America. (After Boutin 1992.)

equilibria (A, C, C') where prey net recruitment is balanced by total predation mortality. The point B is an unstable equilibrium where any perturbation to the system (e.g., from the weather) will result in the prey declining to A or increasing to C. In practice B is never seen and is regarded as a boundary between domains of attraction towards A or C.

Curve (i) illustrates the case where predators can regulate the prey population under the complete range of prey densities and hold the prey at a low density A. One possible example of this occurs where both wolves and grizzly bears (*Ursus arctos*) prey upon moose in Alaska (Ballard *et al.* 1987; Gasaway *et al.* 1992). Wolves appear to keep moose densities at low levels ($<0.4/\text{km}^2$). When wolves were removed in a culling operation, the mortality of juvenile moose caused by bears increased so that moose numbers remained at the low level. Moose are kept at similar low levels by the density-dependent predation from wolves in Quebec (Messier and Crete 1985).

Curve (ii) can occur when prey are regulated by intraspecific competition for food. Predators then kill malnourished animals and the effect on the prey population is depensatory rather than regulatory. This may be occurring on Isle Royale in Lake Superior where wolves cannot increase sufficiently to regulate the moose population (Fig. 10.9), which is regulated by food (Peterson 1992), and wolf predation is merely depensatory (Fig. 10.10) (Messier 1991).

Curve (iii) is the special case where both A and C are present and we have multiple stable states. This situation has been suggested for a few predator–prey systems. One example is that of foxes feeding on rabbits in Australia (Pech *et al.* 1992). Foxes were experimentally removed from two areas and the rabbit populations increased in both, as would be expected from any of the curves in Fig. 10.7, and so by itself the increase in prey tells us little about the nature of predation. However, when foxes were allowed to return to the removal areas there was some evidence that rabbits continued to stay in high numbers rather than return to their original low densities. This result suggests that we have curve (iii) and not (i) or (ii). The interpretation is that rabbit populations, originally at A, were allowed to increase above

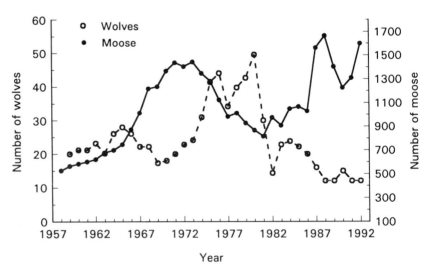

Fig. 10.9 Wolf (○) and moose (●) numbers on Isle Royale, 1959–92, show that the wolf population follows fluctuations in the numbers of moose, which are regulated by food. (After Peterson 1992.)

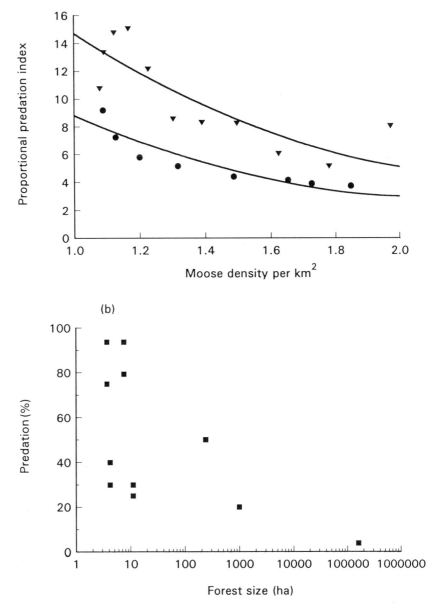

Fig. 10.10 Depensatory total responses of (a) wolves killing moose on Isle Royale, during moose population decline (▼) and increase (●). (After Messier 1991.) (b) Various mammal and bird predators on passerine bird nests as a function of forest patch size. These patches are an index of prey population size. (After Wilcove 1985.)

the boundary density, B, so that when foxes reinvaded the experimental area rabbit numbers continued towards C.

The "forty-mile" caribou herd of Yukon may have exhibited behavior characteristic of multiple stable states (Urquhart and Farnell

1986). Traditionally, this herd, whose range is on the Yukon–Alaska boundary, numbered in the hundreds of thousands – one estimate by Murie in 1920 being 568 000. In the 1920s and 1930s goldminers and hunters killed tens of thousands. After the Second World War, when the Alaska Highway and associated roads were built, hunting increased further. By 1953 numbers were estimated at 55 000 and by 1973 there were only 5000 animals left. Although wolf numbers declined along with their prey, as one might expect, the proportional effect of predation was thought to be high. After 1973 hunting of caribou was restricted and during 1981–3 wolf numbers were reduced from 125 to 60. Thereafter, wolf numbers returned to prereduction levels. Although caribou numbers increased marginally to 14 000 during the wolf reductions, they have remained at approximately this level for the rest of the 1980s. Despite the lack of accurate population estimates the density changes shown by the "forty-mile" herd are so great (almost two orders of magnitude) that it is reasonably clear there has been a change in state from a high level determined by food to a low level determined by predators. The wolves may have been able to take over regulation because hunting could have reduced the caribou population size below the boundary level, B.

Another example of two states may be seen in the wildebeest of Kruger National Park, South Africa (Smuts 1978; Walker et al. 1981). In this case high numbers of wildebeest were reduced by culling. When the culling was stopped numbers continued to decline through lion predation, suggesting the system had been reduced below point B. A herbivore–plant interaction with two stable states is seen in Serengeti woodlands (Dublin et al. 1990b). Woodland density changed from high to low density in the 1950s and 1960s by severe disturbance from fires. In the 1980s elephant browsing was able to hold woodlands at low density.

Figure 10.7b shows a special case where predators have no regulatory effect but can cause the extinction of the prey species if prey numbers are allowed to drop below B. Predation mortality is greater than prey net recruitment below B so that the prey population will decline to extinction. The conditions for this situation occur when there is no switching by predators (i.e., there is a type II functional response), there is no refuge for the prey at low densities, and predators have an alternative prey source to maintain their population when the first prey species is in low numbers. Low densities of the primary prey could be produced by reduction of habitat (as has occurred with many endangered bird species) or by hunting. For example, wolves prey upon mountain caribou in Wells Gray Park, British Columbia, during winter but not in summer when caribou migrate beyond the range of wolves (Seip 1992). Recruitment for this herd in March is 24–39 calves per 100 females. In contrast, caribou in the adjacent Quesnel Lake area experience predation year-round and the average recruitment is 6.9 calves per 100 females. This population suffers an adult mortality of

29% (most of which is caused by wolves), well above the recruitment rate, and so the population is declining. Moose are the alternative prey in this system and maintain the wolf population. However, moose have only recently entered this ecosystem, having spread through British Columbia since the 1900s as a result of logging practices, so that previously wolves would not have had this alternative prey to maintain their populations at low caribou numbers. At the same time the Quesnel Lake caribou do not have a summer refuge from predation, unlike those in the Wells Gray Park. One inference, therefore, is that the caribou of Quesnel Lake are below the boundary density B in Fig. 10.7b and could be declining to extinction (Seip 1992).

Habitat fragmentation for passerine birds breeding in deciduous forests of North America is thought to be the primary reason for the major decline in their populations (Wilcove 1985; Terborgh 1989; 1992). The interior of large patches of forest provide a refuge against nest predation from raccoons (*Procyon lotor*), opossums (*Didelphis virginiana*), and striped skunks (*Mephitis mephitis*), and parasitism from brown-headed cowbirds (*Molothrus ater*). Fragmentation of the forests reduces this refuge because nests are now closer to the edge of the forest where there are more predators and nest parasites. Predation rates are inversely related to forest patch size which must be related to total prey population (Fig. 10.10b). In large forest tracts nest predation is only 2%, in small suburban patches it is close to 100% and well above the recruitment rate. Since small fragmented forest patches are the norm in much of North America, many populations of bird species may be in the situation shown in Fig. 10.7b where the density is left of the boundary B and declining to extinction.

10.8 Behavior of the prey

We have seen how the behavior of predators can influence the nature and degree of predation. We will now examine how the behavior of prey affects predation rates.

10.8.1 *Migration*

If a prey species can migrate beyond the range of its main predators, then their populations can escape predator regulation (Fryxell and Sinclair 1988a). This has been shown theoretically (Fryxell *et al.* 1988) and there are some examples supporting this idea. The explanation for this escape from predator regulation is that predators, with slow-growing, nonprecocial young are obliged to stay within a small area to breed. In contrast, ungulate prey with precocial young do not need to stay in one place because the young can follow the mother within an hour or so of birth. Thus, the prey can follow a changing food supply while the predators cannot. For example, the wildebeest migrations in the Serengeti can follow seasonal changes in food and are regulated by food abundance; meanwhile their lion and hyena predators, although commuting up to 50 km from their territories, cannot move nearly as far as the wildebeest. Other examples from Africa are reported for wildebeest migrations in Kruger Park, South Africa (Smuts 1978), and white-eared kob (*Kobus kob*) in Sudan (Fryxell and Sinclair 1988b). In North America a similar escape from predation is suggested for migrating caribou herds – the George River herd in Quebec (Messier *et al.* 1988), the barren-ground caribou (Heard and Williams 1991), the Wells Gray Park mountain caribou through altitudinal migration (Seip 1992), and possibly the "forty-mile" caribou before hunting reduced the herd (Urquhart and Farnell 1986).

10.8.2 *Herding and spacing*

Theoretical studies propose that animals can reduce their risk of predation by forming groups, herds, or flocks (Hamilton 1971), and that group sizes should increase with increasing predator densities (Alexander 1974). The benefit from avoiding predators, however, is counteracted by the cost of intraspecific competition within the group. There should be some group size where the benefit/cost ratio is optimized (Terborgh and Janson 1986).

The effect of predators on herding behavior is illustrated in Fig. 10.11. The relationship between musk-ox (*Ovibos moschatus*) group size and wolf density suggests that predators are the most likely explanation for differences in group size in different populations (Heard 1992).

The opposite behavior to herding is shown by many female ungulates when they give birth. At that time they leave the herd and become solitary. This is seen in impala and other antelopes in Africa, and in cervids, mountain sheep, and forest caribou in North America

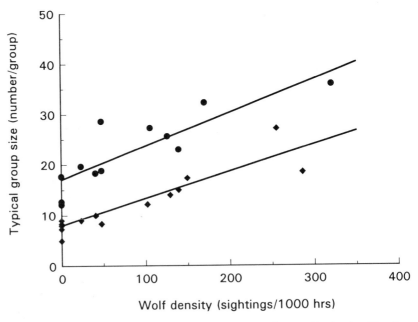

Fig. 10.11 Musk-ox group size on arctic islands in North-West Territories, Canada, is related to wolf density in both summer (◆) and winter (●). (After Heard 1992.)

(Bergerud and Page 1987). This behavior relies on predators spending most of their search time in areas of high prey density, so that solitary prey at low density experience a partial refuge and hence lower predation rates.

Another form of refuge is used by deer in winter when they congregate in loose groups in the small areas between wolf home ranges (Rogers *et al.* 1980; Nelson and Mech 1981). These areas appear to be unused by the wolves, a sort of "no-man's-land," and hence they act as a refuge for the prey.

10.8.3 *Birth synchrony*

Some prey species synchronize their reproduction to reduce the predation rate on their young, a behavior called *predator swamping*. This synchrony is over and above that imposed by the seasons. For example, moose and caribou have highly synchronized birth periods (Leader-Williams 1988), as do wildebeest in Africa (Estes 1976). Other examples of breeding synchrony are seen in lesser snowgeese (Findlay and Cooke 1982) and colonial seabirds (Gochfeld 1980).

Experimental studies, such as that of foxes feeding in breeding colonies of black-headed gulls (*Larus ridibundus*) in England, have shown that those gulls breeding outside the main nesting period are more likely to lose their nests to predators (Patterson 1965). However, synchrony may or may not be adaptive depending on the abilities of the predator and the type of synchrony (Ims *et al.* 1988). Thus, if prey form groups and all groups are synchronized together, then predator swamping can occur. However, if reproduction is synchronized within groups but not between them, then predation rate on juveniles could be increased rather than decreased by this behavior, and this depends on the type of predator (Ims 1990). In general, breeding synchrony should be evaluated not just in terms of predation. Other aspects such as seasonality of the environment should be considered. These aspects are important for conservation because species that rely on grouping behavior and synchrony of breeding will be vulnerable to excessive predation if human disturbances alter either aspect.

10.9 **Summary**

Some of the important points for conservation and management that we can derive from this discussion of predation are as follows.

1 Predator and prey populations usually coexist. Prey may be held at low density by predator regulation or at high density by intraspecific competition for food or other resources, and here predators are depensatory.

2 It is possible that both systems may operate in the same area, leading to multiple stable states. This may be generated by a type III functional response or by a density-dependent numerical response at low prey densities. The system may move from one state to another as a result of disturbance. Such dynamics may occasionally underlie the outbreak of pest species and the decline of species subject to hunting.

3 Conversely, there are situations where the prey population could become extinct, particularly with a type II predator functional response, no refuge for prey, and alternative food sources for the predators. This is important in conservation where habitat changes may reduce refuges, or introduced pests such as rats may provide alternative prey for predators of rare endemic species.

4   Which of the above occurs depends on the ability of the predator to catch prey and the ability of the prey to escape either by using a refuge or by reproducing fast enough to make up the losses. A very efficient predator defined by a high predator/prey ratio will hold the prey at low density.

11 Parasites and pathogens

11.1 Introduction

This chapter briefly introduces parasitism and disease within wildlife populations. It addresses how an infection affects a population's dynamics and how it spreads through a population. The veterinary aspects of infection, special to each parasite and host, are not dealt with here. Instead we look at examples of how parasites and disease affect conservation of endangered species, reduce the potential yield of harvested populations, or are of use in controlling pests.

11.2 Definition and effects

Parasites and pathogens can best be divided into two classes: *microparasites*, which include viruses, fungi, and bacteria, and *macroparasites* such as arthropods (e.g., fleas) and cestodes (e.g., tapeworms). Microparasites and macroparasites have a roughly equivalent effect upon their hosts and so can be lumped together as parasites. The effect of the parasite upon the host is termed disease. Predators have a different effect ecologically and are therefore considered separately.

Parasites may lower the standing crop of a host population. Hence, they are disadvantageous if the host population is to be conserved or harvested and advantageous if the host population is to be controlled.

Yuill (1987) reviewed the role of disease in mammals but his conclusions, summarized below, can apply to all vertebrates. Parasites can be expected in all wildlife species in every ecosystem. Death of the host is unusual and occurs only: (i) if serious illness facilitates transmission, as in rabies; (ii) if the parasite does not depend on the infected host for survival and can complete its life cycle after the host dies; or (iii) if the pathogen moves through host populations over a wide geographic area and over a long period of time.

Disease can have a drastic effect on the survival of wildlife but more commonly its effects are subtle. It can adversely affect natality or normal movement. Brucellosis in caribou has both effects. A caribou cow infected with brucellosis may abort her fetus, and the same disease may also cause lameness from degenerative arthritis in the leg joints. Infective agents can also affect the host's energy balance by reducing energy intake or increasing energy costs through higher body temperature and metabolic rate.

11.3 **Dynamic differences between predation and parasitism**

Both predation and parasitism can reduce the average standing crop of the host species but predation places potential prey into one of two states. Either it is preyed upon in a given time interval or it survives unscathed. Predation is largely an "all or none" process.

Parasitism on the other hand places a susceptible animal into one of three states: over a given time interval it is not affected, or is sick, or is killed by the infection. Predation kills but does not affect the health or fecundity of the survivors. In fact, the level of wellbeing and the fecundity rate of the survivors may be enhanced by the lower density occasioned by the predation. In contrast, parasitism can lower the average health, and hence reproductive rate, of the standing population because of the sick animals within the population. The population of hosts takes longer to bounce back when the parasite load is lifted from it. Note, however, that many parasites have no discernible effect upon the health of their hosts.

11.4 **The basic parameters of epidemiology**

Whether or not an infection will persist within a host population depends on the magnitude of the net reproductive rate R_0 of the disease, which for microparasites may be defined as the average number of secondary infections generated by a primary case. It is better defined for macroparasites as the average number of offspring per parasite that grow to maturity. If the parasite has two sexes it can also be defined as the average number of daughters reaching maturity per adult female.

If R_0 is less than zero the initial inoculum of parasites will decay to extinction. R_0 is not a constant for a parasitic species but is determined by the varying characteristics of both the parasite and the host populations, particularly the density of the host. The conditions leading to persistence of the infection are given by Anderson and May (1986) and Anderson (1991) as:

$$R_0 = \beta X/(\alpha + b + \sigma) > 1$$

where:

X = host density,
β = probability of transmission per contact between susceptible and infected hosts,
α = disease-induced mortality rate,
b = the mortality rate of uninfected hosts,
σ = rate of recovery from infection.

These are not constants for a host–parasite system but vary according to such things as weather affecting the condition of the hosts.

The relationship can be expressed also in terms of a threshold host density X_T below which the infection will die out:

$$X_T = (\alpha + b + \sigma)/\beta > X$$

which neatly makes the point that R_0 is dependent on host density. Note that if the parasite is highly virulent (large α), or if recovery is rapid (large σ), or if the parasite transmits poorly between hosts (small β), then a dense population (large X_T) is needed to stop the infection dying out. That equation can be elaborated to take in the effect of an incubation period and postinfection immunity (both of which increases X_T) and "vertical" transmission of the infection whereby a fraction of the offspring of an infected female are born infected (which lowers X_T).

These equations encapsulate two important concepts of epidemiology: that persistence or extinction of an infection is determined by only a few traits of the host and parasite; and that the density of the hosts must exceed some critical threshold to allow the infection to persist and spread.

11.5 Determinants of spread

The rate of spread (c) of an infection is determined, as is persistence, by traits of both the parasite and the host, particularly the rate of mortality (α) caused by the disease and the net reproductive rate (R_0) of the pathogen. Källén *et al.* (1985) give the relationship as:

$$c = 2\{D\alpha(R_0 - 1)\}^{1/2}$$

where D is a diffusion coefficient more or less measuring the area covered by the wandering of an infected animal over a given period of time. Two recent papers have used this relationship. Dobson and May (1986a) calculated the constants of that equation for rinderpest in Africa from the observed radial spread of 1.4 km/day. Pech and McIlroy (1990) used a more elaborate version the other way around, estimating from a knowledge of the equation's constants a potential spread of foot and mouth disease of 2.8 km/day through a population of feral pigs in Australia.

11.6 Transspecies infection

The long period of natural selection over which a parasite and its obligate host sort out an accommodation with each other ensures that persistent infection has little influence on the density of the host. If, however, the specific characters of the host and parasite are such that usually $R_0 < 1$, then the infection is likely to be sporadic and may have a large but temporary depressing effect on the density of the host. Bubonic plague and (until recently) smallpox acted in this way against people.

Parasites, particularly microparasites, have their greatest effect when they jump from one species of host to another. That process is also a major source of evolutionary opportunities for parasites. For example, the knee worm (*Pelecitus roemeri*) has been "captured" from the parrot family by wallabies and kangaroos. The effect of worm is unknown in parrots but in macropods the worm induces a fibrous capsule up to the size of a cricket ball on the animal's knee. Transspecifics are the parasites and pathogens to watch out for. They can cause significant conservation problems (see Sections 11.7.1 and

15.2.8.) but they can also sometimes be used to control pest species (see Section 11.7.3).

Other transspecies parasites must be guarded against because they cause considerable gratuitous mortality. The myxoviral rinderpest epidemic that swept the length of Africa in the latter years of last century killed huge numbers of wild ungulates, particularly African buffalo. Asian cattle were its original host but it jumped across to wild ungulates when it reached Africa. The decline of moose populations in Nova Scotia and New Brunswick is associated with infestation by the nematode brainworm *Parelaphostrongylus tenuis*, which jumped hosts from white-tailed deer. The infestation in moose is fatal but there is little evidence that the parasite can maintain itself in the moose except by reinfection from white-tailed deer (Anderson 1972). However, as Samuel *et al.* (1992) have noted, the relationship between meningeal worm, white-tailed deer, and moose has not been studied experimentally and not all the circumstantial evidence is consistent.

The translocation of domestic or exotic wildlife may lead to parasites and pathogens jumping to a new suite of species. In Australia, native animals such as kangaroos and wombats became infected with common liver flukes (*Fasciola hepatica*) acquired from sheep and cattle. The liver flukes cause severe lesions in the liver of the wombat (Spratt and Presidente 1981). The process may also operate the other way, with wildlife acting as reservoirs of parasites and pathogens transmitted to domestic stock. The controversies over brucellosis in bison, and its transmission to domestic stock both in the USA and Canada, are obvious examples (Peterson *et al.* 1991).

11.7 **Practical considerations for wildlife management**

Yuill (1987) points out that wildlife management activities may favor transmission of disease and that wildlife managers must continuously keep in mind that infectious and parasitic agents are important components of every biological system. For example, the availability of water limited wildlife densities in the Etosha National Park of Namibia. Waterholes constructed to raise the carrying capacity of wildlife acted as a reservoir for anthrax acquired from cattle in the surrounding areas. This disease caused considerable mortality of wildebeest and zebra.

Parasites and pathogens can be important in all three components of wildlife management. They can cause conservation problems by reducing the densities of species targeted for conservation and they can reduce the potential yield of harvested populations; however, on the positive side, they can be used to control pest species. The next sections provide examples of each to give a feel for the range of effects.

11.7.1 *Parasites and conservation*

Parasites and pathogens can be a factor driving the decline of an endangered species (see Section 15.2.8) and can become an issue in the recovery of endangered species. Parasites and pathogens can hinder or thwart attempts to establish captive breeding populations. Thorne and Williams (1988) have reviewed the well-known example of the first attempts to establish a captive breeding colony of black-footed ferrets (*Mustela nigripes*) in the USA. A previous attempt to establish a breeding colony in the early 1970s failed because canine distemper virus (CDV) killed the only two litters. The source colony also disappeared. The extreme susceptibility of the black-footed ferrets to CDV became apparent when four of six black-footed ferrets died after being vaccinated for CDV in the 1970s. The vaccine had been previously shown to be safe in domestic ferrets. In 1981 the species was rediscovered in Wyoming, USA, and the colony's vulnerability to disease was quickly realized. Precautions were taken to minimize human introductions of disease, especially CDV and influenza. The population declined from an estimated peak population of 128 in 1984 to only 16 in 1985. The decline spurred an attempt to start a captive breeding colony but the first six ferrets captured rapidly succumbed to canine distemper. Despite all precautions CDV infected the colony and most of its members died from CDV. The few surviving eventually formed a breeding population. Nonetheless, as Thorne and Williams (1988) noted, "The captive breeding program went from a carefully planned approach with ideally selected, unrelated founder animals to a crisis situation with related animals, a poor sex ratio, and few mature, experienced breeder males."

Captive-bred animals released into the wild may spread disease or pick up parasites and pathogens from endemic wildlife. A potential example of the former is Jones' (1982) report that the release of Arabian oryx captive-raised in the USA for a national park in Oman was delayed when the animals tested positive for antibodies to blue-tongue disease. The failure of a reintroduction of woodland caribou to an island within their historic range in Ontario, Canada, is an example of lethal transspecies parasites and the problems that can be encountered when reintroduced species become infected with a disease from the endemic wildlife. The area had been colonized by white-tailed deer and the caribou became infected with meningeal worm from the deer via a gastropod secondary host (Anderson 1972). Another example comes from the captive breeding of whooping cranes (*Grus americana*). An eastern equine encephalitis (EEE) virus fatally infected seven of the 39 captive-bred population at the Patuxent Wildlife Research Center in Maryland, USA. At that time in 1985, that captive population accounted for about 25% of the world population. EEE virus causes sporadic outbreaks of disease in mammals and birds in the eastern USA and is spread by mosquitos. No deaths are usually seen in endemic hosts but introduced game birds such as ring-neck pheasants (*Phasianus colchicus*) are vulnerable. Among the 200 sandhill cranes in neighboring pens to the whooping cranes, some birds were serum positive for EEE virus but no clinical signs were found. The discovery of the vulnerability of the whooping cranes to a common pathogen was seen as an unrecognized risk and a obstacle to the species' recovery (Carpenter *et al.* 1989).

11.7.2 *Parasites and harvesting*

The subtle effect of many parasites and pathogens on their hosts frequently makes for difficulties in demonstrating any reductions in yields from harvested wildlife. For example, the nematode *Trichostrongylus tenuis* lives in the cecum of the red grouse and has a direct life cycle with no intermediate hosts. Hudson and Dobson (1988) summarized this host–parasite system in the British Isles, a system that has been of interest and concern in the management of this game bird for almost a century. It was first studied by Edward Wilson, the doctor, artist, and naturalist who died on the way back from the South Pole during Robert Falcon Scott's second expedition to the Antarctic.

The parasite affects both reproduction and survival of its host but there is continuing controversy as to whether it is the cause of the marked year-to-year fluctuations in population density and hence in harvest that is typical of red grouse management. There is further controversy as to whether birds die because of high parasite load or whether birds of low social ranking, and therefore with low survival, are predisposed to heavy parasite burdens. It would appear that rigorously designed field experiments would repay the effort required.

Bighorn sheep populations, whether harvested by sports hunters or protected in national parks, are often affected by a disease complex comprising a lungworm (*Protostrongylus stilesi*) that causes tissue damage leading to a secondary infection of pneumonia. The sheep are infected from accidently eating the lungworm's secondary hosts – terrestrial snails. The scale of the lungworm–pneumonia complex can reach the proportions of substantial die-offs (Uhazy *et al.* 1973). Harvested sheep populations may be less prone to the parasite by being in better condition. However, more information is required to confirm or reject this relationship.

11.7.3 *Parasites and control of pests*

Spratt (1990), in reviewing the possible use of helminths for controlling vertebrate pest species, pointed out the marked contrast between the numerous successes in biological control of insects and the almost universal failure of such methods to control vertebrates. The one unequivocal success has been the use of the myxoma virus to control European rabbits in Australia. A somewhat less successful story is of the attempts to control feral cats on Marion Island in the Southern Ocean off South Africa, to conserve the endemic ground-nesting seabirds (see Section 17.5).

Myxomatosis is a benign disease in *Syvilagus* (cottontail) rabbits in South America which is transmitted mechanically by mosquitoes. In the European rabbit (*Oryctolagus*), which is a pest in Australia and England, the virus from *Sylvilagus* produced a generalized disease that is almost always lethal. Myxomatosis was deliberately introduced into Australia in 1950 and into Europe in 1952. It was first spectacularly successful in controlling the rabbit pest, but biological adjustments occurred in the virulence of the virus and the genetic resistances of the rabbits. After 30 years of interaction, natural selection has resulted in a balance at a fairly high level of viral virulence (Fenner 1983).

The initial annual mortality rates were very high in Australia, >95%, but these dropped progressively over the next few years. There is a widespread perception that the rabbits and the disease accommodated to each other and therefore that myxomatosis provided only a temporary respite. This is not so. The rabbit density at equilibrium with the disease is considerably lower than mean density in the absence of myxomatosis.

Parer *et al.* (1985) tried to measure that difference. They used a relatively benign strain of myxoma to immunize rabbit populations against the more virulent field strains that swept through the study area in most years. Rabbit densities increased by a factor of 10 under this treatment. Even after the rabbits and the virus had reached an accommodation with each other, the disease was apparently holding the mean density of rabbits to about 10% of that prevalent before introduction of the myxoma virus.

Note the use of "apparently" in the previous sentence. We can claim only a weak inference because of an experimental flaw. "Control areas were established on a large adjacent property in 1978. Unfortunately the warrens on these areas were fumigated and ripped by the property owner" (Parer *et al.* 1985). Very droll.

11.8 Summary

Most parasites and pathogens have little effect on their hosts but, if a parasite jumps from one host species to another, virulence may suddenly be increased. The key points from the epidemiology of parasites and pathogens are that the fate of an infection is determined by only a few traits of the host and parasite, and there is a critical density of the host that allows the infection to persist and spread. Efforts to reduce the effects of parasites and pathogens can be at their most important in the management of small populations of endangered species, be they in the wild or in captivity. Diseases of harvested wildlife are more rarely controlled unless they present a potential hazard to people. Few attempts to use parasites and pathogens to control pest wildlife have been successful.

12 **Counting animals**

The trick in obtaining a usable estimate of abundance is to choose the right method. What works in some circumstances is useless in others. Hence, we provide a broad range of methods and indicate the conditions under which each is most effective.

12.2 Estimates

Knowledge of the size or density of a population is often a vital prerequisite to managing it effectively. Is the population too small? Is it too large? Is the size changing and if so in what direction? To answer these questions we may have to count the animals, or we may obtain adequate information by way of an indirect indication of abundance. In any event we need to know when a census is necessary and how it might be done.

Although *census* is strictly the total enumeration of the animals in an area, we use the word in its less restrictive sense of an estimate of population size or density. That estimate may come from a total count, from a sampled count, or by way of an indirect method such as mark–recapture.

Closely related to the census is the index, a number that is not itself an estimate of population size or density but which has a proportional relationship to it. The number of whales seen per cruising hour is an index of whale density. It does not tell us the true density but it allows comparison of density between areas and between years. Indices provide measures of relative density and are used only in comparisons. They are particularly useful in tracking changes in rates of increase and decrease.

Almost all decisions on how a population might best be managed require information on density, on trend in density, or on both. There are many methods to choose from and these differ by orders of magnitude in their accuracy and expense. Hence, before any censusing is attempted the wildlife manager should ask a number of questions.

- Do I need any indication of density and what question will that information answer?
- Is absolute density required or will an index of density suffice?
- Will a rough estimate answer the question or is an accurate estimate required?
- What is the most appropriate method biologically and statistically?
- How much will it cost?

190

- Do we have that kind of money?
- Would that money be better spent on answering another question? This chapter outlines briefly the variety of methods available and their applications, providing references to where each is treated in detail.

12.3 Total counts

The idea of counting every animal in a population, or on a given area, has an attractive simplicity to it. It is the method used by farmers to keep track of the size of their flocks. No arithmetic beyond adding is called for and the results are easily interpreted. That is why total counting was once very popular in wildlife management and why it is still the most popular method for censusing people.

Total counts have two serious drawbacks: they tend to be inaccurate and expensive. Nonetheless they have a place. The hippopotami (*Hippopotamus amphibius*) in a clear-water stretch of river can be counted with reasonable facility from a low-flying aircraft. The number of large mammals in a 1 km² fenced reserve can be determined to a reasonable level of accuracy by a drive count. It takes much organization and many volunteers, but it can be done. Every nesting bird can be counted in an adélie penguin (*Pygoscelis adeliae*) rookery, either from the ground or from an aerial photograph. That is an example of a "total count" providing an index of population size because >50% of the birds will be at sea on any given occasion.

Total counting of large mammals over extended areas was common in North America up to 1950. Gill *et al.* (1983) described the system in Colorado:

> Biologists attempted to count total numbers of deer comprising the most important "herds" in the state. Crews of observers walked each drainage within winter range complexes and counted every deer they encountered. The sum of all counts over every drainage of a winter range was taken as the minimum population size of that herd (McCutchen 1938; Rasmussen and Doman 1943).

Total counting of large mammals from the air was a standard technique in Africa in the 1950s and early 1960s. Witness the total counts of large mammals on the 25 000 km² Serengeti–Mara plains (Talbot and Stewart 1964) and of elephants in parts of East Africa (Buechner *et al.* 1963). These massive exercises were soon replaced by sample counting, only Kruger National Park (20 000 km²) of South Africa retaining the total count as a standard technique:

> ... trends in population totals, spatial distribution and social organization are obtained by means of surveys by fixed-wing aircraft. Due to the size of the Kruger National Park these (total count) surveys require three months to complete and are consequently undertaken only once annually (i.e., during the dry season from May to August) (Joubert 1983).

Similar methods are used to count pronghorn antelope in the USA (Gill *et al.* 1983).

12.4 **Sampled counts: the logic**

There are two important areas in which scientific thinking differs from everyday thinking: the selection of a random or unbiased sample and the choosing of an appropriate experimental control. Knowing how to sample and knowing how to design an experiment that gives an unambiguous answer are the two attributes distinguishing science from ideology. Sampling is the technique of drawing a subset of sampling units from the complete set and then making deductions about the whole from the part. It is used all the time in wildlife research and management, but often incorrectly.

The next section takes you through some of the mystery of what happens when we sample. It explores what actually happens when we sample a population in several different ways, thereby making the point that the true estimate is independent of whatever mathematical calculations are applied to the data.

12.4.1 *Precision and accuracy*

If a large number of repeated estimates of density has a mean that does not differ significantly from the true density then each estimate is said to be accurate or unbiased. Accuracy is a measure of bias error. If that set of estimates has little scatter the estimates are described as precise or repeatable. Precision is a measure of sampling error. A system of estimation may provide very precise estimates that are not accurate, just as a system may provide accurate but imprecise estimates. Ideally both should be maximized, but often we must choose between one or other according to what question is being asked. For example, is density below a critical threshold of one animal per square kilometer? Here we need an accurate measure of density and may be willing to trade off some precision to get it. But if we had asked whether present density is lower than that of last year we would need two estimates each of high precision. Their accuracy would be irrelevant so long as their bias was constant. Most questions require precision more than accuracy. Precision is obtained by rigid standardization of survey methods, by sampling in the most efficient manner, and by taking a large sample.

12.4.2 *Sampling frames*

Before an area is surveyed to estimate the number of animals on it, that area must be divided into *sampling units* that cover the whole area and are nonoverlapping. The sampling units may comprise areas of land if we count deer, or trees if we count nests, or stretches of river if we count beavers or crocodiles. To allow us to sample from this *frame list* of sampling units the list must be complete for the whole area. Hence, the frame of units contains all the animals whose numbers we wish to estimate.

For purposes of explanation we use the first example: sampling units of land. The survey area may be divided up into units in any way the surveyor desires, into quadrats, transects, or irregular sections of land perhaps delimited by fences. The choice is a compromise between what is most efficient statistically and what is most efficient operationally.

12.4.3 *Sampling strategies*

Suppose that we wished to estimate the number of kangaroos or antelopes in a large area by counting animals on a sample of that area. Several strategies are open to us. We could sample quadrats or transects; we could select these sampling units systematically or randomly; and, if the latter, we could ensure that each sampled unit occurred only once in the sample (sampling without replacement) or that the luck of the draw allowed units to be selected more than once (sampling with replacement). The efficiency of these systems will be demonstrated with the hypothetical data of Table 12.1 which may be thought of as the number of kangaroos standing on each square kilometer of an area totalling $144\,km^2$. In all cases one-third of the area will be surveyed. We can test the accuracy of the method by determining whether the mean of a set of repeated estimates is significantly different from the true total of 1737 kangaroos. The precision of a sampling system is indicated by the spread of those repeated and independent estimates, and that spread will be measured by the standard deviation of those estimates:

$$s = \sqrt{(\Sigma x^2 - (\Sigma x)^2/N)/(N - 1)}$$

where x is an independent estimate of total numbers and N is the number of such repeated estimates.

We will first sample $1\text{-}km^2$ quadrats randomly with replacement – *sampling with replacement* (SWR). The quadrats are numbered from one to 144 and a sample of 48 of these is drawn randomly. Quadrat numbers 27, 31, 50, and 53 were drawn two times and quadrat number 7 three times, but since these are independent draws they are included in the sample as many times as they are randomly chosen. The quadrat is replaced in the frame list after each draw, allowing it the chance

Table 12.1 A simulated dispersion of kangaroos on a $1 \times 1\text{-km}$ grid of 144 cells. Marginal totals give numbers on $1 \times 12\text{-km}$ transects oriented across the region and down it

1	2	7	4	7	14	9	18	24	22	19	15	142
0	1	5	6	12	11	9	15	20	21	27	28	155
2	3	5	6	10	13	16	20	160	14	19	21	289
1	4	4	6	9	13	14	17	20	16	25	20	149
2	2	5	7	10	12	16	19	20	16	18	22	149
2	4	5	6	9	12	16	22	18	18	21	23	156
0	2	5	8	4	7	11	13	17	16	21	30	134
1	0	4	9	8	10	11	16	14	20	17	17	127
0	4	2	7	8	11	11	11	12	19	22	21	128
0	2	5	8	8	12	16	20	24	25	23	25	168
1	0	4	9	8	8	8	17	17	14	18	22	126
2	5	7	6	12	12	13	15	20	21	20	23	156
12	29	58	82	105	135	150	203	366	222	250	267	1737

of being drawn again. The number of kangaroos in this sample of quadrats totalled 523, and since we sampled only one-third of the quadrats we multiply the total by three to give an estimate of animals in the study area: 1569.

Note that this answer is wrong in the sense that it differs from the true total known to be 1737 (i.e., it is not accurate). That disparity is called the *sampling error*, and is quite distinct from *errors of measurement* resulting from failure to count all the animals on each sampled quadrat.

We now repeat the exercise by drawing a fresh sample of 48 units and get a sampled count of 493 kangaroos which multiplies up to an estimate of 1479. The third and fourth surveys give estimates of 1836 and 1752. That exercise was repeated 1000 times with the help of a desktop computer. The 1000 independent estimates had a mean of $\bar{x} =$ 1741, very close to the true total of 1737. We can be confident there-fore that this sampling system produces accurate (i.e., unbiased) estimates. The 1000 independent estimates had a standard deviation of $s = 153$, which tells us that there is a 95% chance that any one estimate will fall in the range $\bar{x} \pm 1.96s$ or 1741 ± 300, between 1441 and 2041. The standard deviation of a set of independent estimates is the measure of the efficacy of the sampling system and hence of the precision of any one of the independent estimates. It can be estimated from the quadrat counts of a single survey (see Section 12.5.1), and when estimated in this way it is called the standard error of the estimate. Hence the standard error of an estimate is a calculation of what the standard deviation of a set of independent estimates is likely to be.

With that background we can now compare the efficiency of several sampling systems.

12.4.4 Sampling with or without replacement?

When we use *sampling without replacement* (SWOR) a quadrat may be drawn no more than once, in contrast to the previous system which allowed, by the luck of the draw, a quadrat to be selected any number of times. We draw a unit, check whether that unit has been selected previously, and if so reject it and try again. Having drawn 48 distinct units, we calculate density. The sampling is again repeated 1000 times, yielding 1000 independent estimates, each based on a draw of 48 units, of the number of animals we know to be 1737. Those 1000 estimates had a mean of 1743 and a standard deviation of 131, which is appreciably lower than the $s = 153$ accruing from sampling with replacement.

The gain in precision by SWOR reflects the slightly greater infor-mation on density carried by the 48 distinct quadrats of each survey. SWOR is always more precise than SWR for the same sampling fraction, the relationship being:

$$s(\text{SWOR}) = s(\text{SWR}) \times \sqrt{(1 - f)}$$

where f is the sampling fraction, in this case 0.333. The $s(SWR)$ from the 1000 repeated surveys was 153 and from this we could have estimated, without needing to run the simulation, that the precision of the analogous SWOR system would be about:

$$s = 153 \times \sqrt{0.666} = 125.$$

Our empirical $s(SWOR)$ is 131, which is much the same as the $s = 125$ predicted theoretically.

However, it is not as simple as that. The quadrats chosen more than once in a SWR sample are not surveyed more than once although they are included in the analysis more than once, and so the time taken for the survey is shorter. In the example only about 41 of the 48 units drawn in a SWR sample would be distinct units, the other seven being repeats. To compare the precision of a SWOR sample with that of a SWR sample entailing the same groundwork, we would have to draw by SWR about 58 units. Ten are repeats, "free" units that do not need to be surveyed a second time. Intuitively we would assume that the SWR sample of 48 distinct units and 10 repeats must give a more precise estimate than the SWOR sample with its 48 distinct units, none repeated. Not so. The smaller SWOR samples provide estimates more precise by a factor of $\sqrt{(1 - \frac{1}{2}f)}$. In all circumstances SWOR is more precise than SWR (Raj and Khamis 1958). Precision is increased by rejecting the repeats and cutting the sample size back to that of the analogous SWOR sample.

Why then, if SWOR is always better, is SWR often used. First, when the sampling fraction is low, $<15\%$, the precision of the two systems of sampling is similar. At $f = 0.1$ there is only a 5% difference in precision, reflecting the low likelihood of repeats at low sampling intensity. Most sampling intensity in wildlife management is of this order. Second, it is often convenient to sample with replacement when an area is traversed repeatedly by aerial-survey transects. There is not the same necessity to ensure that no transect crossed another or overlaps it. That is a useful flexibility for an aerial survey in a strong cross-wind or for a ground survey in thick forest.

12.4.5 Transects or quadrats?

A frame of transects is a good or bad sampling system according to how it is oriented with respect to trends in density. The dispersion of Table 12.1 has a marked increase in density from left to right. The precision of the estimate of total numbers would be relatively high if the transects were oriented along this cline but low if oriented at right angles to it. That can be demonstrated empirically by sampling the column totals at one-third sampling intensity. Each column represents a transect and each survey comprises four transects randomly chosen. One thousand independent surveys gives a standard deviation of estimates of 512 for SWR and 427 for SWOR. If these transects had been oriented at right angles so that the rows rather than the columns formed the transects, the standard deviation of estimates of 1000 independent surveys would have been approximately 80 for SWR and 69 for SWOR. In this case precision is increased enormously by swinging the orientation of the transects through 90°.

Transects should go across the grain of the country rather than along it; they should cross a river rather than parallel it; and they should go up a slope rather than hug the contour. They should be oriented such that each one samples as much as possible of the total variability of an area. In essence we must ensure that the variation between transects is minimized and therefore that the precision of the estimate is maximized.

Much the same principle adjudicates between the use of quadrats as against transects. So long as the frame of transects is oriented appropriately, the resultant estimate will be more precise than that from a set of quadrats whose area sums to that of the transects. The more clumped the distribution of the animals the greater will be the gain in precision of transects over quadrats. A quadrat is likely to land in a patch of high density or a patch of low density, whereas a transect is more likely to cut through areas of both. Table 12.2 shows that transects oriented along the cline in density of Table 12.1 provide estimates six times more precise than do quadrats of the same size and number.

12.4.6 Random or nonrandom sampling?

Sampling strategies grade from strictly random to strictly systematic. The region in between is described as restricted random sampling. One might decide, for example, to sample randomly but to reject a unit that

Table 12.2 The effect of sampling system on the precision of an estimate. All systems sample one-third of an area of 144 km² containing the dispersion of kangaroos simulated in Table 12.1. Each sampling system is run 1000 times to provide 1000 independent estimates of the true total of 1737

Sampling system	Mean estimate	Standard deviation of 1000 estimates
Large quadrants (n = 4)		
Random with replacement	1746	487
Random without replacement	1738	414
Small quadrants (n = 48)		
Random with replacement	1741	153
Random without replacement	1743	131
Transects parallel to the density cline (n = 4)		
Random with replacement	1732	80
Random without replacement	1734	69
Restricted random	1730	57
Systematic	1736	48

abuts one previously drawn. Or one might break the area into zones and draw the same number of samples randomly from each zone. These two strategies depart from the requirement of strict random sampling whereby each sampling unit has the same probability of selection. The extreme is systematic sampling in which the choice of units is determined by the position of the first unit selected.

Systematic or restricted random sampling has several practical advantages over strict random sampling. First, it encourages or enforces sampling without replacement which, as we have seen, leads to a more precise estimate. Second, it reduces the disturbance of animals on a sampling unit caused by surveying an adjoining unit. This is particularly important in aerial survey where the noise of the aircraft can move animals off one transect onto another. Third, any deviation from strictly random sampling tends to increase the precision of the estimate because the sampled units together provide a more comprehensive coverage of total variability. Table 12.2 demonstrates this for our example. The standard deviation of 1000 independent surveys is lower for restricted random sampling than for random sampling without replacement, and lower still for systematic sampling.

Statisticians do not like nonrandom sampling because the precision of the estimate cannot be calculated from a single survey. The formulae given in Section 12.5.1 for calculating the standard error of an estimate are correct only when sampling units are drawn at random, and they will tend to overestimate the true standard error when restricted random or systematic sampling is used. But not always. If a systematically drawn set of sampling units tends to align with systematically spaced highs and lows of density, the standard error calculated on the assumption of random sampling will be too low and the estimate of density will be biased.

In practice this tends not to happen. It is entirely appropriate to sample systematically or by some variant of restricted random sampling and to approximate the standard error of the estimate by the equation for random sampling. One can be confident that the estimate is unlikely to be biased and that the true standard error is unlikely to exceed that calculated.

12.4.7 How not to sample

There are a number of traps that sampling can lure one into and which can result in a biased estimate or an erroneous standard error.

Suppose one decided to sample quadrats but, for logistical reasons, laid them out in lines, the distance between lines being considerably greater than the distance between neighboring quadrats within lines. The standard error of the estimate of density cannot then be calculated by the usual formulae because the counts on those quadrats are not independent. Density is correlated between neighboring quadrats and this throws out the simple estimate of the standard error which returns an erroneously low value. There are ways of dealing with the data from this design to yield an appropriate standard error (see

Cochran (1977) for treatment of two-stage sampling) but they are beyond the scope of this book. The simple remedy is to pool the data from all quadrats on each line, the line rather than the quadrat becoming the sampling unit. That procedure may appear to sacrifice information but it does not (Caughley 1977a).

Another common mistake is to throw random points onto a map and to declare them centers of the units to be sampled, the boundary of each being defined by the position of the point. In this case the requirement that sampling units cover the whole area and are non-overlapping is violated and the sampling design becomes a hybrid between sampling with replacement and sampling without replacement, leading to difficulties in calculating a standard error. There is nothing wrong with choosing units to be sampled by throwing random points on a map so long as the frame of units is marked on the map first. The random points define units to be selected. They do not determine where the boundaries of those units lie.

A third trap to watch for is a biased selection of units to be sampled. The most common source of this bias in wildlife management is the so called "road count" in which animals are counted from a vehicle or on foot on either side of a road or track. Roads are not random samples of topography. They tend to run along the grain of the country rather than across it, they go around swamps rather than through them, they tend to run along vegetational ecotones, and they create their own environmental conditions, some of which attract animals while others repel them.

12.5 Sampled counts: methods and arithmetic

Sampled counts of animals fall easily into two categories. There is first the method of counting on sampling units whose boundaries are fixed. For example, we might walk lines and count deer on the area within 100 m each side of the line of march. Or we might count all ducks on a sample of ponds, the shoreline of the pond providing a strict boundary to the sampling unit.

The alternative to fixed boundaries is unbounded sampling units. Instead of restricting the counting to those animals within 100 m of a line of march, those outside the transect being ignored, we might count all the animals that we see. Since the observed density will fall away with distance from the observer the raw counts are no longer an estimate of true density. They must therefore be corrected.

Of these two options (sampling units with boundaries and sampling units without boundaries) the first has immense advantages of simplicity and realism. If the transect width is appropriately chosen, what the observer sees is what the observer gets. The mathematics of such sampling are simple, elegant, and absolutely solid. In contrast, the accuracy of corrected density estimated from unbounded transects depends heavily on which model is chosen for the analysis. There are many to choose from and they give markedly different answers for the same data. The advantage of unbounded transects is in all the sightings being used, none being discarded. Since the precision of an estimate is related tightly to the number of animals actually counted, any sampling scheme increasing the number of sightings also tends to increase the precision of the estimate. That is an advantage if the increased precision is obtained without the sacrifice of too much accuracy.

The choice of one or other system is often determined by density. If the species is rare then one might be tempted to use all the data one can get. If it is common one might be content to use the more dependable sampling units with fixed boundaries, knowing that fewer things can go wrong.

12.5.1 *Fixed boundaries to sampling units*

The appropriate analysis depends on whether the sampling units are of equal or unequal size, and how they are selected.

Notation

y	= the number of animals on a given sampled unit,
a	= the area of a given sampled unit,
A	= the total area of the region being surveyed,
n	= the number of units sampled,
D (or d)	= the estimate of mean density,
SE(D)	= the standard error of estimated mean density,
Y	= the estimate of total numbers in the region of size A,
SE(Y)	= the standard error of the estimate of total numbers.

The simple estimate (for equal-sized sampling units)
The simple estimate is used when sampling units are of constant size, as when the region being surveyed is a rectangle that can be sub-divided into quadrats or transects. It will provide an unbiased, although imprecise, estimate, even when sampling units differ in size, but more appropriate designs are available for that case. We will explore this design at some length because most of the principles are shared with the other designs.

The region to be surveyed, of area A, is divided on a map or in one's head into an exhaustive set of nonoverlapping sampling units, each of constant area a. Let us assume, for illustration, that the region is as given in Table 12.1, and that this region of $A = 144\,km^2$ is to be sampled by $n = 4$ transects each of area $a = 12\,km^2$. Sampling intensity is hence $na/A = 4 \times 12/144 = 0.333$.

In Table 12.1 the rows represent transects and the marginal totals the number of animals on each transect. Numbering the transects from 1 to 12 and selecting at random with replacement from this set we draw transect numbers 4, 8, 1, and 4. On surveying these transects we would obtain counts of:

Transect	1	4	4	8
Count	142	149	149	127

Note that transect 4 has been drawn twice, so in practice we survey only three transects, although the count from transect 4 enters the calculation twice.

Density is estimated as the sum of the transect counts $(142 + 149 + 149 + 127)$ divided by the sum of the transect areas $(12 + 12 + 12 + 12)$. Thus:

$$D = \Sigma y / \Sigma a = 567/48 = 11.81/\text{km}^2$$

The precision of that estimate is indexed by its standard error SE(D), which is itself an estimate of what the standard deviation of many independent estimates of density would be, each estimate derived from four transects drawn at random with replacement:

$$SE(D) = 1/a \cdot \sqrt{(\Sigma y^2 - (\Sigma y)^2/n)/(n(n - 1))}$$

That is a slight approximation. To be exactly unbiased it should be multiplied by a further term $\sqrt{(1 - (\Sigma a)/A)}$, but that usually makes so little difference that it tends to be ignored.

The calculation tells us that this hypothetical distribution of estimates, each of them made in the same way as we made ours, with the same sampling frame and the same sampling intensity, only the draw of sampling units being different, would have a standard deviation in the vicinity of ± 0.43. In fact, that is likely to be an underestimate because it is based on only four sampling units, three degrees of freedom. With samples above 30 sampling units we can form 95% confidence limits of the estimate by multiplying by 1.96, but for smaller samples we must choose a multiplier from a t-table corresponding to a two-tailed probability of 0.05 and the degrees of freedom (d.f.) of our sample. In the case of d.f. = 3, the multiplier is 3.182 and so the 95% confidence limits of our estimate of density are $\pm 3.18 \times 0.43 = \pm 1.37$.

The number of animals (Y) in the surveyed region can now be calculated as the number of square kilometers in that region (A) multiplied by the estimated mean number per square kilo-meter (D):

$$Y = AD = 144 \times 11.81 = 1701$$

It has a standard error of:

$$SE(Y) = \pm A \times SE(D) = \pm 144 \times 0.43 = \pm 62$$

Its 95% confidence limits are calculated as A multiplied by the 95% confidence limits of D: $\pm 144 \times 1.37 = \pm 197$.

We can check that against Table 12.2, which shows that the true total number (Y) is 1737 and so the estimate with 95% confidence of $Y = 1701 \pm 197$ is entirely acceptable.

If the sampling is without replacement the above formula for SE(D) yields an overestimate. The standard error for sampling without re-placement is estimated by the formulation for the standard error with replacement multiplied by the square root of the proportion of the area not surveyed. This *finite population correction* or FPC is:

$$\sqrt{(1 - (\Sigma a)/A)}$$

The simple estimate may be used validly even when sampling units are of unequal size. The constant a is then replaced by the mean area of sampling units. The precision of the estimate will be lower (i.e., the standard error will be higher) than by the ratio method (see next subsection), but the estimate is unbiased and may be precise enough for many purposes.

The simple estimate, with minor modification, can be used when the total area A is unknown. One of us was forced to this exigency while surveying from the air a population of rusa deer (*Cervus timo-rensis*) in Papua New Guinea. The deer lived on a grassed plain, the area of which could not be gauged with any accuracy from the available map. The remedy was to measure the length of the plain by timing the aircraft along it at constant speed, and then to run transects from one side of the plain to the other at right angles to that measured baseline. The area of a sampling unit is entered as $a = 1$, even though areas are of different and unknown size. D then comes out as average numbers per transect rather than per unit area. Total numbers Y on the plain could then be estimated by replacing A by N, where N is the total number of transects that could have been fitted into the area. That is simply the length of the baseline divided by the width of a single transect. A similar approach was used for censusing wildebeest in the Serengeti (Norton-Griffiths 1978).

The ratio estimate (for unequal-sized sampling units)

This is the best method for a frame of sampling units of unequal size, as might be provided by a faunal reserve of irregular shape sampled by transects. Statistical texts warn that the estimate is biased when the number of units sampled is < 30 or so, but the bias is usually so slight as to be of little practical importance. The number of units may be as low as two without generating a bias of more than a few percent.

The appropriate formulae are given in Table 12.3 and in the notation at the beginning of Section 12.5.1. That for the standard error looks quite different from that of the simple estimate but they are mathematical identities when the sampling units are of equal size. The ratio estimate is general, the simple estimate being a special case of it. Hence, if these analyses are to be programmed into a calculator or computer, the ratio method is the only one needed.

The probability proportional-to-size estimate

In the previous two methods all sampling units in the frame have an equal chance of being selected. In the probability proportional-to-size (PPS) method the probability of selection is proportional to the size of

Table 12.3 Estimates and their standard errors for animals counted on transects, quadrants, or sections. The models are described in the text

Model	Density	Numbers
Simple		
Estimate	$D = \Sigma y / \Sigma a$	$Y = A \cdot D$
Standard error of estimate (SWR)	$SE(D)_1 = 1/a \cdot \sqrt{(\Sigma y^2 - (\Sigma y)^2/n)/(n(n-1))}$	$SE(Y) = A \cdot SE(D)_1$
Standard error of estimate (SWOR)	$SE(D)_2 = SE(D)_1 \cdot \sqrt{(1 - (\Sigma a)/A)}$	$SE(Y) = A \cdot SE(D)_2$
Ratio		
Estimate	$D = \Sigma y / \Sigma a$	$Y = A \cdot D$
Standard error of estimate (SWR)	$SE(D)_3 = n/\Sigma a \cdot \sqrt{(1/n(n-1))(\Sigma y^2 + D^2 \Sigma a^2 - 2D\Sigma ay)}$	$SE(Y) = A \cdot SE(D)_3$
Standard error of estimate (SWOR)	$SE(D)_4 = SE(D)_3 \cdot \sqrt{(1 - (\Sigma a)/A)}$	$SE(Y) = A \cdot SE(D)_4$
PPS		
Estimate	$d = 1/n \cdot \Sigma(y/a)$	$Y = A \cdot d$
Standard error of estimate (SWR)	$SE(D) = \sqrt{(\Sigma(y/a)^2 - ((\Sigma(y/a))^2/n)/(n(n-1)))}$	$SE(Y) = A \cdot SE(d)$

SWR, sampling with replacement; SWOR, sampling without replacement. Notation is given in Section 12.5.1.

the sampling unit. Suppose that the area to be surveyed is farmland. We might decide to declare the paddocks (or "pastures" or "fields," depending on which country you are in) as sampling units because the fences provide easily identified boundaries to those units.

If each sampling unit were assigned a number and the sample chosen by lot, we would use the ratio method of analysis. But we might decide instead to choose the sample by throwing random points onto a map. Each strike selects a unit to be sampled, the probability of selection increasing with the size of the unit.

The PPS estimate has the advantages that it is entirely unbiased and that the arithmetic (Table 12.3) is simple. Its disadvantage is that it can be used only when sampling with replacement and so it is not as precise as the ratio method used without replacement. Hence this method should be restricted to surveys whose sampling intensity is < 15%. The PPS estimate is a mathematical identity of the simple estimate and the ratio estimate when units of equal size are sampled with replacement.

12.5.2 *Unbounded transects (line transects)*

The observer walks a line of specified length and counts all animals seen, measuring one or more subsidiary variables at each sighting (angle between the animal and the line of march; radial distance, the distance between the animal and the observer at the moment of sighting; the right-angle distance between the animal and the transect). If we know the shape of the sightability curve relating the probability of seeing an animal on the one hand to its right-angle distance from the line on the other, and if an animal standing on the line will be seen with certainty, it is fairly easy to derive an estimate of density from the number seen and their radial or right-angle distances. We seek a distance from the line where the number of animals missed within that distance equals the number seen beyond it. True density is then the total seen divided by the product of twice that distance and the length of the line.

Therein lies the difficulty. That distance is determined by the shape of the sightability curve, which can seldom be judged from the data themselves. Consequently the shape of the curve must be assumed to some extent and the validity of the assumption determines the accuracy of the method.

We present only two of the many models available, mainly to give some idea of their diversity. The first is the Hayne (1949) estimate which is derived from the assumption that the surveyed animals have a fixed flushing distance and will be detected only when the observer crosses that threshold. If k is the number of animals detected and r the radial distance from a detected animal to the observer,

$$D = (1/2L)\Sigma_k(1/r)$$

where L is the length of the line. Hence density is the sum of the reciprocals of the radial sighting distances divided by twice the length of the line.

It is implicit in Hayne's model that $\sin\theta$, the sine of the sighting angle, is uniformly distributed between 0 and 1, and that the theoretically expected mean sighting angle is 32.7°. Hence the reality of the model can be tested against the data. Eberhardt (1978) recommended tabulating the frequency of $\sin\theta$ in ten intervals of 0.1 (0.0–0.1, 0.1–0.2 . . . 0.9–1.0) and testing the uniformity of the frequencies by χ^2 analysis. He gave a worked example for a survey of the side-blotched lizard (*Uta stansburiana*). Robinette et al. (1974) and Burnham et al. (1980) suggested that most mean sighting distances tended to be around 40° or more, the latter authors being convinced that the Hayne estimate is used far too uncritically in wildlife management. Robinette et al. (1974) compared the accuracy of the Hayne estimate with that of eight other line-transect models, showing that when applied to inanimate objects or to elephants it tended to overestimate considerably. The other methods were seldom much better. They identified only two King's method (Leopold 1933) and a more complicated nonparametric method by Anderson and Pospahala (1970) as worthy of further consideration. Burnham et al. (1980) and Burnham and Anderson (1984) provided a starting point for reading further about line-transect methods.

Our second example is a nonparametric method developed by Eberhardt (1978) from work by Cox (1969). First, we choose arbitrarily a distance, Δ, perpendicular from the line. Eberhardt's estimate of density is:

$$D = (3k_1 - k_2)/4L\Delta$$

where k_1 and k_2 are the number of animals seen on either side of the line-transect at distances that fall within the interval 0 to Δ and Δ to 2Δ, respectively. Eberhardt (1978) considered that the method is most useful as only a cross-check on the results of other methods because its estimate is likely to be imprecise. Precision is enhanced by choosing a large value of Δ but accuracy is enhanced by choosing a small Δ (Seber 1982).

Note that neither of these methods, nor any of the others presently available for treating line-transect data, can be used in aerial survey. They are all anchored by the assumption that all animals on the line of march (equivalent to the inner strip marker of aerial survey) are tallied by the observer. That assumption does not hold for aerial survey because the ground under the inner strip marker is at a distance from the observer, because an animal under a tree on that line may be missed, and because an observer cannot watch all parts of the strip at once and may therefore miss animals in full view on that line.

There was a deluge of mathematics on the theory of line transects around 1980. Eberhardt (1978), Gates (1979), Burnham *et al.* (1980), Seber (1982), and Sen (1982) provided comprehensive reviews of previous work and many further advances. The subject provides a rich lode of theoretical gold which will continue to yield theoretical insights. In contrast, the mathematics of sampling units with fixed boundaries had been cracked 50 years previously and therefore does not generate the same interest.

The biologist must decide whether the theoretical excitement of line transects justifies their practical application. Can the difficulty of measuring sighting distances and the unreliability of the resultant estimates be justified when an alternative with fewer problems is available? The line transect was originally introduced to circumvent the difficulty of counting all animals on a transect or quadrat. It cured that problem by replacing it with several new ones. Perhaps we should give some thought to ways of treating the original problem without introducing new ones. If animals are difficult to see on a transect of fixed width, why not walk two people abreast down the boundaries? If that does not work, put a third person between them. And so on.

Much of the present use of line transects in wildlife management stems from the belief that they are somehow more "scientific" than strip transects, just as there was once a popular belief, by no means extinct, that quadrats are statistically superior to transects. There are rare situations in which transect sampling will not work and where line transect methodology might (e.g., in very thick cover). Faced with such, a wildlife manager is recommended to use the computer program TRANSECT described by Burnham *et al.* (1980). It enables a thorough analysis of both grouped and ungrouped data under the Fourier series estimation procedure. A number of other estimation schemes are available as options within the program. A user's manual is provided by Laake *et al.* (1979). Another useful program for analyzing line transect data is LINETRAN, described by Gates (1980; 1986). The program and users' guide is available from Charles Gates at the Department of Statistics, Texas A & I University, California, USA.

12.5.3 *Stratification*

The precision of an estimate is determined by sampling intensity and by the variability of density among sampling units. Suppose there were two distinct habitats in the survey area and that, from our knowledge of the species, we could be sure that it would occur commonly in one and rarely in the other. If we surveyed those two subareas separately and estimated a separate total of animals for each, the combined estimate for the whole area would be appreciably more precise than if the area had been treated as an undifferentiated whole.

The process is called stratification and the subareas strata. By this strategy we divide an area of uneven density into two or more strata within which density is much more even. The strata are treated as if they were each a total area of survey, the results subsequently being combined. The estimate from each stratum will be called Y_h which has a standard error of SE(Y_h). Total numbers Y are estimated by $Y = \Sigma Y_h$. Its standard error is the square root of the sum of the variances of the contributing stratal estimates. The variance of an estimate is the square of its standard error. Here it is designated Var(est) to distinguish it from the variance of a sample designated s^2. Calculate Var(Y_h) = (SE(Y_h))^2 for each stratum and then:

$$SE(Y) = \sqrt{\Sigma \mathrm{Var}(Y_h)}$$

to give the standard error of the combined estimate of total numbers.

Optimum allocation of sampling effort
If our aim is to obtain the most precise estimate of Y as opposed to a precise estimate of each Y_h, sampling intensity should be allocated between strata according to the expected standard deviation of sampled unit counts in each stratum. That requires a pilot survey or at least knowledge gained on a previous survey. Often we have nothing more than aerial photographs or a vegetation map to give us some idea of the distribution of habitat, and only a knowledge of the animal's ecology to guide us in predicting which habitats will hold many animals and which will hold few. This scant information is in fact sufficient to allow an allocation of sampling effort between strata that will not be too far off the optimum. The trick is to know that for almost all populations the standard deviation of counts on sampling units rises linearly with density. From that can be derived the rule of thumb that the number of sampling units put into a stratum should be directly proportional to what Y_h is likely to be.

At first thought that is a daunting challenge – to guess each Y_h before we have estimated it – but it is easier if we break it down into components. First, guess the density in each stratum. It does not matter too much if this is wrong, even badly wrong, because all we need to get roughly right is the ratio of densities between strata. Multiply each guessed density by the mapped area of its stratum to give a guess at numbers in the stratum. Divide each by the total area to give the proportion of total sampling effort that should be allocated to each stratum. Table 12.4 shows the calculation for a degree block that can be divided into three strata from a vegetation map and to which a total of 10 hours of aerial survey has been allocated.

12.5.4 *Comparing estimates*

If the sampling units are drawn independently of each other, the estimates of density from two surveys may be compared. The surveys may be of two areas, or of the same area in two different years, or the same area surveyed in the same year by two teams or by different methods. A quick and dirty comparison is provided by the normal approximation, which is adequate if each survey covered more than 30 sampling units. The two estimates are significantly different when:

$$(est_1 - est_2)/\sqrt{(Var(est_1) + Var(est_2))} > 1.96$$

If sample sizes are too low, or if more than two surveys are being compared, the determination of significance should be made by one-factor analysis of variance. If the surveys are not independent, as when the same transects are run each year, a comparison may still be made by analysis of variance but with TRANSECTS now declared a factor in a two-factor analysis. Chapter 13 goes further into this and other uses of analysis of variance.

12.5.5 *Merging estimates*

If a comparison shows that two or more independent estimates of the same population are not significantly different, we may wish to merge them to provide an estimate more precise than the individual estimates. This is a procedure quite distinct from stratification where estimates from different populations are combined to give an overall estimate. Merging is restricted to the same population estimated more than once.

Table 12.4 Allocation of $E = 10$ hours of aerial survey among strata to maximize the precision of the estimate of animals in the total area

Stratum (h)	Area (km²) (A_h)	Guessed density (D_h)	Guessed numbers $(Y_h = A_h D_h)$	Proportion of total effort $(P_h = Y_h/\Sigma Y_h)$	Hours allocated $(E_h = P_h \cdot E)$
1	2 000	1	2 000	0.03	0.3
2	7 000	5	35 000	0.52	5.2
3	3 000	10	30 000	0.45	4.5
	12 000		67 000	1.00	10.0

Again there is a quick and dirty method, to be used only when the individual estimates were each made with about the same sampling intensity. The merged estimate Y can then be calculated as:

$$Y = (Y_1 + Y_2 + Y_3 + \ldots + Y_N)/N$$

where there are N surveys. It has a variance of:

$$\text{Var}(Y) = (\text{Var}(Y_1) + \text{Var}(Y_2) + \text{Var}(Y_3) + \ldots + \text{Var}(Y_N))/N^2$$

Thus the merged estimate is simply the mean of the individual estimates, and its variance is the mean of the individual estimate variances divided by their number. $\text{SE}(Y)$ is the square root of $\text{Var}(Y)$. From these, the merged density estimate is $D = Y/A$ which has a standard error of $\text{SE}(D) = \text{SE}(Y)/A$.

A more appropriate method, particularly for surveys utilizing markedly different intensities of sampling, is provided by Cochran (1954), who considers also more complex merging. Here the contribution of an individual estimate to the merged estimate is weighted according to its precision. Letting $w = 1/\text{Var}(Y)$:

$$Y = (w_1 Y_1 + w_2 Y_2 + w_3 Y_3 + \ldots + w_N Y_N)/ \\ (w_1 + w_2 + w_3 + \ldots + w_N)$$

with a variance of:

$$\text{Var}(Y) = 1/(w_1 + w_2 + w_3 + \ldots + w_N)$$

12.6 Indirect estimates of population size

This section outlines some of the methods available for calculating the size of a population by techniques that do not necessarily depend on accurate counts of animals. The line transect method could well come under this head but it is placed in "sampled counts" because it requires accurate counting of animals on the line.

12.6.1 Index-manipulation index method

If we obtain two indices of population size, I_1 and I_2, the first before and the second after a known number of animals C was removed, the population's size can be estimated for the time of the first index by:

$$Y_1 = I_1 C/(I_1 - I_2)$$

The proportion removed is estimated as $p^* = (I_1 - I_2)/I_1$ and the proportion of those remaining as $q^* = 1 - p^*$. Following Eberhardt (1982), the variance of the estimate of Y can be approximated by:

$$\text{Var}(Y) \approx Y^2 (q^*/p^*)^2 (1/I_1 + 1/I_2)$$

from which $\text{SE}(Y) = \sqrt{\text{Var}(Y)}$. Eberhardt (1982) gives three examples from populations of feral horses. The data from his Cold Springs population were:

$$I_1 = 301$$
$$I_2 = 76$$

$$C = 357$$
$$p^* = 0.748$$

Thus, the population at the time of the first index is estimated as:

$$Y = (301 \times 357)/(301 - 76) = 478$$

with a variance of that estimate of:

$$\text{Var}(Y) \approx 478^2(0.252/0.748)^2(1/301 + 1/76) = 428$$

from which $\text{SE}(Y) = \sqrt{428} = 21$.

The index-manipulation index method assumes that the population is closed (no births, deaths, immigration, or emigration) between the estimation of the first and second indices. That assumption is approximated when the entire experiment is run over a short period.

12.6.2 *Change-of-ratio method*

If a population can be divided into two classes, say males and females or juveniles and adults, and one class is significantly reduced or increased by a known number of animals, the size of the population can be estimated from the change in ratio. The method was introduced by Kelker (1940; 1944) for estimating the size of deer populations manipulated by bucks-only hunting.

The two classes are designated x and y. Before the manipulation there was a proportion p_1 of x-individuals in the population, and p_2 after the manipulation which removed or added C_x x-individuals (additions are positive, removals negative) and C_y y-individuals: $C = C_x + C_y$. The size of the population before the manipulation may be estimated as:

$$Y_1 = (C_x - p_2C)/(p_2 - p_1)$$

As with the index-manipulation index method, Kelker's method assumes that the population is closed. Hence the two surveys to estimate the class proportions must be run close together. Additionally, all removals or additions must be recorded and the two classes must be amenable equally to survey.

12.6.3 *Mark–recapture*

Mark–recapture is a special case of the change-of-ratio method. A sample of the population is marked and released, a subsequent sample being taken to estimate the ratio of marked to unmarked animals in the population. From data of this kind we can estimate the size of the population, and with further elaboration (individual markings, multiple recapturing occasions) the rate of gain and loss.

The huge number of mark–recapture models available have been reviewed adequately by Blower *et al.* (1981) and in detail by Seber (1982) and Krebs (1989). Here we outline the range of methods, provide an introduction to the most simple of these, emphasize their pitfalls, and mention some of the recent advances that might circumvent those pitfalls.

Petersen models

A sample of M animals are marked and released. A subsequent sample of n animals are captured of which m are found to be marked. If Y is the unknown size of the population then clearly:

$$M/Y = m/n$$

within the limits of sampling variation. With rearranging, that equation allows an estimate of populations size as:

$$Y = Mn/m$$

Intuitively obvious as that is, it is not quite right because of a statistical property of ratios that leads on average to a slight overestimation. The bias may be corrected (Bailey 1951; 1952) by:

$$Y = (M(n + 1))/(m + 1)$$

which has a standard error of approximately:

$$SE(Y) = \sqrt{(M^2(n + 1)(n - m))/((m + 1)^2(m + 2))}$$

These formulae are for "direct sampling," when the number of animals to be recaptured is not decided upon prior to recapturing. There are further variants for sampling with replacement and for inverse sampling (see Seber 1982).

Except for the unlikely case of half or more of the population being marked, the distribution of repeated independent estimates of population size is always strongly skewed to the right, a positive skew. (The direction of skew is the direction of the longest tail.) Figure 12.1 shows this effect from a computer simulation of 1000 estimates of a population of 500 animals containing 100 marked individuals. Each estimate is derived from a capturing of 50 animals. Apart from demonstrating the skew of estimates, the figure makes the point that only a limited number of estimated values is possible. With $Y = 500$ and $M = 100$, the probability of a given animal being marked is 0.2, and so the expected number of marked animals in a sample of 50 is 10. That would give a population estimate of $Y = 464$. If nine were recaptured the estimate would be $Y = 510$. No estimate between 464 and 510 is possible.

Since the estimates are skewed, the confidence limits of an estimate are also skewed and cannot easily be calculated from the standard error. Blower et al. (1981) recommended an approximating procedure. Let $a = m/n$. In a large sample the 95% confidence limits of a are approximately $\pm 1.96\sqrt{(a(1 - a)/n)}$. Since $Y = M/a$ the upper and lower 95% confidence limit of a can each be divided into M to give upper and lower 95% confidence limits of Y.

The Petersen estimate is the most simple of a family of estimation procedures. If animals are marked on more than one occasion and recaptured on more than one occasion, it is possible to estimate gains and losses from the population as well as its size. Seber's (1982) monumental text describes most of the options.

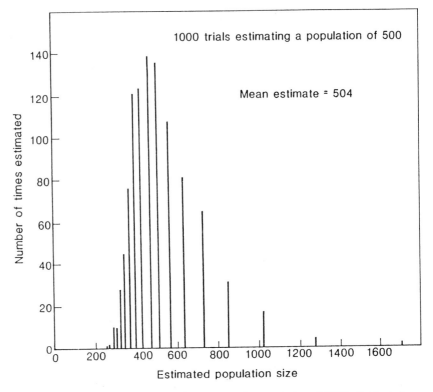

Fig. 12.1 Simulated replications of estimates of a population of 500 individuals by mark–recapture where 100 are marked and 50 captured. Note the positive skew of estimates and the fact that only a limited number of estimated values is possible.

The Petersen estimate depends on these assumptions:

1 all animals are equally catchable,

2 no animal is born or migrates into the population between marking and recapturing,

3 marked and unmarked animals die or leave the area at the same rate,

4 no marks are lost.

Assumption (2) is not needed when marked animals are recaptured on more than one occasion, but the others are common to all elaborations of the Petersen estimate. The least realistic is the assumption of equal catchability, which is routinely violated by almost any population the wildlife manager is called upon to estimate (Eberhardt 1969). For this reason the Petersen estimate and its elaborations (Bailey's triple catch, Schnabel's estimate, the Jolly–Seber estimate, and many others) are of dubious utility in wildlife management.

Frequency-of-capture models

Petersen models work only when all animals in the population are equally catchable. Frequency-of-capture models are not constrained in that way but will work only if the population is closed and if there are no losses from or gains to the population over the interval of the experiment. That is easy enough to approximate by running the exercise over a short period.

Animals are captured on a number of occasions, usually on successive nights, and marked individually at the first capture. At the end of the experiment each individual caught at least once can be scored according to the number of times it was captured. The data come in the form:

No. of times caught (i):	1	2	3	4	5	6	7	8	...	18
No. of animals (f_i):	43	16	8	6	0	2	1	0	...	0

which are from Edwards and Eberhardt (1967) who trapped a penned population of wild cottontail rabbits for 18 days. Of these 43 were caught once only, 16 twice, eight three times, and so on. $\Sigma f_i = 76$ gives the number of rabbits caught at least once and so the population must be at least that large. If we could estimate f_0, the number of rabbits never caught, we would have an estimate of population size $Y = f_0 + 76$.

Traditionally, this has been attempted by fitting a zero-truncated statistical distribution (Poisson, geometric, negative binomial) to the data and thereby estimating the unknown zero frequency. Eberhardt (1969) exemplifies this approach. More recently there has been much interest generated in more complex models by the work of Pollock (1974) and Burnham and Overton (1978), culminating in a comprehensive package of models provided by Otis *et al.* (1978). These can cope with probability of capture varying with time (seasonal trends, changes in weather), with individual animals (sex differences, dominance relationships), with trapping history (capture-shyness and capture-proneness), and with various combinations of these. The fit of each model can be tested against the data and an objective decision made as to which is the most appropriate. The computations are too lengthy to be attempted by hand but a FORTRAN program CAPTURE is available together with a users' manual (White *et al.* 1982).

Estimation of density

All previously reviewed mark–recapture methods yield a population size Y which can be converted to a density D only when the area A relating to Y is known. In most studies Y itself is meaningless because the "population" is not a population in the biological sense but the animals living on and drawn to a trap grid of arbitrary size.

Seber (1982) and Anderson *et al.* (1983) reviewed the methods currently used to estimate A as a prelude to determining density. Most rely on Dice's (1938) notion of a boundary strip around the trapping grid such that the "effective trapping area" A is the grid area plus the area of the boundary strip. Most of these methods are *ad hoc* and subject to numerous problems, or require large quantities of data to produce satisfactory estimates, or require supplementary trapping beyond the trapping grid.

Anderson *et al.* (1983) circumvented this problem with a method of mark–recapture that provides a direct estimate of density. The traps are laid out not in a grid but at equal intervals along the spokes of a wheel. Trap density therefore falls away progressively from the center of the web. The method pivots upon the assumption that the high density of traps at the center guarantees that all animals at the center will be captured. That is analogous to the assumption of line-transect methodology that all animals are tallied on the line of march. The data collected as distance of first capture from the center of the web are analyzed almost exactly as if they were from a line transect. The Fourier series method described by Burnham *et al.* (1980) is the most appropriate and this analysis can be run on the computer program TRANSECT (Laake *et al.* 1979) after slight modification to the input as described in the appendix of Anderson *et al.* (1983).

12.6.4 *Incomplete counts*

The problem of estimating the size of a population from "total counts" known to be inaccurate has been approached from three directions. One family of methods requires a set of replicate estimates, the second requires two estimates, and the third provides an estimate known with confidence to be below true population size.

Many counts

Hanson's (1967) method assumes that all animals have the same probability of being seen but that this probability is less than one. Hence, whether a given animal is seen or not on a given survey is a draw from a binomial distribution. It follows from the mathematics of the binomial distribution that $Y = \bar{x}^2/(\bar{x} - s^2)$, where Y is the population size, $\bar{x}$ the mean of a set of (incomplete) counts and s^2 the variance of those counts.

This method is not recommended because of the restriction that all animals have the same sightability. In practice sightability varies by individuals and between surveys. The variance of a set of replicate counts tends to be greater than their mean (a binomial variance is always less than the mean), indicating that the method is unworkable.

A modification of this method to circumvent that restriction was suggested by Caughley and Goddard (1972). It requires repeated counts made at two levels of survey efficiency (e.g., two sets of aerial surveys, one flown at 50 m and the other at 100 m). However, Routledge (1981) showed by simulation that this method yields a very imprecise estimate unless the number of surveys is prohibitively large, and hence we do not recommend it.

The nonparametric *method of bounded counts* (Robson and Whitlock 1964) provides a population estimate from a set of replicate counts as twice the largest minus the second largest. Routledge (1982) dismissed this method also (as do we) because in most circumstances it greatly underestimates the true number.

Two counts

Caughley (1974) showed that if the counts of two observers of equivalent efficiency were divided into those animals (or groups of animals)
seen by only one observer and those seen by both, the size of the
population could be estimated. Henny *et al.* (1977) and Magnusson *et
al.* (1978) extended the method to allow for the two observers being of
disparate efficiency.

Essentially the method is a Petersen estimate, although animals are
neither marked nor captured. Suppose that the entities being surveyed
are stationary and that their individual positions can be mapped. For
example, Magnusson *et al.* (1978) surveyed crocodile nests and Henny
et al. (1977) the nests of ospreys. If the area is surveyed independently
twice, perhaps once from the ground and once from the air, the entities
can be divided into four categories:

S_1 = the number seen on the first survey but missed on the
second,

S_2 = the number seen on the second survey but missed on the
first,

B = the number tallied by both surveys,

M = the number missed on both.

This is equivalent to a mark–recapture exercise. The first survey maps
(marks) a set of entities, each of which may or may not be seen
(recaptured) on the second survey. But unlike a true mark–recapture
exercise the model is symmetric and the first and second surveys are
interchangeable.

If P_1 is the probability of an entity being seen on the first survey
and P_2 the probability of its being seen on the second survey:

$$P_1 = B/(B + S_2)$$
$$P_2 = B/(B + S_1)$$
$$M = S_1 S_2/B$$
$$Y = ((B + S_1)(B + S_2))/B$$

where Y is an estimate of the size of the population. The last equation
may be corrected for statistical bias (Chapman 1951) to:

$$Y = [((B + S_1 + 1)(B + S_2 + 1))/(B + 1)] - 1$$

which has a variance (Seber 1982) of:

$$\mathrm{Var}(Y) = (S_1 S_2 (B + S_1 + 1)(B + S_2 + 1))/((B + 1)^2 (B + 2))$$

Magnusson *et al.* (1978) reported that, although the method is based on
the assumptions that the two surveys are uncolluded and that there is
a constant probability of seeing an entity on a given survey (equal
catchability), the second assumption is not critical. The population
estimate is close enough even when the probability of being seen
varies greatly between individuals.

Caughley and Grice (1982) extended the method to moving targets, dropping the requirement that the position of stationary entities must be mapped so that they could be identified as seen or not seen at the two surveys. Groups of emus (*Dromaius novaehollandiae*) were tallied simultaneously but independently by two observers seated in tandem on one side of an aircraft. Their counts of $S_1 = 7$, $S_2 = 3$, and $B = 10$ yielded $P_1 = 0.77$ and $P_2 = 0.59$, the population estimate being $Y = 22$ emu groups on the 843 km^2 of transects that they surveyed together, a density of 0.03 groups per square kilometer.

This method of simultaneous but uncolluded tallying carries two dangers, one technical, the other statistical. The two observers must not unconsciously cue each other to the presence of animals in their field of view and ideally should be screened from each other. Second, the chances of "marking" and "recapturing" an entity should be un-correlated, but they are not because marking and recapturing occur at the same instant, the search images transmitted to each observer being nearly identical. Caughley and Grice (1982) showed by simulation that the effect of the close correlation was to underestimate density but that the underestimation became serious only when the mean of P_1 and P_2 was < 0.5.

Known-to-be-alive

Most estimates of population size require that the manager makes a leap of faith. There is seldom any certainty that the population fits the assumptions of the model, nor whether the estimate is wildly inaccurate, nor whether the confidence limits have much to do with reality. The more complex is the model the greater is the uncertainty. Many ecologists, particularly those working on small mammals, have decided that the work needed to achieve an unbiased estimate is not worth the effort. They would prefer an estimate which, although perhaps inaccurate, is inaccurate in a predictable direction and which does not depend on a set of assumptions of dubious reliability. Hence the *known-to-be-alive* estimate, the number of animals that the re-searcher knows with certainty to be in the study area. These estimates for small-mammal populations are usually made by trapping an area at high intensity over a short period. Each animal is marked at first capture, the estimated population size being simply the number of first captures. Such estimates are acknowledged as underestimates but they have the advantage of yielding a real number, not an abstract concept, to work with.

Known-to-be-alive estimates are often the most appropriate in wildlife management. There are several problems of conservation and of harvesting for which an overestimate of density may lead to inap-propriate management action. An underestimate, on the other hand, may simply produce inefficient but entirely safe management. The penalty for a poor estimate is often distributed asymmetrically around the true population size. It is not good to overestimate the number of individuals of an endangered species. It is not safe to apply a harvesting quota, known to be safe for the population size you estimated, to a population that is much smaller than you thought. Where the undesirable consequences of an overestimate are considerably greater than those accruing from an underestimate, the known-to-be-alive number is often the most appropriate estimate to work with.

12.7 **Indices**

An index of density is some attribute that changes in a predictable manner with changes in density. It may be the density of bird nests, or the density of tracks of brown bears, or the number of minke whales (*Balaenoptera acutorostrata*) seen per cruising hour. These indices would reveal something about the density of a bird, of brown bears, and of minke whales. Without knowing anything about the proportional relationship between the index and the abundance of the animal we could be confident that if the index halved or doubled it would reflect roughly a halving or doubling of animal density. Formally, that holds only when the relationship between index and density is a straight line that passes through the point of zero index and zero density.

Indices of density, if comparable, are useful for comparing the density of two populations or for tracking changes in the density of one population from year-to-year. Often a comparison is all we need. The relevant question may be not how large is the population but whether it has declined or increased under a particular regime of management. In such circumstances the accuracy of an index is irrelevant; precision is paramount.

Let us compare an aerial survey designed to yield an estimate of absolute density with one designed to yield an index of density. The first maximizes accuracy, the second precision. The "accurate" survey would probably inspect small quadrats by circling at a low but varying height above the ground. That is a good way to see animals but it is a technique difficult to standardize between pilots. The "precise" survey would sample transects from a fixed height above ground at a constant speed. Since there is no requirement that all the animals be counted on the sampled units, only a fixed proportion being sought, the survey variables are set according to how easily they may be standardized. Ground speed is higher than for an "accurate" survey to allow the pilot to maintain constant ground speed safely even with a strong tail wind. Height above ground is set higher so that the inevitable variations in height will be proportionally less than at low level. Plus or minus 10 m around a height of 30 m results in large variations in search image. The same variation around 90 m has little effect. We might choose a transect width of 50 m per observer for an accurate survey but 200 m for a precise survey. The precision of the estimate is approximately proportional to the square root of the number of animals actually tallied (Eberhardt 1978) and so, although proportionally fewer will be seen on a 200-m strip, we choose the wider strip to increase the absolute number that we see.

Consistency and rigid standardization of techniques are crucial when estimating an index. A good observer is not one who gets a high tally but one who has a consistent level of concentration and who produces results of high repeatability.

All the rules of sampling and of analysis hold as well for indices as for absolute counts of animals. Remember, however, that indices are useful only in comparisons and therefore the quantity to be estimated is the difference between two indices. The variance of an estimate of difference is the sum of the variances of the two estimated indices. As a rule of thumb we should measure the two indices with a precision such that each standard error is less than one-third of the difference we anticipate. Hence, an index must often be estimated much more precisely than a one-off estimate of population size or density.

12.7.1 *Known-to-be-alive used as an index*

Although known-to-be-alive is sometimes used as a one-off estimate of population size, it is more often used to track trends in population size. The operating rules governing these two uses are quite different. In the first exercise we seek the most accurate estimate we can get. In the second we seek consistency of method among several estimates such that their bias is held constant. In the first case we put in as much work as possible. In the second we put in precisely the same amount of sampling on each surveying or capturing occasion. Otherwise the trend in the estimates may reflect no more than variation in capturing effort.

A variant of this aberration, very common in ecological research, is to boost the number known to be alive (because they were caught) on a given occasion by the number of individuals not caught on that occasion but which must have been there because they were caught on both previous and subsequent occasions. Although the accuracy of the estimate of absolute numbers is thereby enhanced the consistency of the string of estimates is thereby lowered. Estimates for the earlier occasions are inflated relative to those of later occasions, the rate of increase being underestimated if density is rising and rate of decrease being overestimated if density is falling.

12.8 Summary

Animal numbers can be estimated by total counts, sampled counts, mark–recapture, or various indirect methods. In each case the usefulness of the method is determined by how closely its underlying assumptions are matched by the realities of what the animals do and how difficult they are to see, trap, or detect. The range of methods provided should allow a wildlife manager to choose one that will be adequate in any given circumstance.

13 Experimental management

13.1 Introduction

We find it convenient to differentiate two modes of wildlife management: that in which management decisions flow from personal experience and received wisdom, and that in which they are based upon data and analysis. For want of better names we will call these the "traditional" and the "experimental" modes of wildlife management. The value of the traditional mode should not be underestimated. Its dominant characteristic is conservatism, a lack of interest in trying out new ideas. That is sometimes a strength rather than a weakness because most new ideas turn out to be wrong (Caughley 1985). However, some new ideas are useful and these are best identified by the experimental approach. In this chapter we explain how a technical judgment can be tested by posing it as a question (hypothesis). A hypothesis is tested by collecting appropriate data using an appropriate design and we provide guidelines for experimental design and analysis. We describe the use of replication to sample the natural range of variability and the use of controls to render the conclusions unambiguous. Finally, we address the conversion of a statistical result back into a biological conclusion.

13.2 Differentiating success from failure

Wildlife management is not like civil engineering. The theory and practice of civil engineering is placed on public display every time a bridge is built. No expertise is needed to interpret that test. If the bridge remains operational for the length of its design life, the engineers got it right. If it collapses they got it wrong, and we look forward to hearing the details of how and why they got it wrong at the subsequent court case.

Wildlife management differs from civil engineering in a number of respects. First, the managers are not erecting something new but acting as custodians of something already there. They are not responsible for the initial conditions but these often constrain their options.

Second, in civil engineering the question is usually obvious. In ecology the appropriate question is seldom obvious. Choosing the appropriate question is the most difficult task, much more difficult than answering that question. Good design does not correct an inadequate grasp of the problem.

Third, criteria for success and failure are seldom tight and often are not available to the public. Compare these two statements.

1 The provision of nest boxes for wood ducks will increase the size of the population.

2 The provision of nest boxes for waterfowl will benefit their overall ecology.

The first is a hypothesis testable against a predicted outcome. The second is of the type that covers everything and probably cannot be disproved. What is an "overall ecology?" How is it measured? What species should it be measured on? Wildlife management objectives that are framed in an unverifiable form (example 2) are not very useful whereas those in the form of testable hypotheses (example 1) allow us to learn more about the system.

Fourth, even when there are such criteria to judge success, the wildlife manager is seldom in complete control of the situation and hence cannot be held fully and personally responsible for the outcome. A failure is usually referable to the acts of many people, often interacting with changes in the environment (sometimes referred to as acts of God).

Fifth, the wildlife and its habitat usually forms a robust ecological system. Within rather wide limits that system will absorb the most inappropriate or irrelevant of management and still look good.

Because the criteria of success are often fuzzy in wildlife management the outcomes of different management systems are sometimes difficult to rank. For example, when managing deer populations do we shoot only bucks, shoot only does, shoot 70% bucks and 30% does, shoot 30% bucks and 70% does, shoot neither. All these schemes have been tried and all have been reported as highly successful. Highly successful with what end in view? How highly successful? Perhaps we ask those questions less often than we should.

Agriculture made a major advance because R.A. Fisher invented the "analysis of variance" (called also A of V or ANOVA) and because a few agriculturalists recognized that here was a technique that could differentiate the effect of different management treatments. More importantly they believed that differentiation was necessary. Wildlife management can learn from the history of agriculture by incorporating more statistical design in management programs.

13.3 Technical judgments can be tested

In contrast to the value judgment discussed above, the wisdom of a technical judgment can be evaluated according to strict criteria. If a manager decides that supplementary feeding will increase the density of quail then that can be tested and the decision rated right or wrong. If a manager decides that elephants must be culled because, if not, they will eliminate *Acacia tortilis* trees from the area, that decision is right or wrong and it can be demonstrated as right or wrong by an appropriate experiment. Note that the decision on whether the local survival of the acacia justifies the proposed reduction of elephants is a value judgment and hence not testable.

So there are value judgments and technical judgments and these must not be confused one with the other. Technical judgments can be tested and should be tested. By this means we learn from our failures as well as from our successes. A recurring theme of this book is that wildlife management advances only when the efficacy of a management treatment is tested. For that to happen the technical decision as to the appropriate treatment must be stated in a form that predicts a verifiable outcome.

13.3.1 *Hypotheses*

Research questions are usually phrased in positive form, e.g., does the bodyweight of black bears (*Ursus americanus*) tend to change as we move from the equator towards the pole? That question is most easily tested statistically if we frame it in negative form, the so-called null hypothesis symbolized H_0: the mean bodyweight of black bears does not change with latitude. If this hypothesis is falsified by data showing that mean weights are not the same at all sampled latitudes, we reject the null hypothesis in favor of an alternative hypothesis H_a. Whereas a question can generate only one null hypothesis, there may be a number of competing alternative hypotheses. In the bear example the alternative to no change of weight with latitude may be an increase with latitude, a decrease with latitude, a peak in the middle latitudes, or a trough in the middle latitudes.

The procedures by which we test hypotheses make up the realm of statistical analysis. They come quite late in the research sequence which goes:

1 pose a research question (usually our best guess or prediction as to what is going on);
2 convert that to a null hypothesis;
3 collect the data that will test the null hypothesis;
4 run the appropriate statistical test;
5 accept or reject the null hypothesis in the light of that testing;
6 convert the statistical conclusion to a biological conclusion.

Most statistical tests estimate the probability that a null hypothesis is false. A probability of say 10% is often interpreted loosely as meaning that there is only a 10% chance of the null hypothesis being true. That is not quite right. Suppose our null hypothesis states that there is no difference in bill length between the females of two populations of a particular species. We draw a sample from each, perform the appropriate statistical test for a difference, and find that the test statistic has a probability of (say) 10% for the sample sizes that we used. That 10% is the estimated probability of drawing two samples as different or even more different in average bill length as those that we drew if the populations from which they were drawn did not differ in that estimated attribute.

If there really is no difference between the two populations in average bill length, then the probability returned by the statistical test will be in the region of 50% (i.e., the chance of drawing samples more disparate than those we drew is the same as the chance of drawing samples less disparate than those we drew). A probability >50% means that the two samples are more similar than we would expect from random sampling of identical populations. If the probability approaches 95% or so, we should investigate whether the sampling procedure was biased.

Statistical tests deal in probabilities, not certainties. There is always a chance that we are wrong. Such errors come in two forms, the *type 1 error* (also known as an α-error) in which the null hypothesis is rejected even though true and the *type 2 error* (β-error) where the null hypothesis is accepted even though false. Following Zar (1974), the relationship between the two kinds of error can be shown as a matrix.

	If H_0 is true	If H_0 is false
If H_0 is rejected	Type 1 error	No error
If H_0 is accepted	No error	Type 2 error

Obviously we are not keen to make either kind of error. The probability of committing a type 1 error is simply the specified significance level. The probability of committing a type 2 error is not immediately specifiable except that we can say that it is inversely related to the significance level for rejecting the null hypothesis. The two kinds of error cannot be minimized simultaneously except by increasing the sample size. Hence, we need a compromise level of significance that will provide an acceptably low chance of rejecting a factual null hypothesis but which is not so low as to generate too large a chance of committing a type 2 error. Experience has indicated that a 5% chance of rejecting the null hypothesis when it is true provides reasonable insurance against both kinds of error. We therefore conventionally specify the 5% probability as our significance level, although that level is essentially arbitrary and little more than a gentlemen's agreement. What is not arbitrary is that the hypothesis to be tested and the level of significance at which the hypothesis is rejected must be decided upon before the data are examined and preferably before they are collected. Otherwise the whole logic of testing is violated.

Our standard statistical tests concentrate on minimizing type 1 errors. The extent to which they minimize type 2 errors is called *power*. Depending on context, avoidance of type 2 errors may be more important than ensuring the warranted rejection of the null hypothesis.

13.3.2 Asymmetry of risk

Converting a statistical result back into a biological conclusion is not at all straightforward. The classical null hypothesis method is at its best when testing whether a treatment has an effect, the treatment representing a cost and the response a benefit. An example might be supplementary feeding to increase the clutch size of a game bird. Here the feeding costs money and time, and we will use it operationally only if an adequate response is clearly demonstrated. First, the null hypothesis must be rejected, and then the difference in response between experimental control and treatment evaluated to determine whether the cost of the treatment is justified by size of the response in fecundity. If the null hypothesis (no effect of treatment) is not rejected we are simply back where we started and no harm is done. Both type 1 and type 2 errors are possible, both are inconvenient, but neither is catastrophic. A type 1 error leads to unnecessary expense until the mistake is identified; a type 2 error results in a small sacrifice in the potential fecundity of the game-bird population.

Null hypothesis testing is less effective and efficient when the treatment itself is a benefit and the lack of treatment is itself a cost. Suppose a marine fish stock appears to be declining although there is considerable year-to-year variation in the index of abundance used: the catch per unit effort. Furthermore, there are good reasons to suspect that the fishing itself is heavy enough to precipitate a decline. The null hypothesis is that fishing has no effect on population size. In this case the failure to reject the "no-effect" null hypothesis is not sufficient reason to operate on the assumption that the fishing is having no effect. At the very least one would first want to know something about the power of the test. In this case the cost of making a type 2 mistake greatly outweighs the benefit of getting it right. The effect of continuing to fish when one should have stopped could be disastrous and irreversible, whereas unnecessary cessation of fishing results only in a temporary cost until fishing resumes. This is an asymmetry of risk. It is particularly prevalent in work on endangered species where an error can result in extinction. Asymmetry of risk demands conservative interpretation of statistical results.

13.4 The nature of the evidence

A management treatment may be successful or it may be a failure. If the first, the manager needs to know whether the success flowed from the treatment itself or whether it would have happened anyway. Otherwise an expensive and unnecessary management scheme might run indefinitely. Alternatively, the management may not achieve its stated aim, in which case the manager must first establish that fact without doubt and then find out why. Was the failure caused by some extraneous factor that formed no part of the treatment? Was the entire management treatment inappropriate or only a part of it? Would a higher intensity of the treatment have been successful whereas the same treatment at a lower intensity was not?

To find out, the management must be run as an experiment. There are rules to designing an experiment that are there for one very important reason: if they are broken the questions the experiment is designed to answer cannot be answered unambiguously.

Suppose a manager wished to increase the density of quail in an area by supplementing their supply of food with wheat. How that is done determines whether anything will be learned from the exercise.

There is a graded series of approaches, ranging from useless in that they yield no verification of the worth of the treatment, through suggestive in that their results allow a cautious choice between alternative interpretations, to definitive where the results can be interpreted without error.

1 Grain is scattered once a month but density is not monitored. The manager assumes that since the treatment *should* increase the density of quail that it *will* increase their density.

This is no test because the outcome of the treatment is assumed rather than observed.

2 The manager measures density on two occasions separated by 1 year, the first before supplementary feeding is instituted. If density were higher on the second occasion the manager might assume that the rise resulted from the feeding.

This is the classic fallacy of "before" being taken as a control on "after." Interpretation of the result rests on an assumption that the density would have remained stable without supplementary feeding, and there is no guarantee of that. For example, it may have been increasing steadily for several years in response to a progressive and general increase of cover.

3 The manager designates two areas, one on which the birds are fed (the treatment) and the other on which they are not (the experimental control). The density of quail is measured before and after supplementary feeding is instituted. If the proportionate increase in density on the treatment area is greater than that on the control area, the difference is ascribed to the effect of feeding.

This design is a radical improvement but still yields ambiguous results: the difference in rate of increase may reflect differences between the two sites rather than between the two treatments. We say that the effect of site and the effect of treatment are *confounded*. Perhaps the soil of one site was heavy and that of the second light, the vegetation on the two sites thereby reacting differently to heavy winter rainfall and the quail reacting to that difference in plant growth.

   Flawed as it is, this design is often the only one available, particularly if the treated area is a national park. In such cases the control must be chosen with great care to ensure that it is in all important respects similar to the treated area, and the response variable should be monitored on each area for some time before the treatment is instituted to establish that it behaves similarly in the two areas. Another way around this problem is to reverse the two treatments and see whether the same result is obtained.

4 The effect of such local and extraneous influences on the results of the experiment is countered by replication. Suppose six sites are designated, three of which are treated by feeding and the other three left as controls. The category of the site is determined by lot. Before and after measurements of density are made at the same time of the year in all six areas. The biological question: does supplementary feeding affect density? is translated into a form reflecting the experimental design: is the difference in quail density between treatments (feeding versus not feeding) greater than the difference between sites (replicates) within treatments?

This is an appropriate experimental design in that the outcome provides an unambiguous test of the hypothesis. Its efficiency and precision could be increased in various ways but its logic is right.

The form of an experimental design is dictated by logic rather than by the special requirements of the arithmetic subsequently performed on the data. This is an immensely important point. If the manager has no intention of applying powerful methods of analysis to the data, that in no way sanctions short-cuts in the basic experimental design.

Another common fallacy is the belief that although a logically designed experiment is necessary for publication in a scientific journal the manager need not bother with all that rigmarole if the only aim is to find out what is going on. The manager might then simply run an "empirical test" like the second or third example given above without realizing that the measurements do not reveal what is going on.

13.4.1 *Why replicate the experiment?*

Suppose we wish to determine whether grazing by deer affects the density of a species of grass. The experimental treatment is grazing by a fixed density of deer and the experimental control is an absence of such grazing. We cannot simply apply the two treatments each to a single area because no two areas are precisely the same. We would not know whether the measured difference in plant density was attributable to the difference in treatment or whether it reflected some intrinsic difference between the two areas. There will always be a measurable difference between areas in the density of any species whatever one does, or does not do, to those areas.

We can postulate that a difference between treated areas is caused by the disparate treatments applied to them only when the difference between the treatments is appreciably greater than the difference within treatments. To determine the scale of variation within the "population" of treatments, we must look at a sample of areas that have received the same treatment. The minimum size of a sample is two. Thus we must designate at least two areas as grazing treatments and two as controls.

The density of a plant species is usually measured within small quadrats scattered over a treatment area. Fifty might be measured in each. Those fifty quadrats are not replicates. They are subsamples of a single treatment and their invalid use as "replicates" is called pseudo-replication. Sampling within a treatment is not treatment replication. Data from such subsamples could be fed into an ANOVA, which would then provide what might appear to be a rigorous test of the hypothesis, but that is an illusion. The arithmetic procedures have been fulfilled but the logic is not satisfied. The result is actually a test of whether the combination of the treatment and the intrinsic characteristics of a single area differ from another treatment, combined with the intrinsic characteristics of another area. We say that area and treatment are *confounded*. Their individual effects cannot be disentangled. No strong test of the effect of the treatments themselves is possible unless those treatments are replicated.

Replicates are not meant to be similar. They are meant to sample the natural range of variability. Consequently, one does not look for six similar sites to provide three treatments and three controls. One picks six sites at random. A common excuse offered for a lack of replication in management experiments (and even in research experiments) is that sites similar enough to act as replicates could not be located. Such an excuse is not valid and points to a lack of understanding of the nature of evidence.

These principles carry over to all other forms of comparison. We cannot conclude from two specimens that parrots of a given species have a higher hemoglobin count near the tops of mountains than at lower altitudes. And we get no farther forward by taking a number of blood samples (subsampling) from the two individuals. Instead, we must test the blood of several parrots from each zone, look at the variation within each group of parrots, and then calculate whether the average difference between groups is greater than the difference within groups. Hence we must replicate. The arithmetic of such a comparison can be extracted from any book on statistical methods. That is the easy part. The difficult part is getting the logic right.

13.5 Experimental and survey design

Experimental design has its own vocabulary. The thing that we monitor, in this case the density or rate of increase of the quail, is the *response variable*. That which affects the response variable, in this case WHEAT, is a *factor*. In our imaginary experiment the factor we examined had two *levels*: no supplementary feeding of wheat and some supplementary feeding of wheat (Fig. 13.1). Equally, its levels could have been set at 0, 30, 70, and 250 kg of grain per hectare per month as in Fig. 13.2. The levels of a factor need not be numbers as in this example. The levels of factor HABITAT, for example, might be pine, oak, and grassland. The levels of factor ORDER might be first, second, third, and fourth. The levels of factor SPECIES might be mule deer, white-tailed deer, and elk.

Suppose we wished to examine the effect of two management treatments simultaneously. Instead of looking at the effect of just wheat on density of quail we might wish also to examine the effect of supplying extra salt. There are now two factors, WHEAT and SALT. The questions now become.

1 Does WHEAT affect density?
2 Does SALT affect density?
3 Is the effect of WHEAT on density influenced by the level of SALT, and vice versa?

In statistical language the last question deals with the interaction

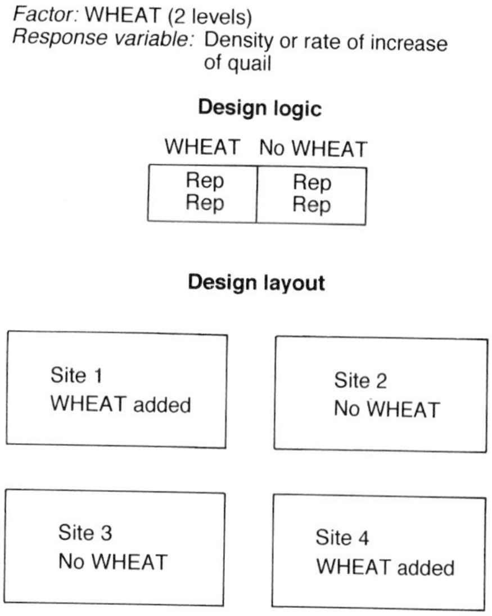

Factor: WHEAT (2 levels)
Response variable: Density or rate of increase of quail

Design logic

WHEAT	No WHEAT
Rep Rep	Rep Rep

Design layout

Fig. 13.1 Minimum one-factor experimental design. The factor is WHEAT (at two levels); the response variable is the density, or rate of increase, of quail.

Factor: WHEAT (4 levels)
Response variable: Density or rate of increase of quail

Design logic

0 WHEAT	30 WHEAT	70 WHEAT	250 (kg/ha) WHEAT
Rep Rep	Rep Rep	Rep Rep	Rep Rep

Design layout

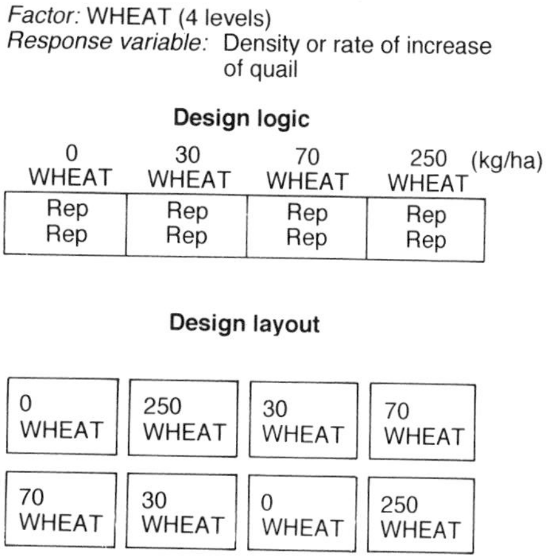

Fig. 13.2 One-factor experimental design where the factor has more than two levels. The factor is WHEAT (at four levels); the response variable is the density, or rate of increase, of quail.

between the two factors, whether their individual effects on density are additive (i.e., independent of each other), or whether the effect of a level of one factor changes according to which level of the other factor is combined with it. Section 13.6.3 considers interactions in greater detail.

Figure 13.3 gives an appropriate experimental design for such a two-factor experiment. Its main features are that each level of the first factor is combined with each level of the second, that there are therefore $2 \times 4 = 8$ cells or treatments, that each treatment is replicated, and that the number of replicates per treatment is the same for all treatments.

13.5.1 Controls

A *control* is that level of a factor subjected to zero treatment. That is not to say that it is necessarily left undisturbed. Everything done to

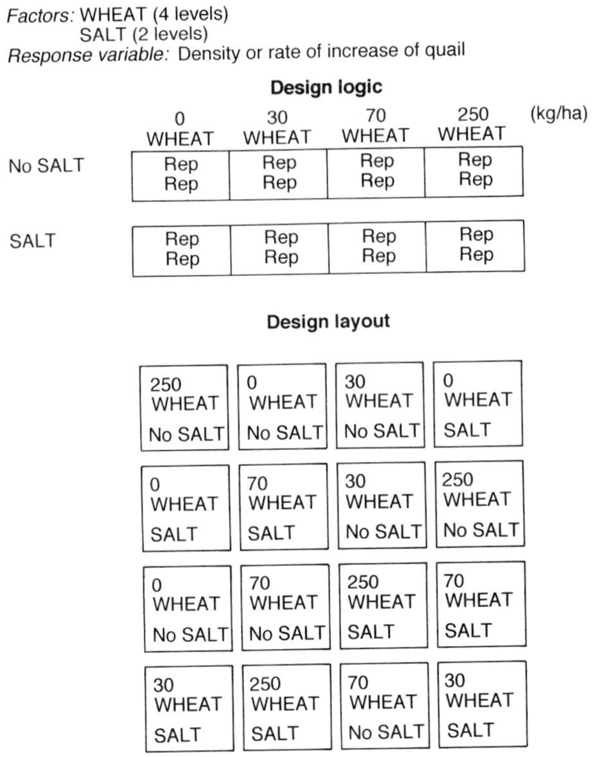

Factors: WHEAT (4 levels)
              SALT (2 levels)
Response variable: Density or rate of increase of quail

Design logic

	0 WHEAT	30 WHEAT	70 WHEAT	250 WHEAT	(kg/ha)
No SALT	Rep Rep	Rep Rep	Rep Rep	Rep Rep	
SALT	Rep Rep	Rep Rep	Rep Rep	Rep Rep	

Design layout

250 WHEAT No SALT	0 WHEAT No SALT	30 WHEAT No SALT	0 WHEAT SALT
0 WHEAT SALT	70 WHEAT SALT	30 WHEAT No SALT	250 WHEAT No SALT
0 WHEAT No SALT	70 WHEAT No SALT	250 WHEAT SALT	70 WHEAT SALT
30 WHEAT SALT	250 WHEAT SALT	70 WHEAT No SALT	30 WHEAT SALT

Fig. 13.3 Two-factor experimental design. The factors are WHEAT (at four levels) and SALT (at two levels); the response variable is the density, or rate of increase, of quail. Questions to be asked are whether the two factors act independently of each other, or whether they interact. If the latter, do they reinforce or oppose each other?

the other levels must also be done to the control, other than for the manipulation that is formally the focus of the treatment. If vehicles are driven over the quail plots to distribute the grain they must also be driven over the control areas.

Controls must obviously be appropriate and often a good deal of thought is needed to ensure that they are. We have previously dealt with the mistake of declaring "before treatment" a control on "after treatment" (see Section 13.4) but there are more subtle traps to keep in mind. If the treatment is an insecticide dissolved in a solvent, then the control plots must be sprayed with the solvent minus the insecticide. If treated birds are banded then control birds must be banded. If animals are removed from the field to the laboratory for treatment and then released, control animals must also be subjected to that disturbance. And so on.

13.5.2 *Sample size*

There is no general answer to the question "How many replicates are necessary?" other than the trite "at least two per treatment." It depends upon the number of treatments to be compared, the average variance among replicates within treatments, and the magnitude of the differences one expects or is attempting to establish. These may be estimated from a pilot experiment or from a previous experiment in the same area.

As a general rule, however, the fewer the treatments the more replicates needed per treatment, but there is little to be gained from increasing replication beyond 30 degrees of freedom for the residual. Suppose the experiment had three factors with i levels in the first, j in the second, and k in the third. There are thus ijk treatments and $ijk(n - 1)$ degrees of freedom in the residual, where n is the number of replicates per treatment.

13.5.3 *Standard experimental designs*

Most questions on the effect of this or that management treatment have a similar logical structure, even though they deal with different animals in different conditions. The most common questions lead to standard experimental designs.

One factor, two levels
Figure 13.1 represents the simplest design that will provide an answer that can be trusted. It evaluates the operational null hypothesis that supplementary feeding with wheat has no effect on the density (or rate of increase) of quail. What is tested, however, is the statistical null hypothesis that the difference between treatments is not significantly greater than the difference between replicates within treatments. If the experiment rejects that null hypothesis we accept as highly likely the alternative hypothesis that supplementary feeding affects the dynamics of quail populations living in conditions similar to those of the populations being studied.

This design tests the effect of only one factor (WHEAT) and evaluates it at only two levels (no wheat and some wheat). Note that the diagram of design logic calls for two replicates at each level. The diagram of design layout shows that the treatments are interspersed: thus we do not have the zero treatments (i.e., controls) bunched together in one region and the wheat-added treatments in a second region.

One factor, several levels
This design (Fig. 13.2) is similar to the last, with the difference that the effect of supplementary feeding with wheat is evaluated at four levels: 0, 30, 70, and 250 kg/ha of wheat distributed each month. It allows an answer to two questions: first, whether supplementary feeding has any effect at all upon the density of the quail; and, second, whether that effect varies according to the level of supplementary feeding. An answer to the second question allows a cost–benefit analysis on the optimum level of supplementary feeding. Treatment replication and interspersion of treatments is maintained.

Two factors, two or several levels per factor

In this design (Fig. 13.3) the effect of supplementary feeding on quail density is evaluated in tandem with evaluation of a second factor, the provision of rock salt.

Although the two factors could have been evaluated by two separate experiments there are large advantages in combining them within the same experiment. It provides an answer to a question that might prove to be of considerable importance: do the two factors interact?

Hypothetically, additional salt in the diet of quail might affect their physiology and hence their dynamics, particularly in sodium depleted areas, and the same may be true of supplementary feeding with wheat. But suppose that supplementary feeding has an effect only when there is adequate salt in the diet. In such circumstances two separate experiments would produce the fallacious conclusion that, whereas salt has an effect, wheat has none. The interactive relationship between the two factors would have been missed and the resultant management would have been inappropriate. One looks for an interaction by calculating whether the effect of the two factors in combination is greater or less than the addition of the two effects when the factors are evaluated separately. That is achieved by ensuring that each level of the first factor is run in combination with each level of the second. The factors are said to be mutually *orthogonal* (at right angles to each other).

The design logic (Fig. 13.3) is seen as a simple extension of the logic of the one-factor design and the design layout continues to adhere to replication and interspersion of treatments. Since there are now eight treatments, each with two replicates, the interspersion of treatments can best be achieved by laying them out either in a systematic manner as with a Latin square or, as in the example, assigning their positions on the ground by random numbers.

13.5.4 Weak-inference designs

Very often a field experiment breaches one or more rules of experimental design and so no longer answers unambiguously the question being posed. Such an occurrence has two causes: an unfortunate mistake or a necessary choice.

Mistakes

Very often there may be no logistical or technical justification for using an inappropriate design. Such a flaw is simply a mistake. One of the most common in ecological and wildlife research is pseudoreplication (or subsampling), used under the misapprehension that it constitutes treatment replication (Hurlbert 1984). In this case site and treatment are confounded (see Section 13.4.1).

A second common mistake is the unbalanced design. Figure 13.4 illustrates an experiment to evaluate the effect of grazing by sheep and rabbits on the density of a species of grass. There are two factors (SHEEP

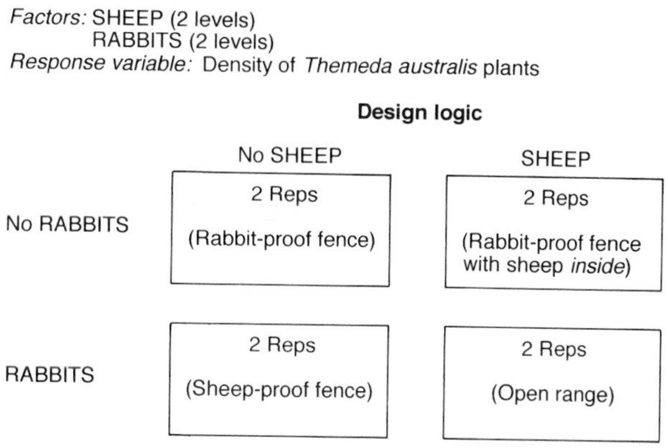

Factors: SHEEP (2 levels)
RABBITS (2 levels)
Response variable: Density of *Themeda australis* plants

Design logic

	No SHEEP	SHEEP
No RABBITS	2 Reps (Rabbit-proof fence)	2 Reps (Rabbit-proof fence with sheep *inside*)
RABBITS	2 Reps (Sheep-proof fence)	2 Reps (Open range)

Fig. 13.4 Design logic for a two-factor experiment on the effect of sheep and rabbits on the biomass of pasture. The factors are SHEEP (at two levels) and RABBITS (at two levels); the response variable is the density of *Themeda australis* plants.

and RABBITS), each with two levels (presence and absence). "Presence" for sheep is taken as the standard stocking rate, and that for rabbits the prevailing density. Variation of rabbit density across the area is taken care of by the replication.

The four treatments may be symbolized by a code in which 1 indicates presence and 0 absence. Most of the practical details of setting up such a trial are simple. A rabbit-proof fence around a quadrat excludes both sheep and rabbits (treatment R0 S0). A sheep-proof fence excludes sheep but allows rabbits in (R1 S0). A quadrat to measure the effect of sheep and rabbits together is simply an unfenced square marked by four pegs (R1 S1). Thus, three of the four treatments are easily arranged. They can be set up and then temporarily forgotten, the experimenter returning after several months or even years to harvest the data.

The final treatment (R0 S1) cannot be managed in this way. No-one has yet invented a fence that acts as a barrier to rabbits while allowing sheep free access to the quadrat. Hence, R0 S1 must be handled differently. It requires a rabbit-proof fence around the quadrat to exclude rabbits (as for R0 S0) but with sheep at standard stocking density within the enclosure. That treatment cannot be set up and then left untended. Sheep need water and husbandry. Hence that treatment is often left out of the experiment. There results a set of data in which the individual effects of sheep on vegetation cannot be disentangled from the effects of rabbits, the total justification of the experiment in the first place.

We have seen many such incomplete experiments set up, often at some expense. They provide estimates of the effect on the vegetation of rabbits alone and of sheep and rabbits together, but not of sheep alone. The effect of sheep alone cannot be obtained indirectly by subtraction because that works only where the two effects are additive (i.e., no interaction). But a significant interaction can quite safely be assumed because each blade of grass eaten by a sheep is no longer available to a rabbit, and vice versa.

Necessary compromises

There are a number of problems that involve passive observation of a pattern or process not under the researcher's manipulative control. In these circumstances a tight experimental design is sometimes not possible, or alternatively the problem may not be open to classic scientific method. In many fields, e.g., astronomy, geology, and economics, such problems are the rule rather than the exception. A common example from ecology is the environmental impact assessment (EIA). As Eberhardt and Thomas (1991) put it: "the basic problem in impact studies is that evaluation of the environmental impact of a single installation of, say, a nuclear power station on a river, cannot very well be formulated in the context of the classical agricultural experimental design, since there is only one "treatment" – the particular power-generating station." In fact, the problem is even more intractable: EIA studies do not test hypotheses. However, environmental impact assessments are still necessary. That they generate only weak inference is no good argument against doing them.

Weak inference results also from a second class of problems: where tight experimental design is theoretically possible but not practicable. In such circumstances we may have an unbalanced design, or poor interspersion of treatments, or insufficient replication or even no replication. Again the results are not useless but they must be treated for what they are: indicating possibilities that may be confirmed by further research.

Weak inference is seldom harmful and can be very useful so long as its unreliability is recognized. Weak inference mistaken for strong inference can be ruinously dangerous.

13.6 Some standard analyses

There are several possible analyses available for any given experimental or survey design. Sometimes they give much the same answer and sometimes different answers. The former reflect only that there is more than one way of doing things; the latter reflect differing assumptions underlying the analyses. Hence, it is important to know what a particular analysis can and cannot do lest one chooses the wrong one. For example, χ^2 tests are used only on frequencies (i.e., counts that come as whole numbers); ANOVA can deal both with frequencies and with continuous measurements. The "t" test is a special case of ANOVA and shares its underlying assumptions.

We will use the ANOVA to introduce a broad class of analyses appropriate for the majority of experimental and survey designs. Any statistical textbook will take this discussion further and present additional analytical options.

13.6.1 One-factor ANOVA

The one-factor ANOVA tests the hypothesis that the response variable does not vary with the level of the factor. The alternative hypothesis is that the response variable differs according to the level of the factor, either generally increasing or decreasing with its level, or going up then down or the reverse, or varying in an unsystematic manner.

Our example (Box 13.1) comprises counts of kangaroos on randomly placed east–west transects, each 90 km in length, on a single degree block in southwest Queensland, Australia. The question of particular concern is whether there is an order effect in days of survey. Did the kangaroos become increasingly irritated by a plane flying backward and forward and therefore sought cover whenever one was heard? Or did they become progressively habituated to the noise such that more were seen each day as the survey progressed? The null hypothesis is that the average seen per transect per day is independent of the day order.

Note that factor DAY contains three levels, the first day, the second day, and the third day. The last contains only six replicates in contrast to the eight of the first 2 days. It will make the point that the arithmetic of one-factor ANOVA does not require that the design is balanced (i.e., the number of replicates is the same for all levels). The analysis can be run without balance although the result must be interpreted more cautiously. Balance should always be sought, if not necessarily always attained.

The analysis of Box 13.1 leads to an *F*-ratio testing the null hypothesis. Appendix 1 gives its critical values. The probability of 20% is too high to call the null hypothesis into serious question. That value is the probability of drawing by chance three daily samples as disparate or more disparate than those we did draw, when there is no difference in density or sightability between days. We would require a probability value of around 10% before we became suspicious of the null hypothesis, and one $< 5\%$ before we rejected the null hypothesis in favor of some alternative explanation.

13.6.2 Two-factor ANOVA

A two-factor ANOVA tests simultaneously for an effect of two separate factors on a response variable and for an interaction between them. Even though the arithmetic is simply a generalization of the one-factor case, the two-factor ANOVA differs in kind from the one-factor because of the interaction term. There are also a number of other differences, and we will reach them after we have considered an example.

Data for a two-factor ANOVA is laid out as a two-dimensional matrix with the rows representing the levels of one factor and the columns the levels of the other. These are interchangable. Each cell of the matrix contains the replicate readings of the response variable, whatever it is. Table 13.1 outlines symbolically and formally the calculation of the sums of squares and degrees of freedom for the four components into which the total sum of squares is split: the effect on the response variable of the factor represented by the rows, the effect

Box 13.1 Red kangaroos counted on the Cunnamulla degree block $(10\,870\,\text{km}^2)$ in August 1986. Each replicate is the number of kangaroos counted on a transect measuring 0.4 km by 90 km

Day 1	Day 2	Day 3
96	71	28
38	45	43
80	45	29
35	67	36
50	31	37
55	28	59
38	84	lost
64	70	lost

$n_1 = 8$	$n_2 = 8$	$n_3 = 6$
$T_1 = 456$	$T_2 = 441$	$T_3 = 232$
$\bar{x}_1 = 57.0$	$\bar{x}_2 = 55.1$	$\bar{x}_3 = 38.7$

k = number of classes = 3
N = number of samples = $n_1 + n_2 + n_3 = 22$
$\Sigma X_{ij} = 96 + 38 + \ldots + 37 + 59 = 1129$
$\Sigma X_{ij}^2 = 96^2 + 38^2 + \ldots + 37^2 + 59^2 = 66\,251$
$\Sigma(T_i^2/n_i) = 456^2/8 + 441^2/8 + 232^2/6 = 59\,273$

Main effects sum of squares (SS):
$\Sigma(T_i^2/n_i) - (\Sigma X_{ij})^2/N = 59\,273 - 1129^2/22 = 1335$

Residual SS:
$\Sigma X_{ij}^2 - \Sigma(T_i^2/n_i) = 66\,251 - 59\,273 = 6978$

Total SS:
$\Sigma X_{ij}^2 - (\Sigma X_{ij})^2/N = 66\,251 - 1129^2/22 = 8313$

ANOVA

Source	SS	d.f.	MS	F
Main effect	1335	$k - 1 = 2$	667.5	$\dfrac{667.5}{367} = 1.8$
Residual	6978	$N - k = 19$	367	
Total	8313	$N - 1 = 21$		

$F = 1.8$ with 2 d.f. in the numerator and 19 in the denominator. The probability is 0.19, too high to argue for rejection of the null hypothesis that observable density does not differ by day of survey.

of the factor represented by the columns, the effect of the interaction between them (of which more soon), and the remaining or residual sum of squares which represents the average intrinsic variation within each treatment cell and which therefore is not ascribable to either the factors or their interaction.

Box 13.2 provides a set of data amenable to a two-factor ANOVA. As with the one-factor example they are real data from an aerial survey

Table 13.1 Calculations of sums of squares for two-factor ANOVA

ROW effect	$(1/nc)\Sigma T_i^2 - (1/nrc)T^2$	d.f. $= r - 1$
COLUMN effect	$(1/nr)\Sigma T_j^2 - (1/nrc)T^2$	d.f. $= c - 1$
ROW × COLUMN effect	$(1/n)\Sigma T_{ij}^2 - (1/nc)\Sigma T_i^2 - (1/nr)\Sigma T_j^2 + (1/nrc)T^2$	d.f. $= (r - 1)(c - 1)$
RESIDUAL	$\Sigma X_{ijk}^2 - (1/n)\Sigma T_{ij}^2$	d.f. $= rc(n - 1)$
Total	$\Sigma X_{ijk}^2 - (1/nrc)T^2$	d.f. $= rcn - 1$

T_{ij}, Total of replicates in the cell at the ith row and jth column; T_i, total of replicates in the ith row; T_j, total of replicates in the jth column; T, grand total; r, number of rows; c, number of columns; n, number of replicates per cell.

whose purpose was to establish whether the counts obtained on a given day were influenced by the disturbance or habituation imparted by the survey flying of previous days. Two species were counted this time, however, the red kangaroo and the eastern gray kangaroo, and since they might well react in differing ways to the sound of a low-flying aircraft their counts are kept separate for purposes of analysis. Red kangaroos and eastern gray kangaroos are now the two levels of the factor SPECIES.

In the ANOVA at the bottom of the box, the sum of squares of each source of variation is divided by the respective degrees of freedom to form a mean square. (Mean square is just another name for variance.) The three sources of variance of interest, those of the two factors and their interaction, are divided (in this case) by the residual mean square to form the F-ratios (named for R.A. Fisher who invented ANOVA) that are our test statistics. That for SPECIES is 5.22 and we check that for significance by looking up the F-table of Appendix 1. It will show that an F-ratio with one degree of freedom in the numerator, and 42 (say 40) in the denominator, will have to exceed 4.08 if the magic 5% or lower probability is to be attained. We therefore conclude that the disparity in observed numbers between reds and grays, 957 as against 1490, is more than a quirk of sampling, that gray kangaroos really were more numerous than reds on the Cunnamulla block at the time of survey.

In like manner we test for a day effect. The trend in day totals: 625, 825, and 997 kangaroos, suggests that the animals are becoming habituated to aircraft noise and hence progressively more visible, day-by-day. The F-tables show however that, with degrees of freedom of 2 and 42, a one-tail probability of 5% or better would require $F = 3.23$. Ours reached only 1.91, equivalent to a probability of 16%, and so we are not tempted to replace our null hypothesis (no day effect) with the alternative explanation suggested by an eye-balling of the data.

The F-ratio for interaction was < 1, indicating that the mean square for interaction is less than the residual mean square. It cannot therefore be significant and we do not even bother to look up the probability associated with it.

Box 13.2 Red kangaroos and gray kangaroos counted on the Cunnamulla degree block ($10\,870\,\text{km}^2$) in June 1987. Each replicate is the number of kangaroos counted on a transect measuring 0.4 km by 90 km

	Day 1	Day 2	Day 3	T_i
Red kangaroos	45	19	18	
	17	51	44	
	8	8	61	
	28	11	35	
	26	72	65	
	48	34	76	
	53	67	52	
	62	27	30	
T_{ij}	287	289	381	957
$\bar{x}$	35.9	36.1	47.6	
Gray kangaroos	66	27	27	
	52	47	66	
	34	13	75	
	8	16	104	
	35	101	109	
	36	150	170	
	42	116	51	
	65	66	14	
T_{ij}	338	536	616	1490
$\bar{x}$	42.3	67.0	77.0	
T_j	625	825	997	$T = 2447$

r = number of rows = 2
c = number of columns = 4
n = number of replicates per cell = 8

$(1/nc)\,\Sigma T_i^2 = (1/24)(957^2 + 1490^2)$ $\qquad = 130\,665$
$(1/nr)\,\Sigma T_j^2 = (1/16)(625^2 + 825^2 + 997^2)$ $\quad = 129\,079$
$(1/n)\,\Sigma T_{ij}^2 = (1/8)(287^2 + 289^2 + \ldots + 616^2) = 136\,506$
$(1/nrc)\,T^2 = (1/48)(5\,957\,809)$ $\qquad = 124\,746$
$\Sigma X_{ijk}^2 = 45^2 + 17^2 + \ldots + 14^2$ $\qquad = 184\,081$

Sum of squares

ROW (species)	$130\,665 - 124\,746$	$= 5919$
COLUMN (days)	$129\,079 - 124\,746$	$= 4333$
ROW × COLUMN	$136\,506 - 130\,665 - 129\,079 + 124\,746$	$= 1508$
Residual	$184\,081 - 136\,506$	$= 47\,575$
Total	$184\,081 - 124\,746$	$= 59\,335$

ANOVA

Source	SS	d.f.	MS	F
ROW (species)	5919	$r - 1 = 1$	5919	5.22
COLUMN (days)	4333	$c - 1 = 2$	2166	1.91
Species × days	1508	$(r - 1)(c - 1) = 2$	754	0.67
Residual	47\,575	$rc(n - 1) = 42$	1133	
Total	59\,335	$rcn - 1 = 47$		

13.6.3 *What is an interaction?*

In the last section we tested an interaction, and found it nonsignificant, without really exploring what question we were answering. A nonsignificant interaction implies that the effect of one factor on the response variable is independent of any effect that may be exerted on it by the other factor, that the two factors are each operating alone. The effect of the two factors acting together is exactly the addition of the effects of the two factors each acting in the absence of the other.

If an analysis produces a significant interaction, the relationship should be examined by graphing the response variable against the levels of the first factor. Figure 13.5 shows the kind of trends most commonly encountered. A significant interaction is telling you that no statement can be made as to the effect on the response variable of a particular level of the first factor unless we know the prevailing level of the second factor. The graph will make that clear.

It is entirely possible for an ANOVA to reveal no main effect of the first factor, no main effect of the second, but a massive interaction between them. A graphing of the response variable will reveal a crossing over pattern, as in the last graph of Fig. 13.5.

13.6.4 *Heterogeneity of variance*

The main assumption underlying ANOVA is that the variance of the response variable is constant across treatments. The means may differ (and that is in fact what we are testing to discover) but the variances remain the same. A violation of this assumption can seriously bias the test. Consequently we need to test for heterogeneity of variance and, if we find it, either transform the data to render the variances homogeneous or use an alternative method such as analysis of deviance that does not employ the assumption of homogeneity.

The most common test for heterogeneity of variance is Bartlett's (Zar 1984). It can be found in almost all statistical tests. Recent work has shown, however, that it is too sensitive. ANOVA is an immensely robust test that performs well even when the assumptions of the analysis are not met in full. It copes well with minor heterogeneity of variance and with deviations from normality. About the only thing that throws it out badly is bimodality of the response variable. A better test is Cochran's C (Zar 1984), where the test statistic is simply the largest variance in a cell divided by the sum of all cell variances. For the two-factor ANOVA given in Box 13.2 the largest cell variance is returned by the replicate counts of gray kangaroos on day 2. It is $s^2 = 2566$. The sum of all six variances is $\Sigma s^2 = 6796.6$ and so Cochran's $C = 2566/6796.6 = 0.378$. Looking up a table of the critical values of Cochran's C (Appendix A2) reveals that the test statistic would have to exceed $C = 0.398$ (7 d.f. per variance and there are six variances) to represent a significant departure from homogeneity of variance. We can thus choose to analyze without transformation.

In many biological cases the variance of the response variable rises with the mean. That is particularly true of counts of animals that tend to fit a negative binomial distribution. A transformation of the counts

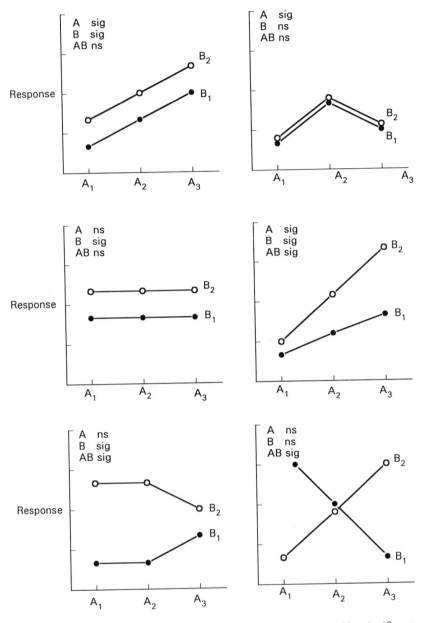

Fig. 13.5 Common forms of interaction in two-factor ANOVAs. ns, Not significant; sig, significant.

to logarithms after adding 0.1 (to knock out the zeros) will usually stabilize the variances. If the animals are solitary rather than gregarious their counts are more likely to fit a Poisson distribution which is characterized by the mean and the variance being identical. Transformation to square roots after adding 3/8 to each will homogenize the variances. The variances of almost all body measurements regress on their means and the data need log transformation. Use an arcsine transformation if data come as percentages.

13.6.5 *Are the factors fixed or random?*

Before any data are collected, let alone analyzed, we must decide precisely what questions are to be asked of them. Take the example of comparing counts of kangaroos on successive days. We were asking whether the act of flying over the study area on one day influenced the counts obtained the next day. The influence could conceivably be negative (disturbance forced the kangaroos into cover in front of the surveying aircraft) or positive (the kangaroos became progressively habituated to aircraft noise).

Now take another question: do viewing conditions differ between days of survey? That question would be answered by counts obtained by days sampled at random. We would want those days to be spaced rather than consecutive as they were in answering the question about the effect of day order. Otherwise the answers to the two questions would be confounded and we would not know which was being answered by a significant day effect. In the question concerning an order effect of days the factor DAY is said to be fixed. No arbitrary selection of any 3 days will do. The days have to follow each other without gaps between them.

Whether a factor is declared fixed or random determines both the question being asked and the denominator of the *F*-ratio that answers it. Table 13.2 shows the appropriate choice of denominators. For two of the three-factor models there is no explicit test for the significance of some of the factors. There are various messy approximations available (see Appendix A in Zar 1984) but it is far better to rephrase the question to one logically answerable from consideration of the data.

Let us generalize the difference between fixed and random factors. A fixed factor is one whose levels cover the range of interest exhaustively. MONTHS therefore usually constitute a fixed factor because they index seasons. YEARS may be fixed or random depending on context. REGIONS may be fixed if the levels of that factor are the only ones of interest. If they are simply a random sample of regions, and any other selection of regions would serve as well, the factor REGIONS is random.

Note that questions change according to whether the factor is fixed or random. Suppose that the response variable is the growth rate of a species of pine and we wish to test for a difference among three soil types (factor SOILS) covering the entire range of soil types of interest. The factor SOILS is thus fixed. If the question concerns the best region in which to plant a plantation of that species, and there are four and only four regions that are possible candidates, the choice of regions is fixed and the appropriate denominator of the *F* testing a difference of growth rate among the soil types of those regions is the residual mean square. However, if we ask the more general question of whether the growth rate differs among soil types across regions in general, any

Table 13.2 The mean square providing the denominator for an F-ratio testing significance of a factor or interaction (i.e., source of variation)

Model	Source	Denominator of F
One-factor ANOVA		
A fixed	A	MS_e
A random	A	MS_e
Two-factor ANOVA		
A fixed, B fixed		
	A	MS_e
	B	MS_e
	AB	MS_e
A fixed, B random		
	A	MS_{AB}
	B	MS_e
	AB	MS_e
A random, B random		
	A	MS_{AB}
	B	MS_{AB}
	AB	MS_e
Three-factor ANOVA		
A fixed, B fixed, C fixed		
	A	MS_e
	B	MS_e
	C	MS_e
	AB	MS_e
	AC	MS_e
	BC	MS_e
	ABC	MS_e
A fixed, B fixed, C random		
	A	MS_{AC}
	B	MS_{BC}
	C	MS_{ABC}
	AB	MS_e
	AC	MS_e
	BC	MS_e
	ABC	MS_e
A fixed, B random, C random		
	A	XXX
	B	MS_{BC}
	C	MS_{BC}
	AB	MS_{ABC}
	AC	MS_{ABC}
	BC	MS_e
	ABC	MS_e
A random, B random, C random		
	A	XXX
	B	XXX
	C	XXX
	AB	MS_{ABC}
	AC	MS_{ABC}
	BC	MS_{ABC}
	ABC	MS_e

XXX, no explicit test is possible; MS_e, mean square of the residual.

random selection of a set of regions will suffice and the denominator of the F testing SOILS is the mean square of the interaction between SOILS and REGIONS (Table 13.2).

13.6.6 *Three-factor ANOVA*

The three-factor ANOVA follows the general lines of the two-factor except that seven rather than three questions are now being addressed simultaneously. Table 13.3 gives the sums of squares and degrees of freedom.

Box 13.3 gives an example, again kangaroos counted by aerial survey, but with the factor YEARS added. YEARS is fixed because we wish to test specifically whether the density changed between 1987 and 1988. We do not wish to test the more general question of whether kangaroo populations remain stable with time. The ANOVA shows that of the three main effects, the three first-order interactions, and the one second-order interaction, only the main effect of kangaroo species is significant. We conclude therefore that the two species certainly differed in density but that there was insufficient evidence to identify a day effect or a change in density between years. Neither did any factor appear to interact with any other.

13.7 **Summary**

Testing the efficacy of wildlife management treatments requires that the expected outcome of the treatment be rigorously defined. Once a verifiable outcome is posed as a hypothesis, the data to test it can be collected by following the logic of experimental design. Insufficient replication of treatments to sample the range of natural variability is a common shortcut but it nullifies the point of the exercise.

The principles illustrated in this chapter can be summarized as the basic rules of experimental design. There are exceptions to several of them but, until the manager or researcher learns how and in what circumstances they may safely be varied, these should be followed in full.

1   To determine whether a factor affects the response variable under study, more than one level of that factor must be examined. The levels may be zero (control) and some nonzero amount, or they may be two

Table 13.3 Calculation of sums of squares and degrees of freedom for three-factor ANOVA

ROW effect	$(1/ncl)\Sigma T_i^2 - (1/nrcl)T^2$	d.f. $= r - 1$
COLUMN effect	$(1/nrl)\Sigma T_j^2 - (1/nrcl)T^2$	d.f. $= c - 1$
LAYER effect	$(1/nrc)\Sigma T_k^2 - (1/nrcl)T^2$	d.f. $= l - 1$
RC interaction	$(1/nl)\Sigma T_{ij}^2 - (1/ncl)\Sigma T_i^2 - (1/nrl)\Sigma T_j^2 + (1/nrcl)T^2$	d.f. $= (r-1)(c-1)$
CL interaction	$(1/nr)\Sigma T_{jk}^2 - (1/nrl)\Sigma T_j^2 - (1/nrc)\Sigma T_k^2 + (1/nrcl)T^2$	d.f. $= (c-1)(l-1)$
RL interaction	$(1/nc)\Sigma T_{ik}^2 - (1/ncl)\Sigma T_i^2 - (1/nrc)\Sigma T_k^2 + (1/nrcl)T^2$	d.f. $= (r-1)(l-1)$
RCL interaction	$(1/n)\Sigma T_{ijk}^2 - (1/nl)\Sigma T_{ij}^2 - (1/nr)\Sigma T_{jk}^2 - (1/nc)\Sigma T_{ik}^2 + (1/ncl)\Sigma T_i^2$ $+ (1/nrl)\Sigma T_j^2 + (1/nrc)\Sigma T_k^2 - (1/nrcl)T^2$	d.f. $= (r-1)(c-1)(l-1)$
Residual	$\Sigma X_{ijkm}^2 - (1/n)\Sigma T_{ijk}^2$	d.f. $= (n-1)rcl$
Total	$\Sigma X_{ijkm}^2 - (1/nrcl)T^2$	d.f. $= nrcl - 1$

Box 13.3 Red kangaroos and gray kangaroos counted on the Cunnamulla degree block ($10\,870\,km^2$) in 1987 and 1988. Each replicate is the number of kangaroos counted on a transect measuring 0.4 km by 90 km

	June 1987			June 1988		
	Day 1	Day 2	Day 3	Day 1	Day 2	Day 3
Red kangaroos	45	19	18	72	31	28
	17	51	44	32	29	9
	8	8	61	94	34	48
	28	11	35	27	47	38
	26	72	65	66	21	91
	48	34	76	55	49	138
	53	67	52	102	29	67
	62	27	30	41	67	106
T_{ijk}	287	289	381	489	307	525
$\bar{x}$	35.9	36.1	47.6	61.1	38.4	65.6
Gray kangaroos	66	27	27	116	65	10
	52	47	66	81	57	22
	34	13	75	63	41	43
	8	16	104	25	33	63
	35	101	109	75	120	74
	36	150	170	77	28	101
	42	116	51	59	62	76
	65	66	14	59	17	82
T_{ijk}	338	536	616	555	423	471
$\bar{x}$	42.3	67.0	77.0	69.4	52.9	58.9

ANOVA

Source	SS	d.f.	MS	F	$F_{0.05}$*
ROW (species)	4 551.3	1	4551.3	4.7	4.0
COLUMN (days)	3 227.3	2	1613.6	1.6	3.1
LAYER (years)	1 086.8	1	1086.8	1.1	4.0
RC interaction	1 018.1	2	509.0	0.5	3.1
CL interaction	4 681.6	2	2340.8	2.3	3.1
RL interaction	1 708.6	1	1708.6	1.7	4.0
RCL interaction	1 444.7	2	722.3	0.7	3.1
Residual	86 359.4	84	1028.1		
Total	104 077.7	95			

* $F_{0.05}$ is the critical level that must be exceeded by the observed F to qualify for a probability $\leqslant 5\%$.

or more categories (e.g., habitat types) or nonzero quantities (e.g., altitudinal bands).

2 "Before" is not a control on "after" because time trends in the response variable are caused by all sorts of influences unrelated to the treatment under study.

3 Treatments must be replicated, not subsampled. (See Hurlbert 1984 for an excellent exposition of the pitfalls of "pseudoreplication" in ecological research.)

4 The number of replications per treatment (including the control treatment) should be as close as possible to equal across treatments.

5 Treatments must be interspersed in time and space. Do not run the replications of treatment A and then the replications of treatment B. Mix up the order. Do not site the replications of treatment A in the north of the study region and the replications of treatment B in the south. Mix them up.

6 If the influence of more than one factor is of interest, each level of each factor is examined in combination with each level of every other factor (factorial design).

7 If an extraneous influence (site in the quail example) is likely to be correlated with one of the designated factors, either it should be declared a factor in its own right and the design modified accordingly or its range should be covered at random by the replication. Thus its influence is factorized out in the first option or randomized across treatments in the second.

8 All these rules may be broken, but that degrades the design to one yielding neither strong inference nor an unambiguous conclusion. Such results are still useful so long as their dubious nature is fully appreciated and declared. Environmental impact assessments are just such examples.

14 Conservation in theory

14.1 Introduction

In this chapter we deal with theory that has been developed to account for why and how populations become extinct. Most of that theory deals with extinction as a consequence of low numbers, the various difficulties that a population can get into when it is too small. A second class of extinction processes – those caused by a permanent and deleterious change in the population's environment – is less well served by theory, but in practice (see Chapter 15) is rather more important.

14.2 Demographic problems contributing to risk of extinction

Demography deals with the probability of individuals living or dying and, if they live, the probability that they will reproduce. Those individual probabilities, accumulated over all individuals in the population, determine what the population as a whole will do next, whether it will increase, decrease, or remain at the same size. Three effects can influence the population outcome underlain by those individual probabilities: individual variation, short-term environmental variation, and environmental change. These will be examined in turn, particularly in the context of the likelihood of the population becoming extinct.

14.2.1 Effect of individual variation

A population's rate of increase is determined by the age-specific fecundity rates interacting with the age-specific mortality rates, but its value is predictable only when the population has a stable age distribution. If it does not have a stable age distribution, or if numbers are low, the actual rate of increase may vary markedly in either direction from that predicted by the life table and fecundity table (see Sections 4.3 and 4.4). This effect is called *demographic stochasticity*.

Take the hypothetical case of a population of 1000 large mammals whose intrinsic rate of increase is $r_m = 0.28$. A female can produce no more than one offspring per year. The population is at a low density of $D = 0.01/km^2$ so there will be little competition for resources and consequently the rate of increase will be close to r_m. On average the probability of an individual surviving 1 year is $p = 0.9$, and the probability that a female will produce an offspring over 1 year is $b = 0.95$. The beginning of the year is defined as immediately after the birth pulse, at which time the population contains 500 males and 500 females. By the end of the year the population will have been reduced by natural mortality to about 900 (i.e., 1000 × 0.9) and these animals produce about 428 offspring (450 × 0.95) at the next birth pulse. The population therefore starts the next year with about 1328 individuals (900 + 428), having registered a net increase over the year at about the rate $r = \log_e(1328/1000) = 0.28$, or 32%. The actual outcome will be very close to those figures because the differences in demographic behavior between individuals tend to cancel out.

Now consider a subset of this population restricted to a reserve of $200\,\text{km}^2$. The density of $0.01/\text{km}^2$ translates to a population size of two individuals. These two obviously cannot increase by 32% to 2.64 individuals as the large-population estimate would imply. They can only increase to 3, or remain at 2, or decline to 1 or even to 0. Table 14.1 gives the probabilities of those outcomes.

Table 14.1 shows that the most likely outcome is three animals and a rate of increase of $r = 0.41$. But even though the population is "trying" to increase, the actual rate of increase may by chance vary between a low of minus infinity to a high of $r = 0.41$. Hence the demographic behavior of a small population is determined by the luck and misfortune of individuals. It is a lottery. That of a large population is ruled by the law of averages. We say that the outcome for a small population is stochastic and for a large population deterministic.

The extent to which actual r is likely to vary from its deterministic value in a constant environment is measured by $\text{Var}(r) = \text{Var}(r)_1/N$ where $\text{Var}(r)_1$ is the component of variance in r attributable to the demographic behavior of an average individual. For a population with a relatively low r_m, as in our hypothetical example, $\text{Var}(r)_1$ will be in the region of 0.5. We adopt that value for purposes of illustration. $\text{Var}(r)$ declines progressively as population size N rises. The variance of r at any population size N can be estimated for this "population" as $\text{Var}(r) = 0.5/N$.

Table 14.2 shows that at a population size of $N = 50$ the effects of small numbers, and hence necessarily unstable age distribution, can result in a rate of increase varying (at 95% confidence) between $r =$

Table 14.1 Probabilities for the population outcome over a year of a population comprising two individuals, one of each sex. The chance of an individual surviving the year is $p = 0.9$ and the chance of the female producing an offspring at the end of that year is $m = 0.95$

N_{t+1}	What happened	Probability of outcome		r
		Symbolic	Numerical	
0	Both die	$(1 - p)^2$	0.01	$-\infty$
1	One dies	$2p(1 - p)$	0.18	-0.69
2	Both live, no offspring	$p^2(1 - m)$	0.0405	0.0
3	Both live, one offspring	$p^2 m$	0.7695	0.41
			1.0000	

Table 14.2 Deviation from expected rate of increase resulting from stochastic variation. The influence of one individual on variance in r is taken as $Var(r)_1 = 0.5$

N	Expected r	$Var(r)$	$SE(r)$	95% Confidence limits of r
10	0.28	0.05	0.224	± 0.500
50	0.28	0.01	0.100	± 0.202
100	0.28	0.005	0.071	± 0.139
500	0.28	0.001	0.032	± 0.063
1000	0.28	0.0005	0.022	± 0.043

0.48 (i.e., 0.28 + 0.202) and $r = 0.08$ (i.e., 0.28 − 0.202). At $N = 10$ the possible outcomes vary between a high rate of increase and a steep decline. In this example the deterministic rate of increase becomes a safe guide to the actual rate of increase only after the population has attained a size of several hundred.

Although the details are special the message is general: populations containing fewer than about 30 individuals can quite easily be walked to extinction by the random demographic variation between individuals, even when those individuals are in the peak of health and the environment is entirely favorable.

14.2.2 The effect of environmental variation

We have seen that the demographic behavior of a population in a constant environment is broadly predictable when it contains several hundred individuals. The larger the population the tighter the correspondency between actual rate of increase and the expected deterministic rate of increase. But that individual variation is by no means the most important source of variation in r. Year-to-year variation in environmental conditions has a more profound effect and, unlike the effect of individual variation, does not decline with increasing population size. It is called environmental stochasticity.

The most important source of environmental variation is yearly fluctuation in weather. Weather has a direct effect on the demography of plants, invertebrates, and cold-blooded vertebrates. Their rates of growth are often a direct function of temperature as measured in degree-days. Wildlife is largely buffered against the direct effect of temperature and humidity, the influence being indirect through food supply.

The variance in r caused by a fluctuating environment is symbolized as $Var(r)_e$. It can be measured as the actual year-to-year variance in r exhibited by a population whose size is large enough to swamp the effect of variance in r due to individual variation. We recommend that such a population should contain at least 5000 individuals. Even so, $Var(r)_e$ will be overestimated because its measurement contains a further component of variance introduced by the sampling variation generated in estimating the year-to-year rates of increase.

The major influence of environmental variation on the probability of extinction is its interaction with the effect of individual variation. Thus, it becomes progressively more important with decreasing population size, even though its average effect on r is independent of population size.

14.3 Genetic problems contributing to risk of extinction

In the next few sections we examine some of the ways in which genetic malfunction may contribute to the extinction of a population. But first we provide a brief introduction to population genetics for those who have not studied it previously. Those who have can skip to Section 14.4.

14.3.1 Heterozygosity

A chromosome may be thought of as a long string of segments, called loci, each locus containing a gene in paired form. The two elements of that pair, one contributed by the individual's mother and the other by its father, are called alleles and they can be the same or different. The chromosomes of vertebrates and vascular plants contain around 100 000 loci.

Suppose the gene pool of a population contains only two alleles for locus A. These will be referred to as A_1 and A_2. Any individual in that population will thus have one of three combinations of alleles at that locus: A_1A_1 or A_1A_2 or A_2A_2. If the first or third combination is obtained, the individual is *homozygous* at that locus; if the second, it is *heterozygous*. The proportions of the three combinations in the population as a whole are called *genotypic frequencies*. Which will be the most common depends on the frequencies (proportions) of the two alleles in the population as a whole. Suppose the frequency of the A_1 allele is $p = 0.1$ and therefore that A_2 is $q = 0.9$ (because the sum of a complete set of proportions must equal 1), then the frequencies of the three genotypes will be:

	Genotypes	A_1A_1	A_1A_2	A_2A_2
	Genotypic frequencies	p^2	$2pq$	q^2
=		0.01	0.18	0.81

Note that the genotypic frequencies (proportions) also sum to one.

That relationship between allelic frequencies and genotypic frequencies is the Hardy–Weinberg equilibrium law. Formally, it holds only when the population is large, its individuals mate at random, and there is no migration, mutation, or selection. In practice, however, it is highly robust to deviations from these assumptions and can be accepted as a close approximation to the actual relationship between allelic frequency and genotypic frequency for two alleles at a single locus. The Hardy–Weinberg equilibrium holds equally for more than two alleles at a locus so long as it is calculated in terms of one allele against all the others. Alleles of the two types A_1 and *not-A_1*

(*not-A₁* = *A₂* + *A₃* + *A₄*, etc.) also take Hardy–Weinberg equilibrium proportions.

The number and frequency of different alleles at a locus can be determined fairly easily by electrophoresis or DNA sequencing. If p_{ij} is the frequency of allele i at locus j in the population as a whole, the proportion of individuals heterozygous at that locus may be estimated as:

$$h_j = 1 - \Sigma p_{ij}^2$$

providing that the number of individuals n_j examined for locus j is > 30. If fewer, h_j is underestimated but can be corrected by multiplying by $2n_j/(2n_j - 1)$. Thus, the more alleles at a locus the higher the value of h_j, and the less diverse the frequencies of the alleles the higher is h_j.

Mean heterozygosity is estimated as:

$$H = (1/L)\Sigma h_j$$

where L is the number of loci examined. H varies considerably between species for reasons that are not understood. As estimated by one-dimensional electrophoresis of loci controlling production of proteins, H ranges between 0.00 and 0.26 for mammals, with an average at about 0.04. Heterozygosity estimated in the same way yields $H = 0.036$ for both white-tailed and mule deer (Gavin and May 1988) and $H = 0.029$ for leopards. Figure 14.1 shows the frequency distribution of H for 169 species of mammals. Note first that the distribution is shaped like a reverse *J*: most species have a low H but a few break out of that pattern to return a high H. Second, a substantial proportion (10.6%) of mammalian species are homozygous at all loci examined by electrophoresis.

Genetic variability can be reported also as the proportion of polymorphic loci in the population (i.e., the proportion of loci for which there is more than one allele within the population as a whole). This is not the same as H above. If all but one individual in the population are homozygous at locus A that locus is nonetheless scored as polymorphic. The proportion of polymorphic loci within the population is therefore higher than the average heterozygosity (the proportion of heterozygous loci in an average individual), usually about three times higher. Furthermore, it is an unstable statistic tending to increase as sample size increases. We recommend against its use.

The level of heterozygosity within a population can be calculated by DNA sequencing or gel electrophoresis. Here we consider only the latter. If the function of a given gene is to direct the synthesis of a protein, an individual with heterozygous alleles at that locus will produce two versions of that protein, but only one if homozygous. Identifying the individual as one or the other is a simple technical exercise. A sample of its blood plasma is placed on a slab of gel and subjected to a weak electrical current which causes the plasma proteins to migrate through the gel at a speed specific to each protein. The protein of interest, say transferrin, can be differentiated from the

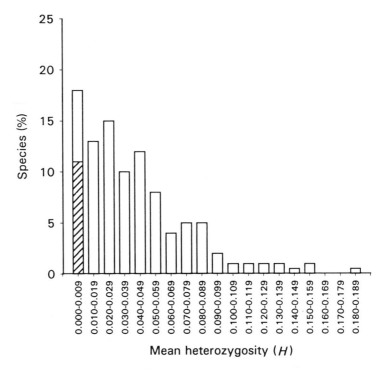

Fig. 14.1 Frequency distribution of mean heterozygosity H for 169 mammalian species. (After Nevo *et al.* 1984.)

other proteins by staining with Nitroso-R salt (1-Nitroso-2 naphthol-3,6-disulfonic acid) and the different versions of that protein, corresponding to different alleles, can be identified by their banding pattern on the gel. By this means we can readily identify which alleles are present at the locus controlling transferrin synthesis and hence whether the individual is heterozygous at that locus. The same applies for other loci and other proteins. A precise value for H can be obtained by examining about 100 loci for each of about 30 individuals, and so H can be monitored fairly easily over time to detect a slump in overall heterozygosity.

14.3.2 *Additive genetic variance*

So far we have dealt only with single genes in isolation, those that either produce an effect or do not. Electrophoretic analysis depends on these: a particular protein is either produced or not according to whether a particular allele is present at the locus of interest. This is the realm of mendelian genetics, which deals with qualitative characters.

Adaptive evolution, however, is based mainly on the selection of quantitative characters, such as foot length or timing of conception. These vary in a continuous fashion. Such a character is influenced by many genes. The variation in a quantitative trait between individuals has two sources, environmental and genetic. The genetic variance can be partitioned into three components: that resulting from the effect of alleles within and between loci (additive genetic variance), that resulting from dominance effects, and that resulting from interaction between loci. The additive genetic variance is by far the most important and for present purposes can be discussed as if it constituted the total genetic variance.

The intricacies of genetic variance are beyond the scope of this book but all the necessary points can be made by using heterozygosity H as an index of additive genetic variance. Those factors that influence additive genetic variance also influence heterozygosity and they do so by similar amounts and in similar ways. The mathematics describing the dynamics of the two are much the same.

At this stage we make an important point on a subject that is widely misunderstood: the nature of "*genetic diversity.*" This is usually conceptualized by conservationists as the number of distinct alleles in the population. The loss of one of those alleles is seen as a reduction in genetic diversity and therefore a bad thing. That notion of genetic diversity is trivial and entirely inappropriate to conservation management. Rather, the important measure is *genetic variance*, which can be conceptualized as a parameter closely akin to heterozygosity H, the proportion of loci that are heterozygous in an average individual. Consequently the amount of heterozygosity carried within a sample comprising a couple of dozen individuals will closely approximate the amount of heterozygosity within the population from which that sample was drawn. By extension, the amount of genetic variance carried by a relatively small sample of the population will closely approximate the magnitude of the genetic variance of that population as a whole. It is the genetic variance of the population that we seek to conserve, not the "genetic variation" or "genetic diversity" represented by the total number of distinct alleles within the population.

That misunderstanding carries over to populations established by liberation. It is often argued that since these were usually started by a nucleus of only a few individuals, and since those individuals must carry only a small fraction of the genetic variability of the population from which they were drawn, that the subsequent robust health of the population that developed by the liberation indicates that a population needs little genetic variability. Those populations tend either to increase rapidly or to go extinct (often apparently by simple demographic stochasticity) within the first few years of liberation. If the former, they are shortly free of any genetic bottleneck. The condition that gets a population into potential genetic trouble is not a history that includes a small population size but a small population size that persists for many generations. That seldom occurs for a population founded by liberation.

14.3.3 Drift and mutation

In the absence of immigration and mutation, the number of different alleles at a locus in the population as a whole can either remain constant or decrease. It cannot increase. In practice it will always decrease because alleles will be lost under the influence of nonrandom mating and unequal reproductive success between individuals. Heterozygosity thus decreases also. Its rate of decline is a function of population size N, the proportion of heterozygous loci in the population as a whole being reduced by the fraction $1/(2N)$ per generation. Over one generation H changes according to:

$$H_1 = H_0(1 - 1/(2N))$$

and over t generations:

$$H_t = H_0(1 - 1/(2N))^t$$

which may alternatively be written:

$$H_t = H_0 e^{-t/2N}$$

After $t = 2N$ generations, the population's heterozygosity will have dropped to 0.37 (i.e., e^{-1}) of its initial value at time $t = 0$. This holds as well for a single locus as it does across all loci. The loss of additive genetic variance is exactly analogous and conforms to the same equations. The process is called *random genetic drift*.

The rate of mutation at a single locus is about 10^{-6} per gamete per generation. However, most evolution is in terms of phenotypic characters that change in very small steps. These quantitative characters are controlled by many genes. Mutation within such a gene complex is much more frequent than at a single locus, closer to 10^3 or even 10^2 per gamete per generation.

Heterozygosity (and hence additive genetic variance) will change over one generation by an amount ΔH according to:

$$\Delta H = -H/2N + m$$

where m is the input of heterozygosity by mutation. Its equilibrium is solved by setting ΔH to zero:

$$H = 2Nm$$

which informs us that for any population size N there will be an equilibrium between mutational input of additive genetic variance and loss of it by drift. What varies, however, is the value of H at equilibrium. It will be higher when N is large and lower when N is small. The population size N must remain constant for many generations for such an equilibrium to establish.

14.3.4 *Selection*

Selection comes in two forms: directional and stabilizing (or normalizing). *Directional selection*, that which moves a trait in one direction, is the stuff of evolution. *Stabilizing selection*, on the other hand, eliminates the extremes of a trait and holds the trait to its optimum for the current environment.

At any particular time most selection will be of the stabilizing type. Any tendency for the breeding season to expand, for example, will be attacked continuously by stabilizing selection. An offspring born during a seasonally inappropriate time of the year has little chance of survival and so that mistake, if it has a genetic basis, will be swiftly corrected. The relaxation of that selection in captivity is presumably the reason why the breeding season of captive populations tends to expand after several generations.

Stabilizing selection is essential to maintaining the fitness of a wild population. Ironically, it is one of the strongest forces reducing additive genetic variance and heterozygosity. We must therefore avoid an obsession with maximizing variance to the exclusion of maintaining fitness by reducing the number of deleterious alleles.

14.3.5 *Inbreeding depression*

Suppose the A_1 is a recessive, its effect being masked by A_2 when the two occur together at the locus. If additionally A_1 is slightly deleterious when its effect is expressed, it can have damaging effects on fecundity and survival of the population if its frequency increases by the statistical luck of random pairing of alleles in each generation. The gene pools of most populations contain many of these sublethal recessives, about enough to kill an individual three times over if by chance they all occurred on its chromosomes in homozygous form and were therefore all expressed in its phenotype. Thus, a decline in heterozygosity tends to lead to a decline in fitness.

Genetic malfunction may follow as a consequence of small population size. The following sequence may be triggered if a population becomes too small.

1 The frequency of mating between close relatives rises and random genetic drift increases,

2 which leads to reduced heterozygosity in the offspring,

3 which exposes the effect of semilethal recessive alleles,

4 which reduces fecundity and increases mortality,

5 which causes the population to become smaller yet, and that trend may continue until extinction. The population must be held at low numbers for several generations before that process is initiated. A short bout of low population size has little effect on heterozygosity.

Loss of fitness during inbreeding can be traced largely to the process of fixation (i.e., reduction of alleles at a locus to one type) of deleterious recessive alleles. In mammals, the mortality rate is 33% higher for the offspring of parent–offspring or full sibling matings than for the offspring of unrelated parents (Ralls *et al.* 1988). Hybrid vigor is largely the reverse (the masking of the effect of those recessives), but it might also contain a component of heterosis where the heterozygote is fitter than either homozygote.

Inbreeding does not automatically lead to inbreeding depression. It is seldom reported for populations larger than a couple of dozen individuals. Nor does low heterozygosity necessarily lead to inbreeding depression. Note that the average individual of most wild populations is heterozygous at < 10% of loci. A population that has survived a bout of inbreeding may come out of it with enhanced fitness because inbreeding exposes deleterious recessives and allows them to be selected out of the population. That is precisely the method used by animal breeders to remove deleterious alleles. Homozygosity causes an immediate problem only when the allele is deleterious. Nonetheless, inbreeding often does produce inbreeding depression. That possibility must always be kept in mind if a population is small. Section 14.6 addresses the question: how small is too small?

14.3.6 *How much genetic variation is needed?*

The cheetah (*Acinonyx jubatus*) has a low level of heterozygosity. O'Brien *et al.* (1985b; 1986) and Cohn (1986) studied the genetic variance of captive populations of that species originating from southern Africa. A standard electrophoretic analysis of 52 loci (in 55 individuals) discovered a heterozygosity value of $H = 0.00$ as compared with $H = 0.063$ for people and 0.037 for lions. A more refined "two-dimensional" electrophoretic analysis, separating the proteins first by electrical charge (as above) and then by molecular weight, uncovered rather more variability and yielded $H = 0.013$ for cheetahs as against 0.024 for people analyzed by the same method. A further sample from East Africa returned $H = 0.014$ (O'Brien *et al.* 1987).

Is the cheetah in peril? It is possible that as a direct consequence of the low heterozygosity the cheetah produces sperm of low viability, its rate of juvenile mortality is abnormally high, and it is particularly susceptible to disease. All these claims have been made but no causal relationship has been established between these putative defects and the peculiarities of the genotype. Alternatively the cheetah may be in no danger of demographic collapse despite its low genetic variability. In support of that is its widespread distribution, which was even wider in the recent past, particularly in Asia. Contraction of range over the past 1000 years has been no greater than that of the lion, another widespread species but with a standard level of heterozygosity. For both species the contraction of range seems to be a result of excessive human predation rather than of a diminished genetic fitness. As far as we know there is no evidence from the wild suggesting that the cheetah is faced by a level of risk beyond those hazards imparted by a rising human population (Caughley 1994).

The cheetah clearly has low genetic variance but well within the range exhibited by mammalian species (see Fig. 14.1). The suggestion that it is in demographic peril as a consequence of that modest genetic variance earns no support from what is known of its ecology. There are two messages transmitted by this example. The first is special: we need more disciplined information on the cheetah in the field to determine whether its diminished genetic variance is associated with demographic malfunction. The second is general: by genetic theory currently followed in conservation biology the cheetah should be in demographic trouble, but there is no convincing evidence for that and considerable but circumstantial evidence to the contrary. A plausible alternative hypothesis is that present genetic theory overestimates the amount of genetic variation needed to sustain an adequate level of individual fitness. One should not jump to the conclusion that a species is in danger simply because it has a low H. There is too much evidence to the contrary.

We cannot yet lay down general rules as to the minimum genetic variance required for adequate demographic fitness. Nor can we define a minimum viable population size (genetic). We need much more research on the incidence of inbreeding depression in the field, on the population size and the period over which that size must be maintained before inbreeding depression becomes a problem, and on the relationship between heterozygosity and fitness.

14.4 Effective population size (genetic)

It can happen that the size of a population appears large enough to avoid genetic malfunction but that the population is acting genetically as if it were much smaller. The proportion of genetic variability lost by random genetic drift may be higher than the computed theoretical $1/(2N)$ per generation because that formulation is correct only for an "ideal population." In this sense "ideal" means that: family size is distributed as a Poisson; sex ratio is $50:50$; generations do not overlap; mating is strictly at random; and the rate of increase is zero. This introduces the notion of *effective population size* in the genetic sense, the size of an ideal population that loses genetic variance at the same rate as does the real population. The population's effective size (genetic) will be less than its census size except in special and unusual circumstances.

Perhaps the greatest source of disparity between census size and effective size is the difference between individuals in the number of offspring they contribute to the next generation. In the ideal population their contribution has a Poisson distribution, the fundamental property of which is that the variance equals the mean. Should the variance of offspring production between individuals exceed the mean number of offspring produced per individual, the effective population size will be smaller than the census size. In the unlikely event of variance being less than the mean the effective population size is greater than the census size and the population is coping better genetically than one might naively have expected. The effective population size N_e corrected for this demographic character can be calculated as:

$$N_e = (NF - 1)/(F + (s^2/F) - 1)$$

where F is the mean lifetime production of offspring per individual and s^2 is the variance of production between individuals. It indicates that when mean and variance are equal N_e approximates to N. Since males and females sometimes differ in mean and variance of offspring production this equation is often solved for each sex separately and the sex-specific N_e values summed.

Genetic drift is minimized when sex ratio is 50:50. Effective population size (genetic) in terms of sex ratio is given by:

$$N_e = 4/((1/N_{em}) + (1/N_{ef}))$$

where N_{em} and N_{ef} are the effective numbers of males and females as corrected above for variation in production of offspring. The series below shows the relationship numerically:

Sex ratio	50:50	60:40	70:30	80:20	90:10
N_e	N	$0.96N$	$0.84N$	$0.64N$	$0.36N$

Further corrections can be made for other sources of disparity between real and ideal populations. These considerations are often important in *Drosophila* research and in managing the very small populations in zoos, but they have little utility in conservation. Rather than attempting to estimate N_e for a threatened population you should simply assume as a rule of thumb that N_e approximates to $0.4N$, and that the censused population is losing genetic variability at a rate appropriate to an ideal population less than half its size.

14.5 Effective population size (demographic)

The *effective population size (demographic)* is the size of a population with an even sex ratio and a stable age distribution that has the same net change in numbers over a year as the population of interest.

14.5.1 *Effect of sex ratio*

If a species is polygamous, as most species of wildlife are, a disparate sex ratio may have a large effect upon net change in numbers over a year and hence effective population size (demographic) at the beginning of that year. Net change in numbers over a year can be calculated as:

$$\Delta N = NpbP_f - N(1 - p)$$

where P_f is the proportion of females in the population, p is the probability of surviving the year averaged over individuals of all ages within a stable age distribution, and b is number of live births produced per female at the birth pulse terminating the year. It indicates that net change in numbers in a population of any given size is a linear function of the proportion of females in the population. The regression of ΔN on P_f has a slope of Npb and an intercept (i.e., the value of ΔN when $P_f = 0$) of $-N(1 - p)$.

The hypothetical example in Section 14.2.1 had a population size of $N = 1000$, a probability of survival per individual per year of $p = 0.9$, and a fecundity rate of $b = 0.95$ live births per female per year. Those values fed into the above equation yield ΔN as 541 when $P_f = 0.75$, as 328 when $P_f = 0.5$, and as 114 when $P_f = 0.25$.

The relationship can be rearranged to estimate N_e, the effective population size (demographic), as:

$$N_e = N(pbP_f + p - 1)/(0.5pb + p - 1)$$

For the above example, N_e is solved as 1653 when $P_f = 0.75$, as 1000 when $P_f = 0.5$, and as 347 when $P_f = 0.25$. Thus, a disparate sex ratio may have a significant effect on a population's ability to increase from low numbers, enhancing that ability when females predominate and depressing it when males dominate. Ungulate populations that have crashed because they have eaten out their food, or because a drought has cut their food from under them, often end the population slide with a preponderance of females. They are thus in better shape demographically to recover from the decline than if parity of sex ratio had been retained. Note here an important point: the sex ratio minimizing genetic drift (50:50) is not that maximizing rate of increase (disparity of females). Hence the appropriate "effective population size" depends on the context. The genetic version is appropriate for a small capped population in a zoo. The demographic version is often more appropriate in the wild where the aim is usually to stimulate the growth of an endangered species.

14.5.2 *Effect of age distribution*

Similarly, an equation for effective population size (demographic) can be written to correct for the effect of an unstable age distribution. It requires a knowledge of the age distribution, population size, age-specific fecundity, and age-specific mortality (see Sections 4.3 and 4.4 for calculating these). However, these estimates are difficult to obtain in practice, particularly for a small population. When faced by the urgent task of diagnosing the cause of the decline of an endangered species, instead of wasting valuable research time on estimating its age-distribution version of effective population size (demographic) you should simply understand the principle and appreciate that, because of instability of age distribution, the population's rate of change may be higher or lower than would be expected from a simple tallying of numbers. Suppose the population experienced a severe drought in the previous year that killed off the vulnerable young and old animals. The age distribution will now be loaded with animals of prime breeding age, and the birth rate, measured as offspring per individual, will therefore be much higher than usual. In consequence, the rate of increase will be higher than usual. Alternatively the population may have experienced a mild winter such that the age distribution is loaded with young animals below breeding age. The birth rate at the next birth pulse is then lower than usual and so is the rate of increase.

14.6 **How small is too small?**

Two values for minimum viable population size (genetic) are commonly quoted: 500 and 50. The difference between them reflects the differing assumptions upon which they are based. The 500 value (Franklin 1980) is the effective population size at which the heritability of a quantitative character stabilizes at 0.5 on average (i.e., 50% of quantitative phenotypic variation is inherited and 50% is environmentally induced). The figure of 0.5 is the heritability coefficient for bristle number in *Drosophila*, and quantitative characters in farm animals often have a heritability coefficient of that general magnitude.

The assumption is that such a level of heritability reflects a genetically healthy population. The genetic variance needed to enforce such a heritability coefficient has an equilibrium (where loss by genetic drift is balanced by gain from mutation) of $N_e = 500$ in the absence of selection, and this is taken as a safe lower limit for population size.

The estimate of 50 comes from the observation of animal breeders that a loss of genetic variance of 1% per generation causes no genetic problems. Since that rate is $1/(2N_e)$ we can write $0.01 = 1/(2N_e)$, rearrange it to $N_e = 1/(2 \times 0.01)$, and solve it as $N_e = 50$.

The arbitrary nature of both estimates of minimum viable population size will be readily apparent. Neither should be accepted as anything more than general speculation.

No single number estimates have been offered for minimum viable population size (demographic). It is clearly a function of $Var(r)_1$ interacting with $Var(r)_e$ (see Sections 14.2.1 and 14.2.2) and so will vary among species and among populations within species. However, for most populations it is likely to be higher than the corresponding minimum viable population size (genetic).

14.7 Extinction caused by environmental change

The most common cause of extinction is a new element introduced into the environment. It is thus distinct from year-to-year environmental fluctuation. We identify the new influence as the driving variable responsible for the population's decline and the population may be driven to extinction by its action. A population seldom "dwindles to extinction," it is pushed. If you can identify the agent imparting the pressure and neutralize that pressure you can save the population.

The most common causes of driven extinctions, roughly in order of importance, are: (i) contraction and modification of habitat; (ii) increased predation or hunting; (iii) competition for resources with a species new to that environment; (iv) a poison in the environment; and (v) a disease, particularly one new to the environment. These are discussed further, with examples, in Chapter 15.

14.8 Summary

A population may, by chance, be forced to extinction by year-to-year variation in weather or other environmental factors. When the population is small it may "random walk" to extinction because its dynamics at low numbers are determined by the unpredictable fortunes of individual members. Genetic drift and inbreeding depression may also operate at low numbers to reduce fitness and thereby lower numbers even further. Their effect increases with declining population size, modified further by the sex and age structure of the population. However, the commonest causes of extinction are habitat modification and the introduction, usually by humans, of a new element into the environment. It is commonly a new predator, competitor, or pathogen. The population is then driven to extinction rather than dropping out by chance.

15 Conservation in practice

15.1 Introduction

In Chapter 14 we examined the ways in which demographics and genetics contribute, at least potentially, to the risk that a population will become extinct. The extinction of a species does not differ in kind. The species goes extinct because the last population of that species goes extinct. Here, we review actual extinctions or near extinctions to show what are the commonest causes of extinction in practice. We then describe how to detect such problems and how to treat a population in danger.

15.2 How populations go extinct

Extinctions may be divided easily into two categories, driven extinctions and stochastic extinctions. They can be characterized as follows.

1 *Driven extinction*: whereby a population's environment changes to its detriment and rate of increase falls below zero. The population declines. Perhaps this lowering of density frees up resources to some extent, or lowers the rate of predation, but this is not sufficient to counteract the force of the driving variable and the population finally goes extinct. Included in this category are extinctions caused by environmental fluctuation and those caused by catastrophes. The latter are viewed here as simply large environmental fluctuations.

2 *Stochastic extinction*: whereby a population fails to solve the small population problem. The effect of chance events, which would be trivial when numbers are high, can have important and sometimes terminal consequences when numbers are low.

(a) *Extinction by demographic malfunction*: whereby a population goes extinct by accident (chance) because it is so small that its dynamics are determined critically by the fortunes of individuals rather than by the law of averages. In those circumstances a population is quite capable, by chance configuration of age distribution or sex ratio, of registering a steep decline to extinction over a couple of years even though its schedules of mortality and fecundity (see Sections 4.3 and 4.4) would result in an increase if the age distribution were stable.

(b) *Extinction by genetic malfunction*: whereby a population at low numbers for several generations loses heterozygosity to the extent that recessive semilethals are exposed, average fitness therefore drops, and the population declines even further and ultimately to extinction. The loss of an allele from the genotype is an event resulting from the lottery of random mating. Although each individual loss is unpredictable, the average rate of loss, as a function of population size, can be predicted fairly accurately.

These mechanisms do not exclude each other entirely but they are sufficiently distinct that we treat them separately. Although the relative contribution of these mechanisms is unknown, enough anecdotal information is available to suggest that the driven extinction is by far the most prevalent. Extinction by demographic malfunction is probably the second most important, but it usually requires that the population is driven to low numbers before demographic stochasticity can operate. Many examples have been documented, particularly for introductions. Extinction by genetic malfunction appears to come a distant third. Genetic problems have a low priority in saving a natural population from extinction. They are more relevant to managing a population in captivity or one whose size is so small and its future so bleak that it should be in captivity.

We will now look at a few examples of species that have become extinct, or came close to extinction, to give us a feel for the range of possibilities.

15.2.1 *Effect of habitat change*

Many extinctions appear to have been caused by habitat changes, but the precise mechanism of population decline is usually difficult to determine retrospectively. There follows examples where extinctions or steep declines were associated with a change in habitat and where that change probably caused the extinction.

The Gull Island vole

The Gull Island vole (*Microtus nesophilus*) was discovered and described in 1889. It was restricted to the 7-ha Gull Island off Long Island, New York. Fort Michie was built there in 1897, its construction requiring that most of the island (and thus the vole's habitat) be coated with concrete. The species has not been seen since.

The hispid hare

The hispid hare (*Caprolagus hispidus*) once ranged along the southern Himalayan foothills from Nepal to Assam but is now restricted to a handful of wildlife sanctuaries and forest reserves in Assam, Bengal, and Nepal. This short-limbed rabbit-like hare depends on tall dense grass formed as a successional stage maintained either by monsoon flooding or by periodic burning (Bell *et al.* 1990). The hare's decline reflects fragmentation of suitable habitat by agricultural encroachment. Most of the surviving populations are now isolated in small pockets of suitable habitats in reserves. Much of the natural grassland has been lost to agriculture, forestry, flood control, and irrigation schemes. What remains is modified, even within the reserves, by unseasonal burning and grass cutting for thatching material.

Like many endangered species, especially those that are small or inconspicuous, neither density nor rate of decline has been measured. The current status was determined from searches of the few remaining pockets of tall grassland. There is some evidence that contraction of the species into pockets of favorable habitat renders individual hares more vulnerable to predation.

Wallabies and kangaroos

Some of the more dramatic examples of driven extinctions involve the ecology of a significant segment of the fauna being disrupted by large-scale and abrupt habitat changes. The substantial depletion of the Macropodoidea (about 50 species of kangaroos, wallabies, and rat-kangaroos) that followed European settlement of Australia is an example, and has been summarized by Calaby and Grigg (1989).

The changes in the macropod fauna included the extinction of six species and the decline of 23, of which two died out completely on the mainland but still occur in Tasmania. The rates of decline are often difficult to estimate because the year in which the decline began is seldom known and indices of population size are seldom available. Land clearing and extensive sheep farming were in full swing in Australia by 1840 and declines in macropods were evident by the late 1800s. Many of the declines followed sweeping changes to habitat as land was cleared for agriculture or the vegetation modified by the grazing of sheep. The smaller macropods (< 5 kg) were the most affected, only one species (*Macropus greyi*) of the larger macropods going extinct. Calaby and Grigg (1989) emphasized the difficulty of determining the cause of declines retrospectively but considered that the evidence strongly suggested that the declines of nine species could be referred to the effects of land clearing, eight to modification of vegetation by sheep, five to introduced predators (foxes and cats), and seven to unknown causes.

The sheep grazing and woodland clearing that led to the decline or extinction of at least 17 species concomitantly benefited five species of the larger macropods, which increased in numbers. A further four large macropods and four of the 11 species of the smaller rock-wallaby (*Petrogale*) have changed little in numbers from the time of the European settlement to the present day.

15.2.2 The effect of introduced predators

The Lord Howe woodhen

The Lord Howe woodhen (*Tricholimnas sylvestris*) is a rail about the size of a chicken. It lives on the 25-km^2 Lord Howe Island in the southwest Pacific, 700 km off the coast of Australia. Lord Howe was one of the few Pacific islands, and the only high one, that was not discovered by Polynesians, Melanesians, or Micronesians before European contact, and which therefore suffered none of the human-induced extinctions common on Pacific islands over the first millennium AD. Humans set foot on it for the first time in 1788, at which time it hosted 13 species of land birds of which nine became extinct over the next 70 years or so.

The story of the Lord Howe woodhen is related by Fullagar (1985). The island was visited regularly for food (which no doubt included woodhen) and water by sailing ships in the late eighteenth and early nineteenth centuries, and finally settled permanently in 1834. Pigs were introduced before 1839, dogs and cats before 1845, domestic goats before 1851, and the black rat (*Rattus rattus*) in 1918. By 1853 the woodhen's range was restricted to the mountainous parts of the island and by 1920 its range had apparently contracted to the summit plateau (25 ha) of Mt Gower, a 825-m (2700 ft) mountain almost surrounded by near vertical cliffs rising out of the sea. The summit plateau is a dreary place: dripping moss forest and perpetual cloud, a rather different place from the coastal flats that used to be the bird's habitat. This was obviously a species on its last legs, but that problem was not recognized until 1969 after which the population was studied intensively. The population was stable at between eight and ten breeding pairs, although in 1 year it went down to six pairs. Apparently no more than ten territories could be fitted into the 25 ha of space, and so we can be confident that the population on Mt Gower, and probably that of the entire species, did not exceed that size over the 60 years between 1920 and 1980.

The obvious candidate for the contraction of population size and range was the black rat, which has been implicated in the extinction of several species of birds on islands. But in this case the rat does not appear to be implicated. The woodhen kills rats, and in any event rats are more common on the summit of Mt Gower than on any other part of the island. The culprit instead appeared to be the feral pig which will kill and eat incubating birds and destroy the nest and eggs. Pigs cannot accomplish the minor feat of mountaineering needed to reach the summit of Mt Gower. The pigs were shot out in the 1970s and the cats severely reduced.

In 1980 a captive breeding center was established on the island at sea level and seeded with three pairs from Mt Gower. Thirteen chicks were reared in the first season of captivity, 19 in the second, and 34 in the third. The birds were released and the captive breeding terminated at the end of 1983. Today (1992) the population appears to be stable at about 180 birds, 50–60 breeding pairs, and that number seems to saturate all the suitable habitat on the island, mainly palm forest. An interesting byproduct of the pig control is the expansion of breeding colonies of petrels and shearwaters.

The Stephen Island wren

The Stephen Island wren (*Xenicus lyalli*), the only known completely flightless passerine, was discovered in 1894. It lived on a 150-ha island in Cook Strait which separates the North Island and South Island of New Zealand. Subfossil remains indicate that it was previously widespread on both main islands but became extinct there several centuries before European settlement, part of the extinction event that followed the colonization of New Zealand by the Polynesians about AD 900. The causal agent of its extinction on the mainland was probably the Polynesian rat (*Rattus exulans*) introduced by the Polynesians.

The wren was extinguished by a single domestic cat, the pet of the lighthouse keeper, Mr Lyall. He was the only European to see the species alive and then on but two occasions, both in the evening. He said it ran like a mouse and did not fly, a fact confirmed subsequently from the structure of the primary feathers. The first one he saw was dead, having been brought in by the cat. Subsequently the cat delivered a further 21, 12 of which eventually found their way to museums. Then it brought in no more. The species became extinct in the same year that it was discovered.

15.2.3 *The side-effects of pest control*

The effects of pest control often exceed the original intentions of the control exercise.

The black-footed ferret

The sinuously elegant black-footed ferret (*Mustela nigrites*) provides an example of a species paying the price for the control of another. This account of its narrow escape from extinction is taken from Seal *et al.* (1989) and particularly from Cohn (1991).

The black-footed ferret once ranged across most of the central plains of North America from southern Canada to Texas. Its life style is closely linked to that of the prairie dog (*Cynomys*), a squirrel-like rodent that lived in huge colonies on the plains. The ferret feeds mainly on prairie dogs but can feed also on mice, ground squirrels, and rabbits. However, 90% of its diet comprises prairie dogs. The ferret lives in the warrens or burrow systems of the prairie dogs and hence this species provided the ferret with both its habitat and a large proportion of its food supply. Around the turn of the century there was a concerted effort to eradicate the prairie dog, which was viewed as vermin by ranchers. It was seen as competing with sheep and cattle for grass and its burrow systems made riding a horse most unsafe. The prairie dogs were poisoned, trapped, and shot in their millions by farmers and by government pest controllers. As the prairie dogs went, so did the ferrets. By the middle of the century they were judged to be extinct, but in 1964 a small population was discovered in South Dakota. That colony died out in 1973. In 1981 a colony was discovered in Wyoming. Careful censusing produced an estimate of 129 individuals in 1984, but by the middle of 1985 the population had declined to 58 animals and within a few months was down to 31. Canine distemper was diagnosed in this population, and might well have been the cause of that decline (see Section 11.7.1).

With the population obviously threatened there was an attempt to capture the remaining animals to add to an already established captive breeding colony. Five were caught in 1985, 12 in 1986, and one in 1987. By February 1987 the last known wild black-footed ferret was in captivity.

15.2.4 *The effect of poorly regulated commercial hunting*

The type of example of serious declines caused by hunting is provided by the history of commercial whaling. It demonstrates the effect of the discount rate (see Section 16.7.2) upon the commercial decision determining whether a sustained yield is taken or whether the stock is driven to commercial extinction.

Market economics will act to conserve a commercially harvested species only when that species has an intrinsic rate of increase r_m (see Section 4.2.1) considerably in excess of the commercial discount rate, the interest a bank charges on a loan to a valued customer. Hence, when a species is harvested commercially, the yield must be regulated by an organization whose existence and funding is independent of the economics of the industry that it regulates; otherwise it will necessarily endorse the quite rational economic decisions of the industry which may well be to drive a stock to very low numbers and then to switch to another stock.

The muskoxen in mainland Canada

Unregulated commercial hunting reduced the muskoxen (*Ovibos moschatus*) on the arctic mainland of Canada to about 500 animals by 1917. In that year the species was protected by the Canadian government. The size of the historical populations will never be known but, ironically, documentation of the purchase of the musk-ox hides from native hunters by the trading companies was detailed. Barr (1991) collated the records and estimated that a minimum of 21 000 muskoxen were taken between 1860 and 1916. Their hides were shipped to Europe as sleigh and carriage robes, replacing bison robes after that species had been reduced almost to extinction.

Commercial hunting appears to be the overriding cause of the virtual extinction of the muskoxen on the Canadian mainland. Legislative protection successfully reversed the trend. Muskoxen now number about 15 000 on the mainland and have reoccupied almost all their historical range. The conservative hunting quotas introduced in the 1970s did not stop that recovery.

15.2.5 *The effect of unregulated recreational hunting*

Recreational hunting is intrinsically safer than commercial hunting because sport hunters operate on an implicit discount rate of zero. Sport hunting hence has an enviable record of conserving hunted stocks. Instances of gross overexploitation are rare but not unknown.

The Arabian oryx

The Arabian oryx (*Oryx leucoryx*) is a spectacular antelope whose demise in the wild and its subsequent reestablishment from captive stock is related by Stanley Price (1989). Its original distribution appears to have included most of the Arabian Peninsula, but by the end of the nineteenth century the remaining Arabian oryx were divided into two populations. A northern group lived in and around the sand desert of north Saudi Arabia known as the Great Nafud and a southern group occupied the Rub' al Khali (the Empty Quarter) of southern Arabia.

The northern population became extinct about 1950. The range of the southern population declined from about $400\,000\,km^2$ in 1930, to $250\,000\,km^2$ in 1950, to $10\,000\,km^2$ in 1970. Within the next couple of years the population was reduced to six animals in a single herd. They were shot out on October 18, 1972.

This extinction was caused by recreational hunting. The countries of the Arabian Peninsula are essentially sea frontages, the inland boundaries being little more than lines on a map. There is little control over activities in the hinterlands. Oil company employees and their followers used company trucks for hunting trips and seemed to have been at least partly responsible for the decline. Then there were the large motorized hunting expeditions originating mainly from Saudi Arabia. These were self-contained convoys that included fuel and water tankers. The vehicles and support facilities allowed large areas to be swept each day with efficient removal of the wildlife. Their main quarry was bustards and hares secured by hawking, but gazelles and Arabian oryx were also chased. One such party crossed into the Aden Protectorate (now the People's Democratic Republic of Yemen) in 1961 and killed 48 Arabian oryx, about half the population of that region. In the 1960s large parties from Qatar would each year capture Arabian oryx with nets in the hinterland of Oman, trucking them the 900 km back to replenish the captive herd of Shaikh Kasim bin Hamid.

15.2.6 *The effect of competition with introduced species*

Hawaiian birds

More bird species have been introduced to the Hawaiian islands than to any other comparable land area. Of 162 species introduced, 45 are fully established and 25 have secured at least a foot-hold (Scott *et al.* 1986). These exotic species have been suggested as one of the causes of the decline and extinction of the native birds.

Mountainspring and Scott (1985) estimated the geographic association within pairs of the more common small to medium-sized insectivorous forest passerines. After statistically removing the effect of habitat they showed that a higher proportion of exotic–native pairs of forest birds were negatively associated than were pairs of indigenous birds. They suggested that these results reflected competition, mainly for food.

The Japanese white-eye (*Zosterops japonicus*) became the most abundant land bird in Hawaiian islands after being introduced to Oahu in 1929 and to the island of Hawaii in 1937. It feeds on a wide variety of foods and is fairly catholic in its choice of habitats. It shares the range of at least three native species with similar food habits. Although causality cannot be demonstrated conclusively, particularly in retrospect, there is strong inference that the Japanese white-eye was implicated in the decline of the Hawaii creeper (*Oreomystis mana*) in the 1940s.

The California condor

The story of the California condor (*Gymnogyps californianus*) reveals a little about the realities of conservation: the gaps between theory and practice and the overwhelming need to determine, not assume, the causes of the decline.

The California condor was probably abundant and widely distributed in southern North America during the Pleistocene. Later, it figured in the ceremonies and myths of the prehistoric and historic Indians and caught the eye (and trigger finger) of the early European explorers. California condors ranged from the Columbia River south into New Mexico in the 1800s but by 1940 their range had contracted to a small area north of Los Angeles. Koford's (1953) estimate, based on sightings, of only 60 individuals surviving by the early 1950s was probably low. Annual surveys by simultaneous observations of known concentrations were begun in 1965 but abandoned in 1981 because they were judged to be subject to unacceptable error. Photographic identifications were then used to generate a total count of 19–21 birds in 1983 (Snyder and Johnson 1985). The decline continued until, in 1985, the last eight wild individuals were caught and added to the captive flock.

The causes of the initial decline was probably shooting and loss of habitat, but the supporting evidence is anecdotal. Low productivity caused by an insufficiency of food was suggested as a cause of the decline during the 1960s. Road-killed deer were cached at feeding stations in 1971–3 to alleviate the perceived shortage of food (Wilbur *et al.* 1974). That program was run for an insufficient time to determine whether supplementary feeding was associated with increased productivity.

The connection between toxic organochlorines and eggshell thinning in birds was established in the late 1960s, but the resulting flurry of studies focused on bird-eating and fish-eating birds because avian scavengers were assumed to be less at risk. The possibility of a causal association between environmental toxins and the later decline of the condor was recognized in the mid-1970s, but determination of the specific role of organochlorines in that decline was delayed (Kiff 1989). Eggshell samples from California condors had been collected in the late 1960s but for various reasons, including mishaps to the samples, analyses were delayed until the mid-1970s. The negative correlation between eggshell thickness and dichlorodiphenyldichoroethane (p,p'-DDD) levels was significant: the shells were thinner and their structure different from shells collected before 1944.

It was known that condor eggs often broke but the cause was open to debate. Even the monitoring activities themselves were suspected as being the cause. The evidence for organochlorines was circumstantial but it led Kiff (1989) to conclude that "p,p'-DDD contamination probably had a very serious impact on the breeding success of the remnant population in the 1960s, leading to a subsequent decline in the number of individuals added to the pool of breeding adults in the 1970s." In 1972 dichlorodiphenyltrichloroethane (DDT) was banned in the USA. The few eggs measured after 1975 had thicker eggshells and this led to guarded optimism. In March 1986, however, an egg laid by the last female to attempt breeding in the wild was found broken. Its thin shell was suspiciously reminiscent of the "p,p'-DDD thin-eggshell syndrome." In the meantime, analysis of tissue from wild condors found dead in the early 1980s revealed that three of the five had died from lead poisoning, probably from ingesting bullet fragments in carrion. Other condors had elevated lead levels in their blood (Wiemeyer et al. 1988). Recognition of the deleterious effects of yet another toxin in the condor's food supply led to provision of "clean" carcasses just before the last condors were taken into captivity.

15.2.8 The effect of introduced diseases

Extinctions caused by disease are particularly difficult to identify in retrospect. Moreover, on theoretical grounds disease is unlikely to be a common agent of extinction. In their review of pathogens and parasites as invaders, Dobson and May (1986b) noted the improbability of a parasite or pathogen driving its host to extinction unless it had access to alternative hosts.

Hawaiian birds

Avian malaria and avian pox have been suggested as contributing to the decline of the Hawaiian birds (Warner 1968). Migratory waterfowl may have provided a reservoir for avian malaria on the Hawaiian islands, and the continuous reintroduction by migration may have maintained a high level of infection in the face of a decline in host numbers (Dobson and May 1986b). Alternatively, avian malaria may have been carried by introduced birds such as the common myna (Acridotheres tristis) which may themselves have maintained the disease at a high level because they are not greatly affected by it.

Originally there were no mosquitos on Hawaii capable of spreading malaria. The accidental introduction of mosquitos in 1826 and their rapid spread throughout the islands coincided with the decline of many species of birds. Six of 11 endemic passerines died out by 1901 on Oahu before their habitats had been modified (Warner 1968). Experiments showed that the Hawaiian passerines, especially the honeycreepers, are highly susceptible to malaria, much more so than are the introduced species (Warner 1968).

Avian malaria is a factor in restricting the present distribution of native birds on Hawaii, lending credence to the suggestion that it is implicated in the extinction of other species. Scott et al. (1986) noted that elevations above 1500 m that were free of mosquitos hosted the highest densities of natives birds, especially of the rarer passerines.

<table>
<tr><td>15.2.9 The effect of multiple causes</td><td>

The heath hen

The history of the heath hen is related by Bent (1932). We use here the summary and interpretation of that history presented by Simberloff (1988).

</td></tr>
</table>

> Probably the best-studied extinction is that of the heath hen (*Tympanuchus cupido cupido*). This bird was originally common in sandy scrub–oak plains throughout much of the northeastern United States, but hunting and habitat destruction had eliminated it everywhere but Martha's Vineyard by 1870. By 1908 there were 50 individuals, for whom a 1600 acre refuge was established. Habitat was improved and by 1915 the population was estimated to be 2000. However, a gale-driven fire in 1916 killed many birds and destroyed habitat. The next winter was unusually harsh and was punctuated by a flight of goshawks; the population fell to 150, mostly males. In addition to the sex ratio imbalance, there was soon evidence of inbreeding depression: declining sexual vigor. In 1920 a disease of poultry killed many birds. By 1927 there were 13 heath hens (11 males); the last one died in 1932. It is apparent that, even though hunting and habitat destruction were minimized, certainly by 1908 and perhaps even earlier, the species was doomed. Catastrophes, inbreeding depression and/or social dysfunction, demographic stochasticity, and environmental stochasticity all played a role in the final demise.

15.2.10 *How to prevent extinction*

The previous sections summarized 12 examples of extinction or steep decline. The decline of the heath hen and the Hawaiian birds may be attributable to several factors but research on those species has not adequately revealed the causes of the declines. The extinction of several species of wallaby seems very likely to reflect habitat modification. The extinction of the Stephen Island wren and the decline of the Lord Howe woodhen were unambiguously caused by an introduced predator. The extinction of the Arabian oryx in the wild (although subsequently reestablished from captive stock) and the near extinction of the muskoxen were also caused by predation, this time by humans. The black-footed ferret went extinct in the wild because the source of its habitat and most of its food supply – the prairie dog – was greatly reduced in density by control operations.

These examples implicate only a few potential causes of decline. Probably the most important is modification or destruction of habitat. The local extinction of several mammals from the sheep rangelands of Australia appears to have been caused by habitat changes caused by the sheep introduced in the mid-1800s. Twelve of the original 38 species of marsupials and six of the original 45 species of eutherians (endemic rodents and bats) no longer live in that region (Robertson *et al.* 1987).

The first step in averting extinction is to recognize the problem. Many species have slid unnoticed to the brink of extinction before their virtual absence was noticed. The smaller mammals and birds, and the frogs and reptiles, are more likely to be overlooked than are the large ungulates and carnivores.

The second step is to discover how the population got into its present mess.

- Is the cause of decline a single factor or a combination of factors?
- Are those factors still operating?
- If so can they be nullified?

The cause of a decline is established by application of the researcher's tools of trade: the listing of possible causes and then the sequential elimination of those individually or in groups according to whether their predicted effects are observed in fact. This is the standard toolkit of hypothesis production and testing.

It is essential that the logic of the exercise is mapped out before the task is begun. The listing of potential causes is followed by a formulation of predictions and then a test of those predictions. The efficiency of the exercise is critically dependent on the order in which the hypotheses are tested. Get that wrong and a 3-month job may become a 3-year project. In the meantime the population may have slid closer to the threshold of extinction.

Box 15.1 gives a specimen protocol for determining the cause of a population's decline. The example comes from the decline of caribou on Banks Island in the Canadian Arctic. The first aerial surveys of the island in 1972 revealed an estimated population of 11 000 caribou. Subsequent surveys in the 1980s traced a dwindling population which numbered barely 900 caribou by 1991. The muskoxen during the same time increased from 3000 to 46 000, leading to fears that there are too many muskoxen for the good of the caribou (Gunn *et al.* 1991). Particularly severe winters restricted foraging for the caribou and caused dieoffs, at least in 1972–3 and 1976–7. The frequency of severe winters with deep snow and freezing rain increased during the 1970s and 1980s. Caribou and muskoxen differ in life styles and responses to winter weather.

An example of how difficult it can be to get the logic of diagnosis right is provided by research and treatment of the endangered Puerto Rican parrot (*Amazona vittata*). This strikingly attired bird has been the focus of some 40 years of intensive conservation efforts, including some 20 000 hours of observations of ecology and behavior (Snyder *et al.* 1987). The parrot may have numbered more than 1 million historically but by the early nineteenth century its distribution had contracted severely, with the clearing of much of the forest of Puerto Rica. By the 1930s it was estimated as 2000 and by the mid-1950s, when the first intensive studies started, its numbers had collapsed to 200. Only

Box 15.1 Hypotheses to be tested to discover the cause of the decline of caribou on Banks Island, Northwest Territories of Canada

Hypotheses to account for the decline:
Either:
- A Food shortage
- B Increased predation

If (A) then mechanisms may be:
- A1 Increase in weather events such as freezing rain that affect availability of food
- A2 Competition for food with muskoxen which are increasing
- A3 Caribou themselves reducing the supply of food

If (B) then mechanisms may be:
- B1 Wolf predation
- B2 Human predation

The food shortage hypotheses (A) may be tested against the predation hypotheses (B) by checking body condition. Hypotheses A predict poor body condition and low fecundity during a population decline; hypotheses B predict good condition and high fecundity during a decline.

If this test identifies the A hypotheses as the more likely, then A1 is separated from A2 and A3 by its predicting a positive rate of increase in some years. A2 and A3 predict negative rates of increase in all years.

A2 (competition with another species) is separated from A3 (competition between caribou) by checking for concomitant decline of caribou where muskoxen are not present in the same climatic zone.

24 parrots were left in 1968 when rescue efforts were resumed. Despite a high profile effort, including a captive breeding program started in 1968, little progress can be reported. The number of parrots in the wild population numbered only 21–23 before the 1992 breeding season (Collar *et al.* 1992). The cause of the decline has not yet been identified unambiguously.

15.3 Rescue and recovery of near-extinctions

Once a decline in a species is recognized and the causes are determined, the problem can be treated. The species accounts in the preceding sections give some idea of the range of management actions available to rescue a species from the risk of extinction. Sometimes, all it takes is a legislative change such as a ban on hunting (as with the Canadian muskoxen). More usually, active management (e.g., predator control and captive breeding for the Lord Howe Island woodhen) is necessary. The management actions needed to reverse the fortunes of a declining species are seldom more than conventional management techniques unless a species is in desperate straits. Then a whole new set of techniques may be called under the heading of *ex situ*. *Ex-situ* techniques preserve and amplify a population of an endangered species outside its natural habitat. Thereafter it can be reintroduced. The Lord Howe Island woodhen and the Arabian oryx are examples of such reintroductions.

Reintroducing a species to the area from which it has died out should not be attempted without some understanding of why the species went extinct there in the first place. Stanley Price (1989) describes the reintroduction of the Arabian oryx with captive stock and details the considerations that should precede a reintroduction. In any event the liberated nucleus should be large enough to avoid demographic malfunction. Twelve individuals are an absolute minimum for an introduction. Twenty are relatively safe.

When the cause of a local extinction is unknown, and when we therefore do not know whether the factor causing the extinction is still operating, a probe liberation should precede any serious attempt to repopulate the area. The 20 or more individuals forming the probe are instrumented where possible (e.g., with radiocollars) and monitored carefully to determine whether they survive and multiply or, if not, the cause of their decline. If the latter, the factor operating against the species can be identified and countermeasures can then be formulated. It is worth noting that a closely related species may be used as a probe when it is too risky to use individuals of the endangered species. For example, a successful probe release of Andean condors (*Vultur gryphus*) cleared the way for the release of two Californian condors from captive breeding population in 1992 (Collar *et al.* 1992).

Short *et al.* (1992) showed the importance of probing for reintroductions of several wallaby species. Of 10 liberations into areas where the species had once been present but had died out, all failed. Of 16 liberations into areas where the species had not previously occurred, about half were successful. Apparently the factors that had caused the original extinctions of the first category were still operating. The authors suggested that exotic predators were probably the dominant factor causing the original extinctions and militating against successful reintroduction.

15.4 Conservation in national parks and reserves

National parks and reserves are preeminently important as instruments of conservation. In these areas alone the conservation of species supposedly takes precedence over all other uses of the land.

15.4.1 What are national parks and reserves for?

On one level that question is trite and leads to the equally trite answer that parks and reserves are to conserve nature. When the question is refined to "what are the precise objectives of *this* park," the answer must be more concrete. But even the general question is not as trite as it might seem. It is instructive to follow the history of ideas about the function of reserves, of which national parks can serve as the type example. Here we summarize those changing perceptions as outlined by Shepherd and Caughley (1987).

The national park idea has two quite separate philosophical springs whose streams did not converge until about 1950. The first is American, exemplified by the US Act of 1872 proclaiming Yellowstone as the world's first national park. The intent was to preserve scenery rather than animals or plants. Public hunting and fishing were at first entirely acceptable.

The second spring is "British colonial," with the Crown asserting ownership over game animals and setting aside large tracts of land for their preservation. The great national parks of Africa grew out of these game reserves, some physically and the others philosophically. Wildlife was the primary concern and scenery came second if at all. The first was Kruger National Park established in 1926 on a game reserve proclaimed in 1898. Kenya's first was established in 1946 on the Nairobi common.

All national parks established for 40 years or more have had their objectives and their management modified several times. The more influential fashions in park theory, listed here roughly in order of appearance over the last 100 years, are not mutually exclusive. They tend to be added to rather than replacing the previous ones.

1 The most important objective is to conserve scenery and "nice" animals. The aim translated into restricting roads and railways and attempting to exterminate the carnivores.

2 The most important objective is the conservation of soil and plants. This aim was a direct consequence of the rise of the discipline of range management in the USA during the 1930s. Its axiom was (and still is) that there is a "proper" plant composition and density. Enough herbivores were to be shot each year to hold the pressure of grazing and browsing at the "correct" level. An ecosystem could not manage itself. If left to its own devices it would do the wrong thing.

3 The most important objective is the conservation of the physical and biological state of the park at some arbitrary date. In the USA, South Africa, and Australia that date marked the arrival of the first European to stand on the land.

4 The fashion shifted to the conservation of representative examples of plant and animal associations. The wording is from Bell's (1981) definition of the function of national parks in Malawi, but the objective underlies the management of many national parks in many other countries.

5 The most important objective is the conservation of "biological diversity" (or biodiversity). This catch phrase had two meanings. It was sometimes used in the sense of "species diversity" (MacArthur 1957; 1960) whereby the information-theory statistic of Shannon and Wiener could be used to estimate the probability that the next animal you saw would differ at the species level from the last. The statistic is maximized for a given number of species when all have the same density. Within park management the idea translated as "the more species the better." The second meaning dealt with associations rather than species: the more diverse a set of plant associations the better the national park. For example, Porter (1977) defined the objectives of the Hluhluwe Game Reserve in Natal as "To maintain, modify and/or improve (where necessary) the habitat diversity presently found in the

area and thus ensure the perpetuation and natural existence of all species of fauna and flora indigenous to the proclaimed area."

6 The most important objective is the conservation of "genetic variability." The phrase can be defined tightly and usefully (e.g., Frankel and Soulé 1981), but within the theory and practice of park management it lacked focus. It was tossed around with little or no attempt to define or understand what it means, whether the variability sought was in heterozygosity, in allelic frequency, or in phenotypic polymorphism. In practice it again translated into "the more species the better."

7 The most recent objective differs in kind from the six above. Frankel and Soulé (1981) express it thus: "the purpose of a nature reserve [in which category they include national parks] is to maintain, hopefully in perpetuity, a highly complex set of ecological, genetic, behavioral, evolutionary and physical processes and the coevolved, compatible populations which participate in these processes." Don Despain (quoted by Schullery 1984) puts it more plainly: "The resource is wildness."

15.4.2 *Processes or states?*

The first six objectives listed above identify biological states as the things to be conserved. The seventh identifies biological processes as the appropriate target of conservation. At first glance Frankel's and Soulé's (1981) purpose of a nature reserve appears also to require the maintenance of states because it refers to the conservation of populations. But populations are not states in the sense that plant associations are states. A plant association has a species composition. Its component populations must have a ratio of densities one to the other that remains within defined limits. If those limits are breached the plant association has changed into another kind of plant association. A population, however, is not defined by ratios. The ratio of numbers in one age class relative to those in another, or the ratio of males to females, has no bearing on its status as a population.

The management of a national park will be determined by whether the aim is to conserve biological and physical states by suppressing processes or whether it is to preserve processes without worrying too much about the resultant states. There are three options.

1 If the aim is to conserve specified animal and plant associations that may be modified or eliminated by wildfire, grazing, or predation, then intervene to reduce the intensity of wildfire, grazing, or predation.

2 If the aim is to give full rein to the processes of the system and to accept the resultant, often transient, states that those processes produce, then do not intervene.

3 A combination of both: if the aim is to allow the processes of the system to proceed unhindered unless they produce "unacceptable" states, then intervene only when unacceptable outcomes appear likely.

15.4.3 *Effects of area*

Within any group of islands (e.g., the Antilles, Indonesia, Micronesia) big islands tend to contain more species than do small islands. Size as such is not the only influence on the number of species – e.g., distance to the mainland plays a part – but area alone provides a close prediction. The relationship between the number of species and the size of the area within which they were surveyed is known as a species–area curve.

Algebraically it takes the form:

$$S = CA^z$$

in which S is the number of species of a given taxon (e.g., lizards, forest birds, vascular plants), A is the area, C is the expected number of species on one unit of area (usually $1\,km^2$), and z indexes the slope of the curve relating the number of species to the number of square kilometers.

Table 15.1 shows the relationship between species number and land area for Tasmania and the islands between it and the Australian mainland (Hope 1972). These were all linked to each other and to the Australian mainland up to about 10 000 years ago, the subsequent fragmentation reflecting rise of sea level at the end of the Pleistocene. The number of marsupial herbivores that they carry therefore reflects differential extinction without reciprocal immigration over the last 10 millennia. The estimated $z = 0.18$ is low for islands, being closer to that expected for areas within continents, and it probably reflects the recent continental nature of those islands. Box 15.2 shows how C and z are calculated from these data.

C, the expected number of species on one unit of area, varies according to latitude, elevation, ecological zone, taxonomic group, and the units in which A is measured. In contrast z tends to be quite

Table 15.1 Relationship between the number of species of herbivorous marsupials and area of land on Tasmania and the islands between it and the Australian mainland.* The "expected" number is calculated as $S = 1.70A^{0.18}$ (see Box 15.2 for calculation). (After Hope 1972)

Island	Area (km²)	Observed species	Expected species
Tasmania	67 900	10	12.6
Flinders	1 330	7	6.3
King	1 100	6	6.0
Cape Barren	445	6	5.1
Clarke	115	4	4.0
Deal	20	5	2.9
Badger	10	2	2.6
Prime Seal	9	2	2.5
Erith-Dove	8	3	2.5
Vansittart	8	2	2.5
West Sister	6	2	2.3

* Number of species as at AD 1800. Only islands > 5 km² are included.

Box 15.2 Estimating the constants of a species–area curve

A species–area curve takes the form $S = CA^z$ where;

  S = number of species
  A = area, in this case always expressed as km^2
  C = expected number of species on an area of 1 km^2
  z = slope of the curve relating species number to area

Taking the data of Table 15.1 as our example, first convert area and species number to logarithms. Any base will do but we will use logs to the base e. Label log area as x and log species number as y:

x	y
11.126	2.3026
7.193	1.9459
7.003	1.7918
6.098	1.7918
4.745	1.3863
2.996	1.6094
2.303	0.6931
2.197	0.6931
2.079	1.0986
2.079	0.6931
1.792	0.6931

We now calculate these:

$$n = 11$$

mean x = 4.510	mean y = 1.336
Σx = 49.61	Σy = 14.70
Σx^2 = 315.2	Σxy = 82.58
$(\Sigma x)^2/n$ = 223.7	$(\Sigma x)(\Sigma y)/n$ = 66.30

$$SS_x = \Sigma x^2 - (\Sigma x)^2/n = 91.5 \qquad SS_{xy} = \Sigma xy - (\Sigma x)(\Sigma x)/n = 16.28$$

The constants of the species–area curve are now solved:

$$z = SS_{xy}/SS_x$$
$$= 16.28/91.5$$
$$= 0.18$$
$$C = \text{antilog}(\Sigma y/n - z\,\Sigma x/n)$$
$$= \exp(1.336 - 0.18 \times 4.51)$$
$$= 1.70$$

Thus $S = 1.7A^{0.18}$

stable. For most taxa and groups of islands it lies between 0.2 and 0.4. At the midpoint, 0.3, an increase or decrease of area by a factor of 10 results in a doubling or halving respectively of the number of species (by virtue of $10^{0.3} = 2$). Thus, when $A = 1$, $S = C$ irrespective of the value of z; and when $A = 10$ and $z = 0.3$, $S = 2C$.

The relationship is the same if we count the number of species on nested areas of progressively larger size on a continent. Here the value of z tends to be lower, usually around 0.15. It implies that a reduction of area by a factor of 10 reduces the number of species by a factor of only 1.4 ($10^{0.15} = 1.41$). The difference between that exponent of 0.3 for islands and 0.15 for continents probably reflects the easier dispersal between contiguous areas of land against between islands. These relationships are particularly important for determining optimum sizes of reserves.

15.4.4 *Is one big national park better than two small national parks?*

Suppose we have the money necessary to acquire 100 km² of land for conversion into national parks. If the aim were to conserve the maximum number of species for a long time, should we go for one park of 100 km² or two each of 50 km²? Obviously a number of factors would influence our choice, but let us assume that the overriding aim is to maximize the number of species of mammals within the single large reserve or the alternative two smaller reserves. Let us assume that 1 km² will on average contain 20 species in this region (i.e., $C = 20$). Furthermore, we know that $z = 0.15$ for mammals in this region. Thus, a national park of 100 km² would contain about 40 species of mammals ($S = CA^z = 20 \times 100^{0.15} = 40$) whereas a park of 50 km² would hold about 36 mammals ($S = 20 \times 50^{0.15} = 36$). Whether we favor one park of 100 km² or two of 50 km² each comes down to how many species are held in common by the two smaller parks. That will depend on the extent to which they differ in habitat and on the distance between them.

The efficacy with which a reserve system conserves species and communities thus depends on the size of the reserves and, more importantly, on where they are – their dispersion relative to the distribution patterns of species. Margules *et al.* (1982) warn against using data-free geometric design strategies (big is better than small, three is better than two, linked is better than unlinked, grouped is better than linear).

15.4.5 *Effects of corridors*

Corridors between reserves provide the benefit of increasing the size of populations and thereby decreasing the chance of demographic malfunction. But the overall benefit of corridors is not at all clearcut and must be decided upon case by case.

Simberloff and Cox (1987) used the Seychelles islands of the Indian Ocean to make the point that corridors are not always beneficial. The Seychelles contained 14 endemic land birds when Europeans arrived in 1770. Land clearing, fires, and the introduction of rats and cats, devastated the archipelago over the subsequent two centuries but resulted in the extinction of only two of those species. Losses were limited partly because no corridors (isthmuses) linked the islands. Introduced predators and fires were unable to reach all the islands.

The potential advantages and disadvantages of conservation corridors as summarized by Noss (1987) are presented in Box 15.3. Saunders and Hobbs (1991) edited a symposium on this subject which contains several useful papers.

Box 15.3 Potential advantages and disadvantages of conservation corridors. (After Noss 1987)

Potential advantages of corridors
1 Increased immigration rate to a reserve, which could:
    (a) increase or maintain species richness (as predicted by island biogeography theory);
    (b) increase population sizes of particular species and decrease probability of extinction (provide a "rescue effect") or permit reestablishment of extinct local populations;
    (c) prevent inbreeding depression and maintain genetic variation within populations.
2 Provide increased foraging area for wide-ranging species.
3 Provide predator-escape cover for movements between patches.
4 Provide a mix of habitats and successional stages accessible to species that require a variety of habitats for different activities or stages of their life cycles.
5 Provide alternative refugia from large disturbances (a "fire escape").
6 Provide "greenbelts" to limit urban sprawl, abate pollution, provide recreational opportunities, and enhance scenery and land values.

Potential disadvantages of corridors
1 Increased immigration rate to a reserve, which could:
    (a) facilitate the spread of endemic diseases, insect pests, exotic species, weeds, and other undesirable species into reserves and across the landscape;
    (b) decrease the level of genetic variation among populations or subpopulations, or disrupt local adaptations and coadapted gene complexes ("outbreeding depression").
2 Facilitate spread of fire and other abiotic disturbances ("contagious catastrophes").
3 Increase exposure of wildlife to hunters, poachers, and other predators.
4 Riparian strips, often recommended as corridor sites, might not enhance dispersal or survival of upland species.
5 Cost, and conflict with conventional land preservation strategy to preserve endangered species habitat (when inherent quality of corridor habitat is low).

15.4.6 *Effects of initial conditions*

Parks are chosen for a number of reasons: great scenery, many species, a cherished plant association, or a set of interesting landforms. Sometimes the area chooses itself, being deemed good for little else.

Most national parks established within the past 30 years (the majority) have been chosen with some care. They are designed to conserve the plant and animal communities and/or their associated ecological processes in a particular climatic zone. Having decided upon the zone the next step is to choose an area within that zone which samples or epitomizes that zone. The decision is determined first by what land is available for conversion to a park. It is then determined by whether a piece of available land is large enough, or can be made large enough by accretion of adjacent areas, to serve as a national park. Finally a choice is made between the various areas of land that meet the above criteria.

At this stage the choice of land is determined mainly by which area contains the greatest internal diversity of habitats and also by which contains the largest number of species. The two tend to be correlated. These two criteria of choice have an effect upon extinction rates within the reserve. They ensure that the park will have an over-diversity of species and habitats. If an area contains a diversity of habitats it will on average contain little of each. A species dependent on a single habitat will therefore be represented on average by a small population within such a reserve.

If the area contains a diversity of species, several of those species will be near the edge of their range and so will be living outside their environmental optima. Such species will be at low density within the reserve. If the reserve becomes an ecological island the number of species it contains will be much higher than that predicted by the species–area curve and it can therefore be expected to lose species. Those are effects of sampling a region by an area containing most of the characteristics of the whole region.

Another consequence of choosing a diversity of habitats is that the main habitat of interest (e.g., rainforest, savanna, taiga) tends to be sampled near the edge of its distribution to allow other habitats to interdigitate with it. If the climate changes then the habitat of interest is likely to be lost.

To summarize, to select an area suitable for conservation, one should:

1   choose an area containing a moderate rather than a high number of species;

2   include only a moderate number of habitats within the same reserve;

3   position a reserve as close as possible to the center of distribution of the habitat of greatest interest.

15.4.7 Culling in parks and reserves

Whether or not the densities of mammals should be controlled artificially in a national park is a matter of some contention, as illustrated by the proceedings of two international conferences (Jewell *et al.* 1981; Owen-Smith 1983). For a specifically North American example we recommend the debate on management in Yellowstone National Park where the case for culling is given by Chase (1987) and that against by Houston (1982). Our own prejudices are to avoid culling in parks and reserves except in rare, special, and well-defined circumstances.

15.5 Conservation outside national parks and reserves

The principles of conservation discussed above with reference to parks and reserves hold also for conservation outside those reserves. There are, however, a few important differences.

Some species or associations of species occur only rarely in reserves because parks and reserves do not capture a representative sample of the biota. For example, in Australia few reserves contain forest types that grow on sites of high fertility. Most such sites were incorporated into state forests or alienated from common ownership before the reserve system was established. The koala (*Phascolarctos cinereus*) is dependent on such sites and so almost all attempts to conserve koalas must be made outside the reserve network where the manager does not have the same control over land-use practices.

Legislation is the main means by which conservation is advanced outside reserves. Various practices, such as the killing of nominated species, are banned. Less commonly there are controls over land clearing, thereby protecting the habitat of species that dwell in forest and woodland. Activities on land owned by the people as a whole, even though that land is not designated as a conservation reserve, may be subject to environmental impact assessment (EIA). Laws governing conservation outside reserves should take legal precedence over forestry and mining law.

15.6 **International conservation**

Conservation is the responsibility of sovereign nations unless the issue is subject to international treaty (polar bears, ivory trade, migratory birds) or unless the problem occurs on the high seas (whales and pelagic fish stocks), on essentially unclaimed land (Antarctica) or on land under disputed sovereignty (parts of the high Arctic).

15.6.1 *IUCN red data books*

The International Union of Nature and Natural Resources (IUCN) issues "red data books" listing threatened species. Four categories are recognized, their exact wording varying according to the taxon. What follows is generalized.

Extinct (Ex)
Species not definitely located in the wild during the past 50 years.

Endangered (E)
Taxa in danger of extinction and whose survival is unlikely if the causal factors continue operating.

Included are taxa whose numbers have been reduced to a critical level or whose habitats have been so drastically reduced that they are deemed to be in immediate danger of extinction. Also included are taxa that are possibly already extinct but have definitely been seen in the wild in the past 50 years.

Vulnerable (V)
Taxa believed likely to move into the "endangered" category in the near future if the causal factors continue to operate.

Included are taxa of which most or all populations are decreasing because of overexploitation, extensive destruction of habitat, or other environmental disturbance; taxa with populations that have been seriously depleted and whose ultimate security has not yet been assured; and taxa with populations which are still abundant but under threat from severe adverse factors throughout their range.

Rare (R)

Taxa with small world populations that are not at present "endangered" or "vulnerable," but are at risk.

These taxa are usually localized within restricted geographic areas or habitats or are thinly scattered over a more extensive range.

Indeterminate (I)

Taxa known to be "endangered," "vulnerable," or "rare" but where there is not enough information to say which of the three categories is appropriate.

Of these categories the "endangered" and "vulnerable" are the most important and there is widespread agreement on what the terms mean. "Rare" is not a particularly useful category of extinction risk and probably should not be used as such. If rarity itself is the cause of the risk, in the sense that the population size is at a level low enough to place it in danger of demographic or genetic malfunction, then it should be placed in one of the categories of threat.

The information from which the red data books are produced is extracted largely by the Species Survival Commission (SSC) of the IUCN, which is a network of the world's most qualified specialists in species conservation that serve on a voluntary basis. The various groups and their membership are listed in the SSC Membership Directory published in 1990 by the IUCN.

15.6.2 *The role of CITES*

CITES is the acronym for Convention on International Trade in Endangered Species of Wild Fauna and Flora. The convention regulates trade in species of wildlife that are perceived to be at risk from commercial exploitation. There are 99 countries that are party to the convention.

The teeth of the convention are contained in its appendices listing the species covered by CITES. Article II of the convention decrees that:

1 Appendix I shall include all species threatened with extinction which are or may be affected by trade. Trade in specimens of these species must be subject to particularly strict regulation in order not to endanger further their survival and must only be authorized in exceptional circumstances.

2 Appendix II shall include:

(a) all species which although not necessarily now threatened with extinction may become so unless trade in specimens of such species is subject to strict regulation in order to avoid utilization incompatible with their survival; and

(b) other species which must be subject to regulation in order that trade in specimens of certain species referred to in the above sub-paragraph may be brought under effective control.

Table 15.2 Number of species covered by Appendix I and II of the Convention on International Trade in Endangered Species of Wild Fauna and Flora (CITES). (After Woodruff 1989)

	Appendix I (endangered)	Appendix II (threatened)
Mammals	179	303
Birds	133	618
Reptiles	52	340
Amphibians	4	7
Fish	7	15
Insects	0	49
Molluscs	26	5
Plants	87	Approximately 27 000

3   Appendix III shall include all species which any Party identifies as being subject to regulation within its jurisdiction for the purposes of preventing or restricting exploitation, and as needing the cooperation of other parties in the control of trade.
4   The Parties shall not allow trade in specimens of species included in Appendices I, II, and III except in accordance with the provisions of the present Convention.

Table 15.2 indicates the number of species covered by Appendices I and II of CITES.

15.7 Summary

Extinctions can be driven by a permanent change to a species' environment (e.g., a new predator, disease or competitor, or modification of its habitat) or can result from stochastic events. Driven extinctions are the most common. Stochastic extinctions are the chance fate of small populations: factors that would be swamped in a large population can have serious consequences for the individuals of a small population. The critical step in averting extinction is to follow the logical pathway of hypothesis testing to diagnose the cause of the decline. The species can seldom be rescued until the factors driving the decline have been identified and removed. Rescue and recovery operations are standard wildlife management practices (e.g., regulation of harvest, predator control) but sometimes more elaborate steps such as captive breeding and translocations are called for. Reserves and national parks play a key role in the nurturing and recovery of endangered species.

16 Wildlife harvesting

16.1 Introduction

In this chapter we consider how to estimate an appropriate offtake for a wildlife population. It differs according to whether the population is increasing, whether it is stable, and whether or not the environment fluctuates from year-to-year.

16.2 Kinds of harvesting

Wildlife is harvested for many different purposes. Sports hunting usually takes a sample of the population during a restricted season and often with a restriction placed on the sex and age of the harvest. Harvesting for sport is a complex activity whose product is as much a quality of experience as it is meat or trophies. On the other hand the purpose of commercial hunting or pot hunting is simply to harvest a product such as meat and skins.

Both recreational and commercial wildlife harvesting are controversial but it is not the purpose of this chapter to delve into that controversy. Whether or not one considers it is appropriate and ethical to harvest a population of a given species depends more on one's view of life than on what may be happening on the ground. But there is an ethical aspect that is fundamental to wildlife harvesting: the operation, be it for recreation or profit, must result in a sustainable offtake, a yield that can be taken year after year without jeopardizing future yields.

16.3 What harvesting does to a population

The details of sustained yield harvesting differ according to whether a population's key resources are renewable or not, how the population uses those resources, and various interactions between the resources and the population. However, a few generalizations are possible.

16.3.1 A population is harvested at its rate of increase

In all but special circumstances the strategy of harvesting is simple; it is to harvest the population at the same rate as it seeks to increase. Hence, a population increasing at 20% per year can be harvested at around 20% per year. That proportion of the population can be taken each year, and year after year, with the result that the population is held to an induced rate of increase of zero. The use of "rate of increase" as the appropriate harvesting rate is uninfluenced by whether the population is actively spreading, whether it is subject to predation or not, and whether sources of mortality are additive or compensatory.

Rate of increase must not be confused with various ratios calculated from the age distribution of the population. "Recruitment rate," for example, calculated as the proportion of first-year or second-year animals in the population, has no necessary correspondency with rate of increase.

Figure 16.1 illustrates harvesting at a rate identical to the rate at which the population would increase if not harvested. The population is deemed to be increasing at its intrinsic rate which is $r_m = 0.2$, represented by the horizontal dashed line. Thus, irrespective of a population's size, it may be harvested at that rate to hold the population stable. The sloping line represents the sustained yield (population size multiplied by harvesting rate) that accrues from the harvesting. It is a simple multiple of the population's size. The more the population is allowed to increase before harvesting the larger will be the sustained yield.

Surprisingly, this simple model of harvesting is sometimes applicable in reality. A species introduced into a region from which it had previously been extinguished, or where it had never previously occurred, may increase at its intrinsic rate for some time after liberation. It can be harvested at that rate. Muskoxen introduced to the west coast of Greenland, and reintroduced to the high Arctic of Canada and Alaska, fit this model well.

16.3.2 *A population must be stimulated to produce a yield*

Most unharvested populations are not increasing and so the strategy outlined above cannot be used. Usually the population has a rate of increase which, when averaged over several years, is close to zero.

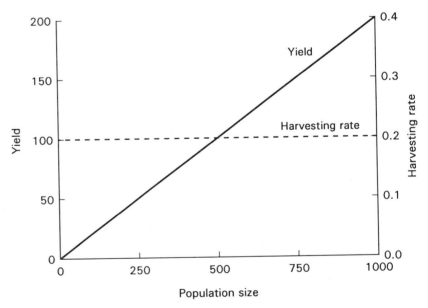

Fig. 16.1 Sustained rate of harvesting and sustained yield in relation to population size when no resource is limiting. In the absence of harvesting the population would increase at its intrinsic rate of 0.2/year.

Hence, the sustained yield for such a population is also zero. Before such a population can be harvested for a sustained yield it must be stimulated to increase.

A population can most easily be stimulated into a burst of growth by increasing the level of a limiting resource or, much more rarely, reducing the level of predation to which it is subject. The key resource may be nest sites or cover, but in most cases it is food. For example, red kangaroos increase if the pasture biomass is > 200 kg/ha dry weight and decline if it is below that level (Bayliss 1987). The easiest way to increase the amount of food available to an individual animal is to reduce the number of animals competing with it for that food. The standing crop of food then rises and the amount available to an individual animal is thereby increased. As a direct consequence the fecundity of individuals is enhanced and mortality, particularly juvenile mortality, is reduced. The population enters a regime of increase as it climbs back towards its unharvested density.

16.3.3 *Harvesting trades off yield against density*

The trade-off between yield and density is the most important thing to know about sustained-yield harvesting. In general, the further the density is reduced the higher is the offtake as a percentage of population size. (We use yield and offtake interchangably.) The maximum rate of sustainable offtake is the population's intrinsic rate of increase, but that rate of population growth is obtained only when food (or whatever other resource is limiting) is at maximum, which in turn usually occurs only when the population is at minimum density. It is important to realize that density itself is having no causal affect upon sustained yield. It is the food supply, influenced by the number of animals eating it, that sets the level of sustained yield.

Whereas sustained rate of offtake (yield divided by population size) tends to increase as density is reduced, the same is not true of the absolute sustained yield. If the population is reduced just a little, the induced rate of increase will be small and the sustained yield will be a small proportion of a relatively large population. The absolute yield will be modest. If the population is drastically reduced the induced rate of increase will be large and the sustained yield a large proportion of what is now a relatively small population. Again the absolute yield will be modest. The highest yield is taken from a density at which the induced rate of increase multiplied by the density is at a maximum. It tends to be at intermediate density levels. Figure 16.2 shows these relationships.

16.3.4 *The two levels of sustained yield*

For a population conforming to the relationship between sustained yield and population size or density given in Fig. 16.2, a sustained yield of a given size may be taken from either of two densities. They comprise what is known as a sustained-yield pair. The member of the pair taken from the lower density is to be avoided because its harvesting requires more effort than is required to harvest the same

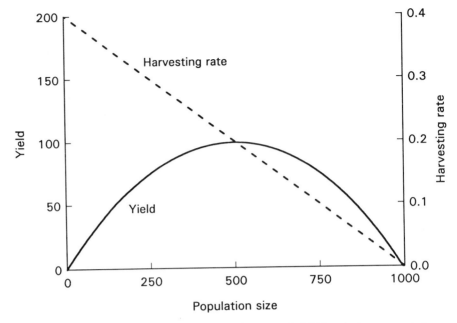

Fig. 16.2 Sustained rate of harvesting and sustained yield in relation to population size when density is limited by a renewable resource such as grass.

yield from the higher density. Figure 16.3 shows how maximum sustained yield and any pair of identical sustained yield values are related to the size of the population from which they are harvested.

When a constant number of animals is taken each year from a previously unharvested population, the population will decline and stabilize at the upper density for which that harvest is a sustained yield. Should that number exceed the maximum sustained yield, the population will decline to extinction (Fig. 16.4).

If a constant proportion of animals is taken each year from a previously unharvested population, the population will decline and stabilize, depending on the rate of harvesting, at any level between unharvested density and the threshold of extinction (Fig. 16.5). Harvesting a proportion of the population can, in contrast to harvesting a constant number, settle the population upon density, generating either an upper or lower sustained yield.

16.4 **The maximum sustained yield**

Harvesting a population at maximum sustained yield (MSY) should never be contemplated. It imparts an instability to the population's dynamics. The MSY can be taken only from the unique MSY density. If the population density has, for reasons such as drought or crusted snow, dropped below that value then the MSY represents an overharvest and the population's density is reduced further. A further harvesting of the MSY will make the problem worse.

Harvesting of the MSY from a density only slightly below that for

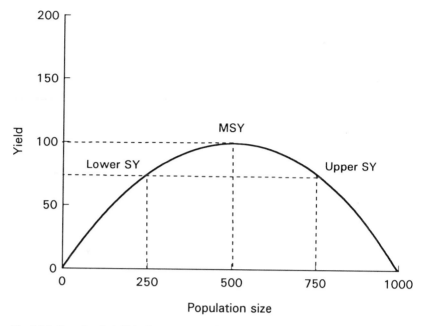

Fig. 16.3 Sustained yield (SY) against population size, showing that for any level of sustained yield other than the maximum sustained yield (MSY) there are two population sizes from which it can be harvested.

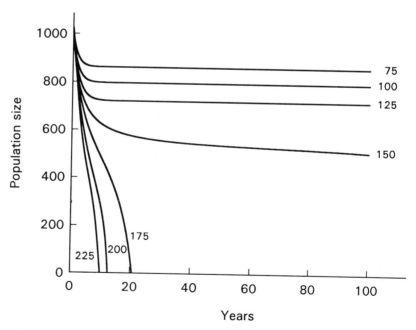

Fig. 16.4 Trend in the size of a population (logistic growth assumed) when a constant number of animals is harvested each year. The MSY is 150/year. (After Caughley 1977b.)

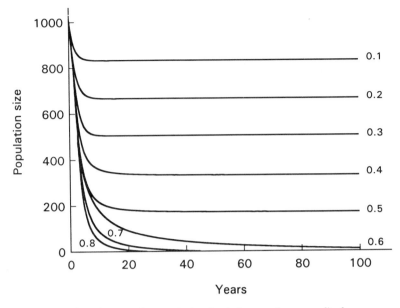

Fig. 16.5 Trend in the size of a population (logistic growth assumed) when a constant proportion of the population is harvested each year. The MSY is removed at a rate of 0.3/year. (After Caughley 1977b.)

which the MSY is appropriate triggers a positive feedback loop that drives the population into a progressively accelerating decline. For this reason a harvesting project should keep well away from the MSY. A margin of error of about 25% below the estimated maximum sustained yield is appropriate; more where year-to-year variation in weather is above average.

16.4.1 *Should we control yield or effort?*

A harvest can be controlled either by placing a quota on offtake or by controlling harvesting effort. The latter can be regulated by setting a hunting season or by limiting the number of people harvesting the population. The essence of controlling effort is that there is no direct attempt to control the number of animals harvested.

The control of harvest by quotas has an intuitive appeal because there is a direct relationship between the prescription and the result. With the harvest regulated by control of harvesting effort, an intermediate step has been inserted between prescription and outcome. Administrators tend to favor regulation by quotas because the size of the offtake is directly under their control.

In fact the disadvantages of regulating effort are more conceptual than real. It is usually a safer and more efficient means of regulating a harvest than is regulation by quota. Harvesting a constant number of animals each year is inefficient when the population is subjected to large, environmentally induced, swings in density. The quota must be set low enough to be safe at the lowest anticipated density, or alternatively the size of the population must be censused each year before the harvesting season, the quota being adjusted according to the estimate. And regulation by quota is unsafe when the quota is near the MSY. As mentioned above, the density at equilibrium with that yield is unstable such that a small environmental perturbation may trigger a density slide.

If yield is controlled indirectly by limiting harvesting effort (e.g., by limiting the number of hunters), but with no further restriction on yield per unit of effort, those dangerous sources of instability are eliminated. A fixed-effort system will, within limits, harvest the same proportion of the population at high and low density. Yield tracks density, the system automatically producing a higher yield when animals are abundant and a lower yield when they are scarce. A regulatory mechanism is built into the harvesting system itself and it is thus fairly safe so long as the appropriate harvesting effort has been calculated correctly. That is not difficult because fine tuning of the appropriate effort does not destabilize the system in the way that fine tuning a quota can. And because of the built-in regulation there is not the same need for frequent monitoring.

16.5 Offtake, resources, and population growth pattern

Offtake is set by a population's pattern of growth, which is determined by the relationship between the population and resources. The latter relationship is manipulated by changing the population's density.

In Chapter 4 we considered the pattern of population growth to be expected for four relationships between a population and its limiting resource. Each of these requires its own estimate of sustained yield.

1 If no resource is limiting, the pattern of population growth is exponential. In these circumstances the relationship between harvesting rate, yield, and population size is as given in Fig. 16.1. The sustained yield is $SY = hN$ where h is harvesting rate equal to the population's rate of increase when unharvested and N is the size of the population.

2 If the population is limited by a consumable resource whose rate of renewal is unaffected by the animals, the trajectory of growth will be a sigmoid curve similar to a logistic. Figure 16.2 shows the relationships. MSY is taken at a harvesting rate of approximately $h_{MSY} = r_m/2$ from a population size of $N_{MSY} = K/2$, K being the average size of the population when it is not harvested; hence MSY $= r_m K/4$.

3 If the population is limited by a consumable resource whose rate of renewal is influenced by the animals, the trajectory of growth will vary according to the parameter values of the two components of the system, the animals and the resource, and the interaction between them. As a general rule the N_{MSY} will be higher than $K/2$ and the MSY greater than the $r_m K/4$ of the logistic model discussed above. Thus, one can estimate the MSY by the logistic model and be confident that the calculation will supply a margin of error. Figure 16.6 shows the relationships for a population of herbivores limited by food supply.

4 If the population is limited by a nonconsumable resource such as space or nesting sites, the trajectory of growth will be a ramp in the

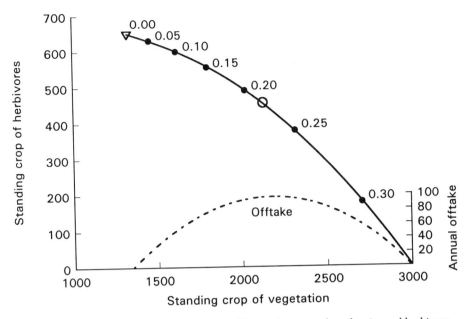

Fig. 16.6 The isocline (———) of equilibrium between plant density and herbivore density, the rate of harvesting (- - - - -) needed to enforce the equilibria, and the sustained annual yield accruing therefrom. ∇, Ecological carrying capacity; ○, economic carrying capacity. (After Caughley 1976a.)

form of an exponential curve truncated by an asymptote. Figure 16.7 shows the relationships among population size, harvesting rate, and sustained yield. The yield curve is asymmetric, the trend of SY on population size reaching a maximum when taken from a population size of $N_{MSY} = K/(r_m + 1)$. The MSY $= r_m K/(r_m + 1)$. The N_{MSY} will be below $K/2$ only when the intrinsic rate of increase is greater than $r_m = 1$ on a yearly basis (i.e., an increase by a factor of more than 2.7 over 1 year). Vertebrate populations seldom have that reproductive capacity and hence the N_{MSY} for wildlife populations will tend to be above $K/2$.

16.6 Harvesting in practice: recreational

Most harvesting of wildlife for recreational hunting has been managed largely by trial and error. This approach works well when populations and their habitats are good at looking after themselves, when intrinsic rates of increase are high, when rate of increase and density are related by tight negative feedback, and when the population size is kept on the right side of N_{MSY}. That describes what generally has happened in traditional wildlife management. The populations and their habitats have been resilient because it is only such animals that evolve into game species.

In practice one cannot have both high density and high yield, except while density is temporarily being reduced. It cannot stay there for long because it will track rapidly towards its equilibrium. Most managers seek to maximize offtake and so would like to raise yield

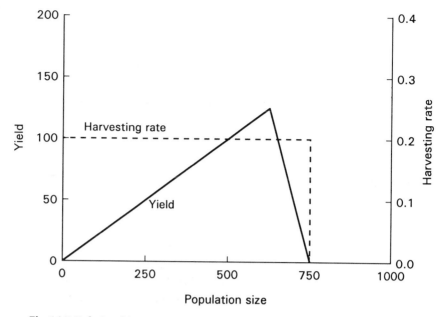

Fig. 16.7 Relationships among population size, harvesting rate, and sustained yield for a population limited by a nonrenewable resource such as nesting sites.

and lower density. This is the cause of the frequent clashes between hunters and managers over whether females and young should be harvested. In their review of harvest management for game birds, Robertson and Rosenberg (1988) explain that "In America the potential harvest is assessed on an annual basis and the activity of sportsmen controlled by bag and season limits. These restrictions rarely aim to achieve MSY, partly due to the reluctance of sportsmen and managers to reduce breeding populations, a situation often wrongly referred to as over-shooting."

The results of this misunderstanding mostly kept offtake well below the maximum sustained yield and game populations were seldom threatened by the harvesting. However, the majority of sustained yield estimates popular during the past few decades were over-estimates and would have resulted in extinction of populations had they been applied rigorously.

The trick with managing a population for sustained yield is to play it safe. You estimate the MSY on what information is available to you (usually the trend of population indices under a known constant offtake or constant effort), refine that estimate of the MSY as often as you can or at least as often as your monitoring system allows, but keep the harvest well below the MSY. Make certain that your estimate of population size remains well above your estimate of N_{MSY}. Remember that in the early stages of managing a population for sustained yield your estimate of both the current population size N and the N_{MSY} may be wildly inaccurate. Unless you have done this sort of thing experimentally several times before you may not appreciate how inaccurate your estimates are likely to be: allow yourself a wide margin of error. Remember that the standard error of an estimate tells nothing about the accuracy of that estimate. Your monitoring of population size will let you know in plenty of time when you need to ease off harvesting effort.

16.7 **Harvesting in practice: commercial**

There is no difference in principle between harvesting for commercial benefit and harvesting for recreational benefit. Both are based on the MSY concept, suitably cushioned by a margin of error. However, in practice there are a number of pitfalls to the management of sustained commercial offtake.

Game harvesting comes in two forms: game ranching (or game farming) and game cropping. The difference is in degree rather than kind, but essentially game ranching seeks to bring the animals under human control, as in the farming of domesticants, whereas game cropping is the harvesting of wild populations.

16.7.1 *Game ranching*

Game ranching is a spectrum of activities overlapping conventional animal husbandry at one end (e.g., the reindeer industry in Finland and Russia) and game cropping at the other. It is beyond the scope of this book. The interested reader might try papers in Beasom and Roberson (1985) and Hudson *et al.* (1989) for a general overview of game ranching, and the volume edited by Bothma (1989) for a thorough treatment of game ranching in southern Africa.

16.7.2 *Game cropping and the discount rate*

There are two quite distinct phases to a cropping operation: first, the population must be reduced below its unharvested density (capital reduction), and then it must be harvested at precisely the rate it seeks to bounce back (sustained-yield harvesting). Biologists tend not to think too much about the capital-reduction phase because they look forward to the prospect of a yield sustainable into the indefinite future.

If you were offered $1000 now as against $1000 in 10 years' time you would take the money now. But if you were offered $400 now as against $1000 in 10 years' time, the decision is no longer clearcut. Against money in the hand you are offered a guarantee of sure but unquantified future benefit. How much is $1000 in 10 years actually worth? A simple answer is that it is worth a present sum which, when prudently invested, yields $1000 10 years hence. Capital expands at about 10% per year, and so $1000 in 10 years is worth $385 now, or even less if the currency is inflating. Hence the answer to $400 now or $1000 in 10 years' time is simple. Take the $400 now; it is worth more.

By the same reasoning a game animal harvested in 10 years' time is worth nothing like an animal harvested now. All future earnings must be discounted by the time it takes to get the money, and the economics of the harvesting operation are thus dictated by the ratio of present to future earnings.

Biologically, the rational scheme for harvesting is to reduce the population to a little above the density allowing MSY and then to take the appropriate yield year after year. But it transpires that the "obvious" biological strategy is not necessarily that leading to maximum economic gain. Clark's (1976) book on the economics of harvesting natural resources shows unambiguously that the best biological strategy and the best economic strategy coincide only when a population's maximum rate of increase is relatively high. Rabbits and herrings go into that category. When the maximum rate of increase is somewhat lower, the real money is made more by capital reduction than by sustained yield. Discounted net revenue is maximized for the total operation when the population is taken by capital reduction to a level below that generating the MSY. A lower sustained yield in the future is thereby traded off advantageously against a higher immediate gain. When the maximum rate of increase is lower still, as with elephants and whales, it may be economically clear sighted to make a total trade-off, taking all revenue by capital reduction and sacrificing all future sustained yield. This strategy maximizes net revenue, discounted to present value, when the population's maximum rate of increase is below about 10% per year. That describes elephants, rhinos, and all species of whales. This is the economic justification for the extinction of a population (and maybe a species).

To summarize, free market trading in a privately owned renewable resource results in overharvesting, particularly (and perhaps paradoxically) when the participants in the market have perfect knowledge. That happens less often for a publicly owned resource because the public's discount rate is much lower. However, publicly owned resources take on the character of privately owned resources when the persons managing the resource and the persons harvesting the resource imagine that they, and not the persons as a whole, own the resource.

Symbiotic relationship in management

Any scheme to harvest a publicly owned renewable resource necessarily involves three parties: the owners of the resource (the persons), the harvester of the resource (usually a private company), and the manager of the resource (a government agency) that regulates the harvesting. According to constitutional theory, the persons in the agency are supposed to act for the owners of the resource, but often they become locked into a symbiotic relationship with the harvesters as if those two groups were themselves joint owners of the resource. The technical advice on sustainable yield offered by the agency's own research branch is commonly ignored by its policy and planning branch when it conflicts with the short-term requirements of the industry. Thus, the ecological aberrations that necessarily follow from the economic implications of the discount rate, and against which the persons in the managing agency are employed to guard, often dominate the harvesting operation. Forestry and fisheries provide numerous examples, and commercial wildlife harvesting is not necessarily immune.

16.8 **Summary**

The way in which a safe sustained yield is estimated depends on the population's growth pattern which, in turn, is determined by the relationship between the population and its resources. The yield can be estimated either in terms of a numerical offtake or in terms of an appropriate harvesting effort, the latter being safer. Although from an ecological viewpoint recreational harvesting and commercial harvesting do not differ in principle, in practice commercial harvesting always has the potential to exceed MSY. The "discount rate" of economic analysis encourages overharvesting by imparting a greater value to present yields than to future ones. Regulating agencies often go along with it.

17 Wildlife control

17.1 Introduction

We show that a control operation is similar to a sustained yield exercise but is conceptually more complex. The objective must be defined precisely, not in terms of the number of pest animals removed but according to the benefit derived therefrom. Methods include mortality control, fertility control, and various indirect manipulations.

17.2 Definitions

Control has three meanings in wildlife research and management. The first two deal with manipulating animal numbers, the third with experimentation.

"Control" is used first in the sense of a management action designed to restore an errant system to its previously stable state by reducing animal numbers. We speak of controlling an outbreak of mice in a grain store or wheat-growing district. The action is temporary.

The second use of "control" has to do with moving a system away from its stable state to another which is more desirable. The animals are reduced in density and the new density enforced by continuous control operations. The word is here used in a somewhat different sense than its use in engineering. There, a "control" (e.g., a governor on an engine) stops an intrinsically unstable system from shaking itself apart. It is a regulator. That connotation is inappropriate to wildlife management (although it has been so employed on occasion) because, except in special circumstances, the original state is more stable than that created by the control operation.

"Control" is used in a third sense within the parlance of experimental design. As Chapter 13 explains at length, an experimental control is the absence of an experimental treatment. That meaning of the word is usually obvious from the context except when the experiment tests the efficacy of a control program (i.e., "control" in one or other of the first two senses). The control operation is then the treatment and the control is the absence of control.

The obvious ambiguity in the previous sentence can easily lead to misunderstandings. For example, in an experiment testing the effect on riverside vegetation of controlling (reducing, i.e., the second meaning) hippopotamuses, they were shot (controlled) periodically in one stretch of river. The vegetation along the bank was compared with that of another stretch of river where the animals were protected (the control stretch, i.e., the third meaning). However, a change of hunting staff led inevitably to the control (protected) stretch being controlled (hunted) one sunny Sunday morning. We have seen similar mistakes (discovered at the last minute) in the testing of rabbit control methods. There is no sure remedy but the chance of a disaster can be reduced somewhat by always linking "experimental" to "control" when discussing experimental design.

17.3 Effects of control

If the density of a population is lowered by control measures, the standing crop of renewable resources (e.g., grass needed by a herbivore) will increase because of the lowered use. Nonrenewable resources such as nesting holes will be easier for an individual to find. Hence, control, like harvesting, increases the resources available to the survivors of the operation. Their fecundity, and their survival in the face of other mortality agents, is thereby enhanced. The reduced density therefore generates a potential increase which will become manifest if the control or harvesting is terminated. Table 17.1 shows just such an effect generated by control operations against feral donkeys in Australia.

The enhanced demographic vigor following reduction in density is a desirable outcome of a harvesting operation, and in fact the success of the harvesting is determined by such an effect, but it acts against the success of a control operation. The further density is reduced the more the population seeks to increase. Thus control, in the sense of enforcing a permanently lowered density, is simply a sustained-yield operation that seldom utilizes the harvest. It is an attempt to drive a negative feedback loop in the opposite direction.

17.4 Objectives of control

More than the other two areas of wildlife management (conservation and sustained-yield harvesting), control is often flawed by a lack of appropriate and clearly stated objectives.

Table 17.1 Differences between donkey populations on two 225-km^2 blocks in the Northern Territory of Australia, 3–4 years after one population was reduced by 80%. (After Choquenot 1991)

Measurement	High-density block	Low-density block
Initial density (1982) (donkeys/km^2)	>10	>10
Treatment (1983)	None	80% shot
Density (1986)	3.3	1.5
Density (1987)	3.2	1.8
Trend	Nonsignificant decrease	Significant increase (20%)
Sexual maturity male (%) at 2.5 years age	43	100
Female fecundity rate (%) at 2.5 years age	30	50
Juvenile mortality rate (%) at 0.5 years age	62	21

Control, in contrast to conservation and sustained-yield harvesting, is not itself an objective. It is simply a management action. Its use must be legitimized by a technical objective such as increasing the density of a food plant of a particular species of bird, say, from 1/ha to 3/ha. The control operations would be aimed at a herbivore for which that plant was a preferred food. The success of the operation would be measured by the density of plants, not by the density of the herbivore or by the number of herbivores killed.

Control campaigns in many countries share a common characteristic. Very often the original reason for the management action is forgotten and the control itself (lowering density) becomes the objective. The means become the end.

A good example is provided by the history of deer control in New Zealand. It is one of the largest and longest running control operations against vertebrates in any country. Table 17.2 lists the sequence of official justifications for government-funded control of deer from 1920 onward (Caughley 1983).

Whereas the stated justification for the control operations changed with time, those changes had virtually no effect on the management action. There were certainly changes in control techniques but, with the exception of the change of 1967, these were evolutionary adjustments in the management action itself. They were not driven by changes in policy. The means themselves were the end.

Up until 1980 the reasons given for the control operations were that deer and other species caused erosion of the higher slopes and silting of lower rivers (Table 17.2). However, in 1978, new meteorologic, hydrologic, geomorphologic and stratigraphic research showed that deer, chamois (*Rupicapra rupicapra*), and thar (*Hemitragus jemlahicus*) had little or no effect on the rate at which river beds silted up or on the frequency and size of floods. Despite these data, deer control continued after 1980 for no verifiable reason. All that changed were the stated objectives which were variously for "esthetics," for "proper land use," to "ensure the continuing health of the forest," to "protect intrinsic natural values," and to "maintain the distinctive New Zealand character of our landscapes." These are not open to scientific testing.

Table 17.2 Published official justification for government control operations against deer in New Zealand. (Source: Annual Reports of Department of Internal Affairs and New Zealand Forest Service)

Years	Official objectives of deer control
1920–29	Increase the size of antlers
1930–31	Reduce competition with sheep
1932–66	Prevent accelerated erosion generally
1967–80	Prevent accelerated erosion in the heads of rivers that may flood cities
1981–92	No verifiable reason offered

Many similar examples could be cited from other countries. Control operations must have clear objectives framed in terms of damage mitigation. Their success must be measured by how closely those objectives are met, not by the number of animals killed. The operations must be costed carefully to ensure that their benefit exceeds their cost. And their success or failure must be capable of independent verification. Table 1.1 gives a matrix of possible objectives and actions. It can be filled in to ensure that the management action is appropriate to the chosen objective.

17.5 Determining whether control is appropriate

There are three circumstances in which control may be an inappropriate management action: where the cost exceeds the benefit, where the "pest" is not in fact the cause of the perceived problem, and where the control has an unacceptable effect upon nontarget species. These are best investigated experimentally before a control program is instituted. We give two examples.

Cats were introduced to the subantarctic Marion Island in 1947 to deal with house mice marooned by shipwrecks. They increased rapidly to 3000 by 1977 and fed mostly on ground-nesting petrels. The breeding success of the petrels, particularly the great-winged petrel (*Pterodroma macroptera*), seemed to be declining and the cats were suspected to be the cause. The neighboring island of Prince Edward was conveniently free of cats and became the experimental control. The objective of reducing the cats was to increase the breeding success of the petrels. Hence, the success must be defined in terms of the birds' breeding success, not in terms of reduced numbers of cats. An introduced disease, shooting, and trapping reduced the cats. The petrel breeding success increased from 0% to 23% (1979–84) to 100%; the chick mortality rate decreased from 60% in 1979–84 to 0% in 1990. Comparisons with breeding on Prince Edward Island, and within a cat-free enclosure on Marion Island, identified the cats as the cause of the initial high mortality and the reduction in cat numbers as the reason for the increase in recruitment (Cooper and Fourie 1991).

The next example deals with nontarget species. The insecticide fenitrothion is a well-known organophosphorus pesticide, but its effects on song birds and other nontarget animals are little known. The Forestry Commission in Scotland wanted to use fenitrothion to control the pine beauty moth (*Panolis flammea*) and was required to undertake environmental assessment of the effects of spraying on nontarget species. For 3 years Spray *et al.* (1987) monitored the effect on the forest birds.

Their design comprised two pairs of plots, each plot measuring about 70 ha. The elements of each pair were matched by soil type, age of planting, and tree composition. One element of each pair was sprayed, all plots being monitored before and after spraying to detect annual variation of the density of breeding birds, short-term changes in abundance within 5 days of spraying, and breeding performance of the coal tit (*Parus ater*). They detected no significant difference in these variables between the insecticide-treated plots and the experimental control plots.

17.6 **Methods of control**

Animal welfare is an important consideration in any control operation. An animal has the right to be treated in a humane manner whether it is to be protected or controlled. Unfortunately the notion of humane treatment is often the first casualty of turning a species into a pest. That is particularly noticeable when the species is an exotic. The wildlife manager's paramount responsibility in any control operation is to ethical conduct rather than to operational efficiency.

Control methods can be divided into those aimed at directly increasing mortality, those aimed at directly reducing fertility, and those that act indirectly to manipulate mortality, fertility, or both. The success of an operation is not gauged by the reduction in the density of the target species but by the reduction in the deleterious effects of the target species. In all cases the prime responsibility of the wildlife manager is to determine whether the control adequately reduces deleterious effects and whether its benefit exceeds its cost.

17.6.1 *Control by manipulating mortality*

Control by increasing mortality may be direct, as in poisoning, trapping, or shooting, or it may be indirect as in biological control through pathogens.

Direct killing

Four simple principles guide the control of a target population living in an environment that remains reasonably constant from year-to-year. These are largely independent of a population's pattern of growth and they emphasize the conceptual similarities between control and sustained yield harvesting.

1 When a constant number of animals is removed from the population each year the size of the population will be stabilized by the control operations unless the annual offtake exceeds the population's maximum sustained yield (MSY).

2 The level at which the population is stabilized by the removal of a constant number each year is equal to or greater than the density from which the MSY is harvested.

3 Density is stabilized by the removal each year of a constant proportion of the population, provided the proportion is lower than the intrinsic rate of increase r_m.

4 The level at which a population is stabilized by removing a constant proportion of the population each year can be at any density above the threshold of extinction.

5 If animals are removed at an annual rate greater than r_m, the population will decline to extinction.

These simple rules can be sharpened for those populations whose pattern of population growth is approximated by a logistic curve

(Caughley 1977b). In general it will serve for populations of large mammals (i.e., low r_m) feeding on vegetation that recovers rapidly from grazing.

1 When the constant number C removed each year is less than the MSY of $r_m K/4$, the population is stabilized at a size of:

$$N = (r_m + \sqrt{(r_m^2 - 4Cr_m/K)})/(2r_m/K)$$

where K is the ecological carrying-capacity density, corresponding to the asymptote of the logistic curve.

2 When a constant proportion of the population is removed each year at a rate less than r_m, the population is stabilized at a size of:

$$N = K - (KH/r_m)$$

where H is the instantaneous rate of removal.

3 If a constant number of animals greater than $r_m K/4$ is removed each year, the population will eventually become extinct.

Biological control

Biological control, so effective against insects, has a poor record against pest wildlife. One of the few successes is the use of myxoma virus against rabbits. It holds the density of rabbits in Australia to about 20% of their uncontrolled density despite a decline in the virulence of the virus and of susceptibility of the rabbit, both a product of massive natural selection.

The chances of finding a biological agent to control vertebrates is always low, largely because the pathogen must be highly host specific and highly contagious.

17.6.2 Control by manipulating fertility

Population control by manipulating fecundity has several advantages over simply killing animals, but it also has problems of its own. It was first suggested as a control method by E.F. Knipling in 1938 (Marsh 1988) but was not applied for another 20 years. Its first use was against the screwworm fly (*Cochliomyia hominivorax*), a serious pest of livestock in the southeast of the USA. Subsequently it has been used against a number of insect pests in various parts of the world.

The use of contraceptive techniques for population control has been reviewed by Marsh (1988) with respect to rodents and lagomorphs, by Turner and Kirkpatrick (1991) with respect to horses, and by Bomford (1990) for vertebrates in general. Bomford showed that although contraception has often been advocated as a useful control method against vertebrates, and tried from time to time, there is no clear and well-documented example of unqualified success. "Many tests of fertility control have not been robust enough to allow clear conclusions. Experiments have often failed to include treatment replicates, or have relied on small samples. These results cannot be analyzed statistically to estimate the probability of a treatment effect" (Bomford 1990).

The usual method of use against insects – flooding the population with sterile males – is dependent on the females mating only once. That is common behavior among insects that live for only 1 year, but is rare among vertebrates.

Most attempts at control by contraception or sterilization have utilized chemicals such as bromocriptine, quinestrol, mestranol, and cyprosterone. Table 17.3 gives Marsh's (1988) criteria for an ideal rodent chemosterilant.

The effect of a contraceptive or sterilizing agent upon the population's dynamics depends on the breeding system of the species and particularly upon the form of dominance. In general, a vertebrate population will seldom be controlled adequately by a contraceptive or sterilant specific to males (Bomford 1990) and so the target should be either the female segment of the population or both sexes.

Caughley et al. (1992) explored the theoretical effect on productivity of three forms of behavioral dominance, two effects of sterilization on dominance, and four modes of transmission. Seventeen of the 24 combinations are feasible but lead to only four possible outcomes. Three of these result in lowered productivity. The fourth, where the breeding of a dominant female suppresses breeding in the subordinate females of her social group, leads to a perverse outcome. Productivity *increases* with sterilization unless the proportion of females sterilized exceeds $(n - 2)/(n - 1)$, where n is the average number of females in the social group (Fig. 17.1). Hence, a knowledge of social structure and

Table 17.3 Characteristics of an ideal chemosterilant for rodents. (After Marsh 1988)

1 Orally effective, preferably in a single feeding
2 Effective in very low doses (not exceeding 10 mg/kg)
3 Permanent or long-lasting sterility (preferably) lasting 6 months or longer, or at least through the major breeding period of the pest species
4 Effective for both sexes or preferably for females if only one sex
5 Rodent-specific or genus-specific
6 Relatively inexpensive
7 Wide margin between chemosterilant effects and lethal doses. (If high in specificity, this may be unimportant or the narrow margin be of value)
8 Well accepted (i.e., highly palatable) in baits at effective concentrations
9 Biodegradable after a few days in the environment
10 If not highly specific, rapid elimination from the body of the primary target to avoid secondary effects
11 No acquired tolerance or genetic or behavioral resistance
12 Free of behavioral modification (such as altering libido, aggression, or territoriality)
13 Free from producing discomfort or ill feelings that could suppress consumption (i.e., bait shyness) on repeat or subsequent feedings
14 Humane (i.e., produces no stressful symptoms)
15 Easy to formulate into various kinds of baits
16 Sufficiently stable when prepared in baits (i.e., adequate shelf-life)
17 Not translocated into plants (or at a very low level), thus permitting use on crops

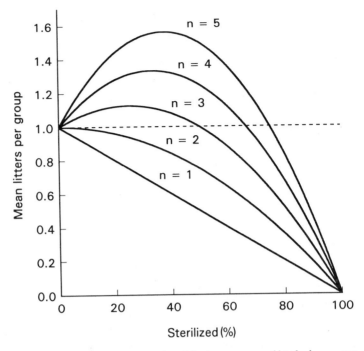

Fig. 17.1 Mean number of litters produced during a season of births by a group of females of size *n* subject to varying rates of sterilization. One female is dominant in each group and the other females subordinate to her. Only the dominant female breeds. She relinquishes dominance if sterilized and the subordinates are then free to breed. (After Caughley *et al.* 1992.)

mating system is desirable before population control by suppressing female fertility is attempted.

Those theoretically derived examples of reducing litter production exclude the effect of increased fertility consequent upon lowered density. It cannot be modeled from first principles in the same way as the expected reduction in litter production because its effect is specific to particular species and habitats. It must be examined by way of carefully designed field experiments. Such experiments should follow the effect of the treatment on the population's dynamics rather than simply on reproduction, because compensatory changes to mortality may also occur and should always be expected.

Two additional methods of fertility control have been suggested but which must await further research to demonstrate their general applicability: immunocontraception and genetic engineering.

Immunocontraception

Antibodies can be raised in an individual against some protein or peptide involved in reproduction, the antibodies hindering the reproductive process. The vaccine must usually be administered by injection or implant because most are broken down by digestion. Bomford

(1990) reviewed the work to date (mainly on horses) and concluded that although it has considerable appeal, mainly because it is potentially highly species specific, the technology is at too early a stage to gauge its usefulness against populations of wild or feral animals.

Genetic engineering

Tyndale-Biscoe (1989) suggested using a pathogen of low virulence to vector a foreign gene that would disrupt reproduction. He suggested particularly that the myxoma virus could be used to carry an inserted gene that would reduce the birth rate of the European rabbit (*Oryctolagus cuniculus*) in Australia.

17.6.3 *Control by indirect methods*

Exclusion

The most obvious way of reducing the deleterious effect of wildlife is to exclude animals from the area. That can be achieved by fencing, by chemical repellents, or by deterrents of one sort or another. Exclusion is often necessary for only part of the time, e.g., damage by deer to regenerating pines may be limited to only the first few years after establishment.

Exclosures can be as small as 1 ha or of mind-boggling proportions. The first of the latter was the Great Wall of China erected at the instigation of Shih Huang Ti, the first Emperor of the Ch'in Dynasty, between 228 BC and 210 BC. It protected his northern and western frontiers, the direction from which he was most frequently attacked. The wall traverses about 2400 km of rough country.

Another big one is the Australian barrier fence built to keep dingoes out of sheep country (Breckwoldt 1988). It runs from the south coast, divides South Australia into two, skirts around the top inland corner of New South Wales, and then loops to enclose all of central Queensland. It was started in 1914 and built in sections, often as an upgrading of previous state border fences and rabbit fences. At its greatest extent it spanned 8614 km, 3.5 times longer than the Great Wall of China. In 1980 the loop up through inland Queensland was fenced off about halfway up and the upper fence abandoned. The present exclosure has a perimeter of 5614 km. In contrast to the numerous rabbit fences that have been built in Australia, the dingo barrier fence has been relatively successful in reducing the spread of a pest species.

Sonic deterrents

The modern forms of the scarecrow comprising sonic devices (bangers, clangers, alarm calls, ultrasonics) have been reviewed by Bomford and O'Brien (1990), who suggested that, at best, these achieve only a short-term reduction in damage. They were particularly critical of the claims made for commercially produced ultrasonic devices and of the standard of experimental testing in this field.

Habitat manipulation

This is certainly the most elegant of control techniques because it does not operate against a negative feedback loop. The key habitat elements are water and shelter. Red squirrels (*Tamiasciurus hudsonicus*) in British Columbia lodgepole pine forests can be dissuaded from feeding on the stems of very young trees by aerial spreading of sunflower seeds. This alternative food source is preferred over the pines (Sullivan and Klenner 1993).

Plant and animal resistance

Plants and animals can be selected to increase their immunity to disease, thereby reducing the need for control of the disease. Since plants generate secondary compounds naturally, probably as a consequence of selection being imposed by grazing and browsing, this capacity may become the basis for further artificial selection.

17.7 **Summary**

The control of a pest species, in the sense of holding its density at a reduced level, is essentially a sustained-yield operation where the yield is not used. Reduction in density is not an end in itself: the success of the operation is measured not by the number of animals removed but by whether the objective was attained, be it the increase in density of an endangered species, an increase in grass biomass, or the reduction of damage to fences. The logic of experimental design must be utilized to determine whether benefits exceed costs, whether the treatment has a deleterious effect on nontarget species, and whether the targeted "pest" is really the cause of the perceived problem.

Appendix

Table A1 Cumulative F distribution. The body of the table contains critical values of F for both "one-sided" and "two-sided" significance probabilities

d.f. in denominator	One-sided tests	d.f. in numerator 1	2	3	4	5	6	7	8	9	10	12	15	20	30	60	∞	Two-sided tests
1	0.10	39.9	49.5	53.6	55.8	57.2	58.2	58.9	59.4	59.9	60.2	60.7	61.2	61.7	62.3	62.8	63.3	0.20
	0.05	161	200	216	225	230	234	237	239	241	242	244	246	248	250	252	254	0.10
	0.01	4050	5000	5400	5620	5760	5860	5930	5980	6020	6060	6110	6160	6210	6260	6310	6370	0.02
2	0.10	8.53	9.00	9.16	9.24	9.29	9.33	9.35	9.37	9.38	9.39	9.41	9.42	9.44	9.46	9.47	9.49	0.20
	0.05	18.5	19.0	19.2	19.2	19.3	19.3	19.4	19.4	19.4	19.4	19.4	19.4	19.5	19.5	19.5	19.5	0.10
	0.01	98.5	99.0	99.2	99.2	99.3	99.3	99.4	99.4	99.4	99.4	99.4	99.4	99.4	99.5	99.5	99.5	0.02
3	0.10	5.54	5.46	5.39	5.34	5.31	5.28	5.27	5.25	5.24	5.23	5.22	5.20	5.18	5.17	5.15	5.13	0.20
	0.05	10.1	9.55	9.28	9.12	9.01	8.94	8.89	8.85	8.81	8.79	8.74	8.70	8.66	8.62	8.57	8.53	0.10
	0.01	34.1	30.8	29.5	28.7	28.2	27.9	27.7	27.5	27.3	27.2	27.1	26.9	26.7	26.5	26.3	26.1	0.02
4	0.10	4.54	4.32	4.19	4.11	4.05	4.01	3.98	3.95	3.94	3.92	3.90	3.87	3.84	3.82	3.79	3.76	0.20
	0.05	7.71	6.94	6.59	6.39	6.26	6.16	6.09	6.04	6.00	5.96	5.91	5.86	5.80	5.75	5.69	5.63	0.10
	0.01	21.2	18.0	16.7	16.0	15.5	15.2	15.0	14.8	14.7	14.5	14.4	14.2	14.0	13.8	13.7	13.5	0.02
5	0.10	4.06	3.78	3.62	3.52	3.45	3.40	3.37	3.34	3.32	3.30	3.27	3.24	3.21	3.17	3.14	3.10	0.20
	0.05	6.61	5.79	5.41	5.19	5.05	4.95	4.88	4.82	4.77	4.74	4.68	4.62	4.56	4.50	4.43	4.37	0.10
	0.01	16.3	13.3	12.1	11.4	11.0	10.7	10.5	10.3	10.2	10.1	9.89	9.72	9.55	9.38	9.20	9.02	0.02
6	0.10	3.78	3.46	3.29	3.18	3.11	3.05	3.01	2.98	2.96	2.94	2.90	2.87	2.84	2.80	2.76	2.72	0.20
	0.05	5.99	5.14	4.76	4.53	4.39	4.28	4.21	4.15	4.10	4.06	4.00	3.94	3.87	3.81	3.74	3.67	0.10
	0.01	13.7	10.9	9.78	9.15	8.75	8.47	8.26	8.10	7.98	7.87	7.72	7.56	7.40	7.23	7.06	6.88	0.02
7	0.10	3.59	3.26	3.07	2.96	2.88	2.83	2.78	2.75	2.72	2.70	2.67	2.63	2.59	2.56	2.51	2.47	0.20
	0.05	5.59	4.74	4.35	4.12	3.97	3.87	3.79	3.73	3.68	3.64	3.57	3.51	3.44	3.38	3.30	3.23	0.10
	0.01	12.2	9.55	8.45	7.85	7.46	7.19	6.99	6.84	6.72	6.62	6.47	6.31	6.16	5.99	5.82	5.65	0.02
8	0.10	3.46	3.11	2.92	2.81	2.73	2.67	2.62	2.59	2.56	2.54	2.50	2.46	2.42	2.38	2.34	2.29	0.20
	0.05	5.32	4.46	4.07	3.84	3.69	3.58	3.50	3.44	3.39	3.35	3.28	3.22	3.15	3.08	3.01	2.93	0.10
	0.01	11.3	8.65	7.59	7.01	6.63	6.37	6.18	6.03	5.91	5.81	5.67	5.52	5.36	5.20	5.03	4.86	0.02
9	0.10	3.36	3.01	2.81	2.69	2.61	2.55	2.51	2.47	2.44	2.42	2.38	2.34	2.30	2.25	2.21	2.16	0.20
	0.05	5.12	4.26	3.86	3.63	3.48	3.37	3.29	3.23	3.18	3.14	3.07	3.01	2.94	2.86	2.79	2.71	0.10
	0.01	10.6	8.02	6.99	6.42	6.06	5.80	5.61	5.47	5.35	5.26	5.11	4.96	4.81	4.65	4.48	4.31	0.02

d.f.	p																	p
10	0.10	3.28	2.92	2.73	2.61	2.52	2.46	2.41	2.38	2.35	2.32	2.28	2.24	2.20	2.16	2.11	2.06	0.20
	0.05	4.96	4.10	3.71	3.48	3.33	3.22	3.14	3.07	3.02	2.98	2.91	2.84	2.77	2.70	2.62	2.54	0.10
	0.01	10.0	7.56	6.55	5.99	5.64	5.39	5.20	5.06	4.94	4.85	4.71	4.56	4.41	4.25	4.08	3.91	0.02
12	0.10	3.18	2.81	2.61	2.48	2.39	2.33	2.28	2.24	2.21	2.19	2.15	2.10	2.06	2.01	1.96	1.90	0.20
	0.05	4.75	3.89	3.49	3.26	3.11	3.00	2.91	2.85	2.80	2.75	2.69	2.62	2.54	2.47	2.38	2.30	0.10
	0.01	9.33	6.93	5.95	5.41	5.06	4.82	4.64	4.50	4.39	4.30	4.16	4.01	3.86	3.70	3.54	3.36	0.02
15	0.10	3.07	2.70	2.49	2.36	2.27	2.21	2.16	2.12	2.09	2.06	2.02	1.97	1.92	1.87	1.82	1.76	0.20
	0.05	4.54	3.68	3.29	3.06	2.90	2.79	2.71	2.64	2.59	2.54	2.48	2.40	2.33	2.25	2.16	2.07	0.10
	0.01	8.68	6.36	5.42	4.89	4.56	4.32	4.14	4.00	3.89	3.80	3.67	3.52	3.37	3.21	3.05	2.87	0.02
20	0.10	2.97	2.59	2.38	2.25	2.16	2.09	2.04	2.00	1.96	1.94	1.89	1.84	1.79	1.74	1.68	1.61	0.20
	0.05	4.35	3.49	3.10	2.87	2.71	2.60	2.51	2.45	2.39	2.35	2.28	2.20	2.12	2.04	1.95	1.84	0.10
	0.01	8.10	5.85	4.94	4.43	4.10	3.87	3.70	3.56	3.46	3.37	3.23	3.09	2.94	2.78	2.61	2.42	0.02
30	0.10	2.88	2.49	2.28	2.14	2.05	1.98	1.93	1.88	1.85	1.82	1.77	1.72	1.67	1.61	1.54	1.46	0.20
	0.05	4.17	3.32	2.92	2.69	2.53	2.42	2.33	2.27	2.21	2.16	2.09	2.01	1.93	1.84	1.74	1.62	0.10
	0.01	7.56	5.39	4.51	4.02	3.70	3.47	3.30	3.17	3.07	2.98	2.84	2.70	2.55	2.39	2.21	2.01	0.02
60	0.10	2.79	2.39	2.18	2.04	1.95	1.87	1.82	1.77	1.74	1.71	1.66	1.60	1.54	1.48	1.40	1.29	0.20
	0.05	4.00	3.15	2.76	2.53	2.37	2.25	2.17	2.10	2.04	1.99	1.92	1.84	1.75	1.65	1.53	1.39	0.10
	0.01	7.08	4.98	4.13	3.65	3.34	3.12	2.95	2.82	2.72	2.63	2.50	2.35	2.20	2.03	1.84	1.60	0.02
120	0.10	2.75	2.35	2.13	1.99	1.90	1.82	1.77	1.72	1.68	1.65	1.60	1.54	1.48	1.41	1.32	1.19	0.20
	0.05	3.92	3.07	2.68	2.45	2.29	2.18	2.09	2.02	1.96	1.91	1.83	1.75	1.66	1.55	1.43	1.25	0.10
	0.01	6.85	4.79	3.95	3.48	3.17	2.96	2.79	2.66	2.56	2.47	2.34	2.19	2.03	1.86	1.66	1.38	0.02
∞	0.10	2.71	2.30	2.08	1.94	1.85	1.77	1.72	1.67	1.63	1.60	1.55	1.49	1.42	1.34	1.24	1.00	0.20
	0.05	3.84	3.00	2.60	2.37	2.21	2.10	2.01	1.94	1.88	1.83	1.75	1.67	1.57	1.46	1.32	1.00	0.10
	0.01	6.63	4.61	3.78	3.32	3.02	2.80	2.64	2.51	2.41	2.32	2.18	2.04	1.88	1.70	1.47	1.00	0.02

d.f., degrees of freedom.

Table A2 Critical 5% values for Cochran's test for homogeneity of variance

d.f. for s^2_j	k = number of variances										
	2	3	4	5	6	7	8	9	10	15	20
1	0.9985	0.9669	0.9065	0.8412	0.7808	0.7271	0.6798	0.6385	0.6020	0.4709	0.3894
2	0.9750	0.8709	0.7679	0.6838	0.6161	0.5612	0.5157	0.4775	0.4450	0.3346	0.2705
3	0.9392	0.7977	0.6841	0.5981	0.5321	0.4800	0.4377	0.4027	0.3733	0.2758	0.2205
4	0.9057	0.7457	0.6287	0.5441	0.4803	0.4307	0.3910	0.3584	0.3311	0.2419	0.1921
5	0.8772	0.7071	0.5895	0.5065	0.4447	0.3974	0.3595	0.3286	0.3029	0.2195	0.1735
6	0.8534	0.6771	0.5598	0.4783	0.4184	0.3726	0.3362	0.3067	0.2823	0.2034	0.1602
7	0.8332	0.6530	0.5365	0.4564	0.3980	0.3535	0.3185	0.2901	0.2666	0.1911	0.1501
8	0.8159	0.6333	0.5175	0.4387	0.3817	0.3384	0.3043	0.2768	0.2541	0.1815	0.1422
9	0.8010	0.6167	0.5017	0.4241	0.3682	0.3259	0.2926	0.2659	0.2439	0.1736	0.1357
10	0.7880	0.6025	0.4884	0.4118	0.3568	0.3154	0.2829	0.2568	0.2353	0.1671	0.1303

C = (Largest s^2)/(Σs^2_j); d.f., degrees of freedom.

References

Abaturov, B.D. & Magomedov, M.-R.D. 1988. Food value and dynamics of food resources as a factor characterizing the state of populations of herbivorous mammals. *Zoologicheskii Zhurnal* 76:223–234.

Abramsky, Z. 1981. Habitat relationships and competition in two Mediterranean *Apodemus* spp. *Oikos* 36:219–225.

Abramsky, Z. & Sellah, C. 1982. Competition and the role of habitat selection in *Gerbillus allenbyi* and *Meriones tristrami*: a removal experiment. *Ecology* 63:1242–1247.

Abramsky, Z., Bowers, M.A. & Rosenzweig, M.L. 1986. Detecting interspecific competition in the field: testing the regression method. *Oikos* 47:199–204.

Abramsky, Z., Dyer, M.I. & Harrison, P.D. 1979. Competition among small mammals in experimentally perturbed areas of the shortgrass prairie. *Ecology* 60:530–536.

Alexander, R.D. 1974. The evolution of social behavior. *Annual Review of Ecology and Systematics* 5:325–383.

Anderson, A. 1989. *Prodigious Birds: Moas and Moa-hunting in Prehistoric New Zealand*. Cambridge University Press, Cambridge.

Anderson, A.E., Bowden, D.C. & Medin, D.E. 1990. Indexing the annual fat cycle in a mule deer population. *Journal of Wildlife Management* 54:550–556.

Anderson, A.E., Medin, D.E. & Bowden, D.C. 1972. Indices of carcass fat in a Colorado mule deer population. *Journal of Wildlife Management* 36:579–594.

Anderson, A.E., Medin, D.E. & Ochs, D.P. 1969. Relationships of carcass fat indices in 18 wintering mule deer. *Proceedings of the Western Association of State Game and Fish Commissioners* 49:329–340.

Anderson, D.R. & Pospahala, R.S. 1970. Correction of bias in belt transect studies of immotile objects. *Journal of Wildlife Management* 34:141–146.

Anderson, D.R., Burnham, K.P., White, G.C. & Otis, D.L. 1983. Density estimation of small-mammal populations using a trapping web and distance sampling methods. *Ecology* 64:674–680.

Anderson, E.W. & Sherzinger, R.J. 1975. Improving quality of winter forage for elk by cattle grazing. *Journal of Range Management* 28:120–125.

Anderson, R.C. 1972. The ecological relationships of meningeal worm and native cervids in North America. *Journal of Wildlife Diseases* 8:304–310.

Anderson, R.M. 1991. Populations and infectious diseases: ecology or epidemiology? *Journal of Animal Ecology* 60:1–50.

Anderson, R.M. & May, R.M. 1986. The invasion, persistence and spread of infectious diseases within animal and plant communities. *Philosophical Transactions of the Royal Society of London. Series B: Biological Sciences* 314:533–570.

Andrewartha, H.G. & Birch, L.C. 1984. *The Ecological Web: More on the Distribution and Abundance of Animals*. University of Chicago Press, Chicago.

Arcese, P. & Smith, J.N.M. 1988. Effects of population density and supplemental food on reproduction in song sparrows. *Journal of Animal Ecology* 57:119–136.

Bailey, J.A. 1968. A weight–length relationship for evaluating physical condition of cottontails. *Journal of Wildlife Management* 32:835–841.

Bailey, N.T.J. 1951. On estimating the size of mobile populations from recapture data. *Biometrika* 38:293–306.

Bailey, N.T.J. 1952. Improvements in the interpretation of recapture data. *Journal of Animal Ecology* 21:120–127.

Baker, C.S., Perry, A., Bannister, J.L., *et al.* 1993. Abundant mitochondrial variation and world-wide population structure in humpback whales. *Proceedings of the Natural Academy of Science USA* 90:8239–8243.

Baker, J.R. 1938. The evolution of breeding seasons. In: de Beer, G.R., ed. *Evolution: Essays on Aspects of Evolutionary Biology Presented to Professor E.S. Goodrich on his Seventieth Birthday*, pp. 161–177. Clarendon Press, Oxford.

Ballard, W.B., Whitman, J.S. & Gardner, C.L. 1987. Ecology of an exploited wolf population in south–central Alaska. *Wildlife Monographs* 98:1–54.

Bamford, J. 1970. Estimating fat reserves in the brush-tailed possum, *Trichosurus vulpecula* Kerr (Marsupialia: Phalangeridae). *Australian Journal of Zoology* 18: 415–425.

Barker, S.C., Singleton, G.R. & Spratt, D.M. 1991. Can the nematode *Capillaria hepatica* regulate abundance of wild house mice?: Results of enclosure experiments in southeastern Australia. *Parasitology* 103:439–449.

Barr, W. 1991. *Back from the Brink: The Road to Muskox Conservation in the Northwest Territories*. Komatik Series, no. 3. The Arctic Institute of North America, University of Calgary.

Bayliss, P. 1987. Kangaroo dynamics. In: Caughley, G., Shepherd, N. & Short, J., eds. *Kangaroos: Their Ecology and Management in the Sheep Rangelands of Australia*, pp. 119–134. Cambridge University Press, Cambridge.

Bazely, D.R. & Jefferies, R.L. 1989. Lesser snow geese and the nitrogen economy of a grazed salt marsh. *Journal of Ecology* 77:24–34.

Beasom, S.L. & Roberson, S.F. (eds) 1985. *Game Harvest Management*. Caesar Kleberg Wildlife Research Institute, Kingsville, Texas.

Bell, D.J., Oliver, W.L.R. & Ghose, R.K. 1990. The hispid hare *Caprolagus hispidus*. In: Chapman, J.A. & Flux, J.E.C., eds. *Rabbits, Hares and Pikas: Status Survey and Conservation Action Plan*, pp. 128–136. IUCN, Gland, Switzerland.

Bell, R.H.V. 1970. The use of the herb layer by grazing ungulates in the Serengeti. In: Watson, A., ed. *Animal Populations in Relation to their Food Resources*, pp. 111–124. The Tenth Symposium of the British Ecological Society. Blackwell, Oxford.

Bell, R.H.V. 1971. A grazing ecosystem in the Serengeti. *Scientific American* 225: 86–93.

Bell, R.H.V. 1981. An outline of a management plan for Kasungu National Park, Malawi. In: Jewell, P.A., Holt, S. & Hart, D., eds. *Problems in Management of Locally Abundant Wild Mammals*, pp. 69–89. Academic Press, New York.

Belsky, A.J. 1987. The effects of grazing: confounding of ecosystem, community and organism scales. *American Naturalist* 129:777–783.

Bender, E.A., Case, T.J. & Gilpin, M.E. 1984. Perturbation experiments in community ecology: theory and practice. *Ecology* 65:1–13.

Bent, A.C. 1932. *Life Histories of North American Gallinaceous Birds*. Smithsonian Institution, US National Museum Bulletin no. 162, Washington, DC.

Bergerud, A.T. 1971. The population dynamics of Newfoundland caribou. *Wildlife Monographs* 25:1–55.

Bergerud, A.T. & Page, R.E. 1987. Displacement and dispersion of parturient caribou at calving as antipredator tactics. *Canadian Journal of Zoology* 65:1597–1606.

Birch, L.C. 1957. The meanings of competition. *American Naturalist* 91:5–18.

Björnhag, G. 1987. Comparative aspects of digestion in the hindgut of mammals. The colonic separation mechanism (CSM) (a review). *Deutsche Tieraerztliche Wochenschrift* 94:33–36.

Blank, T.H., Southwood, T.R.E. & Cross, D.J. 1967. The ecology of the partridge: I. Outline of the population processes with particular reference to chick mortality and nest density. *Journal of Animal Ecology* 36:549–556.

Blower, J.G., Cook, L.M. & Bishop, J.A. 1981. *Estimating the Size of Animal Populations*. George Allen and Unwin, London.

Bomford, M. 1988. Effect of wild ducks on rice production. In: Norton, G.A. & Pech, R.P., eds. *Vertebrate Pest Management in Australia: A Decision Analysis/Systems Analysis Approach*, pp. 53–57. Project report no. 5. CSIRO, Melbourne, Australia.

Bomford, M. 1990. *A Role for Fertility Control in Wildlife Management?* Bureau of Rural Resources, Bulletin no. 7, Canberra.

Bomford, M. & O'Brien, P.H. 1990. Sonic deterrents in animal damage control: a review of device tests and effectiveness. *Wildlife Society Bulletin* 18:411–422.

Bothma, J. du P. (ed.) 1989. *Game Ranch Management.* J.L. van Schaik, Pretoria.

Boutin, S. 1992. Predation and moose population dynamics: a critique. *Journal of Wildlife Management* 56:116–127.

Boutin, S., Krebs, C.J., Sinclair, A.R.E. & Smith, J.N.M. 1986. Proximate causes of losses in a snowshoe hare population. *Canadian Journal of Zoology* 64:606–610.

Boyce, M.S. 1989. *The Jackson Elk Herd: Intensive Wildlife Management in North America.* Cambridge University Press, Cambridge.

Breckwoldt, R. 1988. *A Very Elegant Animal: The Dingo.* Angus & Robertson, North Ryde, Australia.

Bredon, R.M., Harker, K.W. & Marshall, B. 1963. The nutritive value of grasses grown in Uganda when fed to Zebu cattle: I. The relation between the percentage of crude protein and other nutrients. *Journal of Agricultural Science (Cambridge)* 61:101–104.

Briggs, S.V. 1991. Effects of breeding and environment on body condition of maned ducks *Chenonetta jubata. Wildlife Research* 18:577–588.

Brightwell, L.R. 1951. Some experiments with the common hermit crab (*Eupagurus bernhardus*) Linn, and transparent univalve shells. *Proceedings of the Zoological Society of London* 121:279–283.

Brower, L.P. 1984. Chemical defense in butterflies. In: Vane-Wright, R.I. & Ackery, P.R., eds. *The Biology of Butterflies*, pp. 109–134. Academic Press, London.

Brown, K.S., Jr. 1987. Conclusions, synthesis, and alternative hypotheses. In: Whitmore, T.C. & Prance, G.T., eds. *Biogeography and Quaternary History in Tropical America*, pp. 175–191. Oxford University Press, New York.

Bryant, D.M. 1989. Determination of respiration rates of free-living animals by the double-labelling technique. In: Grubb, P.J. & Whittaker, J.B., eds. *Toward a More Exact Ecology*, pp. 85–109. 30th Symposium of the British Ecological Society. Blackwell Scientific Publication, Boston.

Bryant, J.P. 1981. Phytochemical deterrence of snowshore hare browsing by adventitious shoots of four Alaskan trees. *Science (Washington)* 213:889–890.

Bryant, J.P. & Kuropat, P.J. 1980. Selection of winter forage by subarctic browsing vertebrates: the role of plant chemistry. *Annual Review of Ecology and Systematics* 11:261–285.

Buckner, C.H. & Turnock, W.J. 1965. Avian predation on the larch sawfly, *Pristiphora erichsonii* (Htg.), (Hymenoptera: Tenthredinidae). *Ecology* 46:223–236.

Buechner, H.K., Buss, I.O., Longhurst, W.M. & Brooks, A.C. 1963. Numbers and migration of elephants in Murchison Falls National Park, Uganda. *Journal of Wildlife Management* 27:36–53.

Burke, T. 1989. DNA fingerprinting and other methods for the study of mating success. *Trends in Ecology and Evolution* 4:139–144.

Burke, T. & Bruford, M.W. 1987. DNA fingerprinting in birds. *Nature (London)* 327:149–152.

Burnham, K.P. & Anderson, D.R. 1984. The need for distance data in transect counts. *Journal of Wildlife Management* 48:1248–1254.

Burnham, K.P. & Overton, W.S. 1978. Estimation of the size of a closed population when capture probabilities vary among animals. *Biometrika* 65:625–633.

Burnham, K.P., Anderson, D.R. & Laake, J.L. 1980. Estimation of density from line transect sampling of biological populations. *Wildlife Monographs* 72:1–202.

Calaby, J.H. & Grigg, G.C. 1989. Changes in macropodoid communities and populations in the past 200 years, and the future. In: Grigg, G., Jarman, P. & Hume, I., eds. *Kangaroos, Wallabies and Rat-kangaroos*, pp. 813–820. Surrey Beatty & Sons, Chipping Norton, New South Wales.

Carpenter, J.W., Clark, G.G. & Watts, D.M. 1989. The impact of eastern equine encephalitis virus on efforts to recover the endangered whooping crane. In: Cooper, J.E., ed. *Disease and Threatened Birds*, pp. 115–120. International Council for Bird Preservation, Technical Publication no. 10. ICBP, Cambridge.

Carr, S.M., Ballinger, S.W., Derr, J.N., Blankenship, L.H. & Bickham, J.W. 1986.

Mitochondrial DNA analysis of hybridization between sympatric white-tailed deer and mule deer in west Texas, USA. *Proceedings of the National Academy of Science of the United States of America* 83:9576–9580.

Caughley, G. 1970. Population statistics of chamois. *Mammalia* 34:194–199.

Caughley, G. 1974. Bias in aerial survey. *Journal of Wildlife Management* 38: 921–933.

Caughley, G. 1976a. Wildlife management and the dynamics of ungulate populations. In: Coaker, T.H., ed. *Applied Biology*, Vol. 1, pp. 183–246. Academic Press, New York.

Caughley, G. 1976b. The elephant problem – an alternative hypothesis. *East African Wildlife Journal* 14:265–283.

Caughley, G. 1977a. Sampling in aerial survey. *Journal of Wildlife Management* 41:605–615.

Caughley, G. 1977b. *Analysis of Vertebrate Population.* John Wiley, London.

Caughley, G. 1981. Overpopulation. In: Jewell, P.A., Holt, S. & Hart, D., eds. *Problems in Management of Locally Abundant Wild Mammals*, pp. 7–19. Academic Press, New York.

Caughley, G. 1983. *The Deer Wars: The Story of Deer in New Zealand.* Heinemann, Auckland.

Caughley, G. 1985. Harvesting of wildlife: past, present and future. In: Beasom, S.L. & Roberson, S.F., eds. *Game Harvest Management*, pp. 3–14. Caesar Kleberg Wildlife Research Institute, Kingsville, Texas.

Caughley, G. 1987. Ecological relationships. In: Caughley, G., Shepherd, N. & Short, J., eds. *Kangaroos: Their Ecology and Management in the Sheep Rangelands of Australia*, pp. 159–187. Cambridge University Press, Cambridge.

Caughley, G. 1994. Directions in conservation biology. *Journal of Animal Ecology* (in press).

Caughley, G. & Goddard, J. 1972. Improving the estimates from inaccurate censuses. *Journal of Wildlife Management* 36:135–140.

Caughley, G. & Grice, D. 1982. A correction factor for counting emus from the air, and its application to counts in Western Australia. *Australian Wildlife Research* 9:253–259.

Caughley, G. & Gunn, A. 1993. Dynamics of large herbivores in deserts: kangaroos and caribou. *Oikos* 67:47–55.

Caughley, G. & Krebs, C.J. 1983. Are big mammals simply little mammals writ large. *Oecologia* 59:7–17.

Caughley, G., Pech, R. & Grice, D. 1992. Effect of fertility control on a population's productivity. *Wildlife Research* 19:623–627.

Caughley, G., Grice, D., Barker, R. & Brown, B. 1988. The edge of the range. *Journal of Animal Ecology* 57:771–785.

Chapman, D.G. 1951. Some properties of the hypergeometric distribution with application to zoological sample censuses. *University of California Publications in Statistics* 1:131–160.

Chase, A. 1987. *Playing God in Yellowstone.* Harcourt Brace Jovanovich, San Diego, CA.

Cheatum, E.L. 1949. Bone marrow as an index of malnutrition in deer. *New York State Conservationist* 3:19–22.

Chepko-Sade, B.D., Shields, W.M., Berger, J., *et al.* 1987. The effects of dispersal and social structure on effective population size. In: Chepko-Sade, B.D., & Halpin, Z.T., eds. *Mammalian Dispersal Patterns: the Effects of Social Structure on Population Genetics*, pp. 287–321. University of Chicago Press, Chicago.

Cherfas, J. 1988. *The Hunting of the Whale.* The Bodley Head, London.

Chilcott, M.J. & Hume, I.D. 1985. Coprophagy and selective retention of fluid digesta: their role in the nutrition of the common ringtail possum, *Pseudocheirus peregrinus*. *Australian Journal of Zoology* 33:1–15.

Child, G. & Von Richter, W. 1969. Observations on ecology and behaviour of lechwe, puku, and waterbuck along the Chobe River, Botswana. *Zeitschrift für Säugetierkunde* 34:275–295.

Choquenot, D. 1991. Density-dependent growth, body condition, and demography in feral donkeys: testing the food hypothesis. *Ecology* 72:805–813.

Clark, C.W. 1976. *Mathematical Bioeconomics: The Optimal Management of Renewable Resources.* John Wiley, New York.

Clark, W.C., Jones, D.D. & Holling, C.S. 1979. Lessons for ecological policy design: a case study of ecosystem management. *Ecological Modelling* 7:1–53.

Clutton-Brock, T.H. & Harvey, P.H. 1983. The functional significance of variation in body size among mammals. In: Eisenberg, J.F. & Kleiman, D.G., eds. *Advances in the Study of Mammalian Behavior*, pp. 632–663. Special Publication no. 7. The American Society of Mammalogists, Shippensburg State College, Shippensburg, PA.

Clutton-Brock, T.H., Guinness, F.E. & Albon, S.D. 1982. Red deer: behaviour and ecology of two sexes. University of Chicago Press, Chicago.

Clutton-Brock, T.H., Major, M. & Guinness, F.E. 1985. Population regulation in male and female red deer. *Journal of Animal Ecology* 54:831–846.

Clutton-Brock, T.H., Price, O.F., Albon, S.D. & Jewell, P.A. 1991. Persistent instability and population regulation in Soay sheep. *Journal of Animal Ecology* 60:593–608.

Cochran, W.G. 1954. The combination of estimates from different experiments. *Biometrics* 10:101–129.

Cochran, W.G. 1977. *Sampling Techniques*, 3rd edn. John Wiley, New York.

Cohn, J.P. 1986. Surprising cheetah genetics. *Bioscience* 36:358–362.

Cohn, J.P. 1991. Ferrets return from near extinction. *Bioscience* 41:132–135.

Collar, N.J., Gonzaga, L.P., Krabbe, N., *et al.* 1992. *Threatened Birds of the North Americas.* International Council for Bird Preservation, Cambridge.

Collet, C. 1991. Genetic profiling in wildlife law enforcement. In: Hungerford, T.G., ed. *Avian Medicine. Refresher Course for Veterinarians, Proceedings 178*, pp. 99–110. Post Graduate Committee in Veterinary Science, University of Sydney.

Cooke, F. 1988. Genetic studies of birds – the goose with blue genes. In: Quellet, H., ed. *Acta 19th Congressus Internationalis Ornithologicus*, pp. 189–214. University of Ottawa Press, Ottawa.

Cooper, J. & Fourie, A. 1991. Improved breeding success of great-winged petrels *Pterodroma macroptera* following control of feral cats *Felis catus* at subantarctic Marion Island. *Bird Conservation International* 1:171–175.

Cooper, S.M., Owen-Smith, N. & Bryant, J.P. 1988. Foliage acceptability to browsing ruminants in relation to seasonal changes in the leaf chemistry of woody plants in a South African savanna. *Oecologia* 75:336–342.

Coppock, D.L., Ellis, J.E., Detling, J.K. & Dyer, M.I. 1983. Plant–herbivore interactions in a North American mixed-grass prairie: II. Responses of bison to modification of vegetation by prairie dogs. *Oecologia* 56:10–15.

Corfield, T.F. 1973. Elephant mortality in Tsavo National Park, Kenya. *East African Wildlife Journal* 11:339–368.

Cork, S.J. & Warner, A.C.I. 1983. The passage of digesta markers through the gut of a folivorous marsupial, the koala *Phascolarctos cinereus*. *Journal of Comparative Physiology B* 152:43–51.

Cox, D.R. 1969. Some sampling problems in technology. In: Johnson, N.L. & Smith, H., eds. *New Developments in Survey Sampling*, pp. 506–527. John Wiley, New York.

Crawley, M.J. 1987. Benevolent herbivores? *Trends in Ecology and Evolution* 2:167–168.

Cronin, M.A., Vyse, E.R. & Cameron, D.G. 1988. Genetic relationships between mule deer and white-tailed deer in Montana. *Journal of Wildlife Management* 52:320–328.

Crowell, K.L. & Pimm, S.L. 1976. Competition and niche shifts of mice introduced onto small islands. *Oikos* 27:251–258.

Dailey, T.V., Thompson Hobbs, N. & Woodard, T.N. 1984. Experimental comparisons of diet selection by mountain goats and mountain sheep in Colorado. *Journal of Wildlife Management* 48:799–806.

Dark, J., Forger, N.G. & Zucker, I. 1986. Regulation and function of lipid mass during the annual cycle of the golden-mantled ground squirrel. In: Heller, H.C., Musacchia, X.J. & Wang, L.C.H., eds. *Living in the Cold: Physiological and Biochemical Adaptations*, pp. 445–451. Elsevier Scientific, New York.

Dasmann, R.F. 1956. Fluctuations in a deer population in California chaparral. *Transactions of the North American Wildlife Conference* 21:487–499.

Davidson, N.C. & Evans, P.R. 1988. Prebreeding accumulation of fat and muscle protein by arctic-breeding shorebirds. *Proceedings of the International Ornithological Congress* 19:342–363.

Davies, S.J.J.F. 1976. Environmental variables and the biology of Australian arid zone birds. In: Frith, H.J. & Calaby, J.H., eds. *Proceedings of the 16th International Ornithological Congress, Canberra*, pp. 481–488. Australian Academy of Science, Canberra.

Davis, M.B. 1986. Climatic instability, time lags, and community disequilibrium. In: Diamond, J. & Case, T.J., eds. *Community Ecology*, pp. 269–284. Harper & Row, New York.

Dawson, T.J. & Ellis, B.A. 1979. Comparison of the diets of yellow-footed rock-wallabies and sympatric herbivores in western New South Wales. *Australian Wildlife Research* 6:245–254.

Dawson, T.J. & Hulbert, A.J. 1970. Standard metabolism, body temperature, and surface areas of Australian marsupials. *American Journal of Physiology* 218: 1233–1238.

DeAngelis, D.L. & Waterhouse, J.C. 1987. Equilibrium and nonequilibrium concepts in ecological models. *Ecological Monographs* 57:1–21.

DelGiudice, G.D. & Seal, U.S. 1988. Classifying winter undernutrition in deer via serum and urinary urea nitrogen. *Wildlife Society Bulletin* 16:27–32.

DelGiudice, G.D., Mech, L.D. & Seal, U.S. 1990. Effects of winter undernutrition on body composition and physiological profiles of white-tailed deer. *Journal of Wildlife Management* 54:539–550.

Derr, J.N. 1991. Genetic interactions between white-tailed and mule deer in the southwestern United States. *Journal of Wildlife Management* 55:228–237.

Diamond, J.M. 1975. Assembly of species communities. In: Cody, M.L. & Diamond, J.M., eds. *Ecology and Evolution of Communities*, pp. 342–444. Belknap, Cambridge, MA.

Diamond, J.M. 1983. Taxonomy by nucleotides. *Nature (London)* 305:17–18.

Diamond, J. & Case, T.J. 1986. Overview: introductions, extinctions, exterminations and invasions. In: Diamond, J. & Case, T.J., eds. *Community Ecology*, pp. 65–79. Harper & Row, New York.

Dice, L.R. 1938. Some census methods for mammals. *Journal of Wildlife Management* 2:119–130.

Dinius, D.A. & Baumgardt, B.R. 1970. Regulation of food intake in ruminants: 6. Influence of caloric density of pelleted rations. *Journal of Dairy Science* 53:311–316.

Dobson, A.P. & May, R.M. 1986a. Disease and conservation. In: Soulé, M.E., ed. *Conservation Biology: The Science of Scarcity and Diversity*, pp. 345–365. Sinauer Associates, Sunderland, MA.

Dobson, A.P. & May, R.M. 1986b. Patterns of invasions by pathogens and parasites. In: Mooney, H.A. & Drake, J.A., eds. *Ecology of Biological Invasions of North America and Hawaii*, pp. 58–76. Ecological Studies 58. Springer, New York.

Dublin, H.T., Sinclair, A.R.E., Boutin, S., Anderson, E., Jago, M. & Arcese, P. 1990a. Does competition regulate ungulate populations?: further evidence from Serengeti, Tanzania. *Oecologia* 82:283–288.

Dublin, H.T., Sinclair, A.R.E. & McGlade, J. 1990b. Elephants and fire as causes of multiple stable states in the Serengeti–Mara woodlands. *Journal of Animal Ecology* 59:1147–1164.

DuBowy, P.J. 1988. Waterfowl communities and seasonal environments: temporal variability in interspecific competition. *Ecology* 69:1439–1453.

Eberhardt, L.L. 1969. Population estimates from recapture frequencies. *Journal of Wildlife Management* 33:28–39.

Eberhardt, L.L. 1978. Transect methods for population studies. *Journal of Wildlife Management* 42:1–31.

Eberhardt, L.L. 1982. Calibrating an index by using removal data. *Journal of Wildlife Management* 46:734–740.

Eberhardt, L.L. & Thomas, J.M. 1991. Designing environmental field studies. *Ecological Monographs* 6:53–73.

Edwards, W.R. & Eberhardt, L. 1967. Estimating cottontail abundance from live trapping data. *Journal of Wildlife Management* 31:87–96.

Elton, C. 1927. *Animal Ecology.* Macmillan, New York.

Emlen, J.T., Dejong, M.J., Jaeger, M.J., Moermond, T.C., Rusterholz, K.A. & White, R.P. 1986. Density trends and range boundary constraints of forest birds along a latitudinal gradient. *Auk* 103:791–803.

Estes, R.D. 1976. The significance of breeding synchrony in the wildebeest. *East African Wildlife Journal* 14:135–152.

Fancy, S.G. & White, R.G. 1985. Energy expenditures by caribou while cratering in snow. *Journal of Wildlife Management* 49:987–993.

Farentinos, R.C., Capretta, P.J., Kepner, R.E. & Littlefield, V.M. 1981. Selective herbivory in tassle-eared squirrels: role of monoterpenes in ponderosa pines chosen as feeding trees. *Science (Washington)* 213:1273–1275.

Fedak, M.A. & Seeherman, H.J. 1979. Reappraisal of energetics of locomotion shows identical cost in bipeds and quadrupeds including ostrich and horse. *Nature (London)* 282:713–716.

Feeny, P.P. & Bostock, H. 1968. Seasonal changes in the tannin content of oak leaves. *Phytochemistry (Oxford)* 7:871–880.

Fenner, F. 1983. The Florey Lecture, 1983: biological control, as exemplified by smallpox eradication and myxomatosis. *Proceedings of the Royal Society of London. Series B: Biological Sciences* 218:259–286.

Ferrar, A.A. & Walker, B.H. 1974. An analysis of herbivore/habitat relationships in Kyle National Park, Rhodesia. *Journal of the Southern African Wildlife Management Association* 4:137–147.

Finch, V.A. 1972. Energy exchanges with the environment of two East African antelopes, the eland and the hartebeest. In: Maloiy, G.M.O., ed. *Comparative Physiology of Desert Animals. Symposia of the Zoological Society of London, no. 31*, pp. 315–326. Academic Press, London.

Findlay, C.S. & Cooke, F. 1982. Synchrony in the lesser snow goose (*Anser caerulescens*): II. The adaptive value of reproductive synchrony. *Evolution* 36:786–799.

Finger, S.E., Brisbin, I.L., Jr., Smith, M.H. & Urbston, D.F. 1981. Kidney fat as a predictor of body condition in white-tailed deer. *Journal of Wildlife Management* 45:964–968.

Fisher, R.A. 1930. *The Genetical Theory of Natural Selection.* Clarendon Press, Oxford.

Fogden, M.P.L. & Fogden, P.M. 1979. The role of fat and protein reserves in the annual cycle of the grey-backed camaroptera in Uganda (Aves: Sylvidae) *Journal of Zoology (London)* 189:233–258.

Foley, W.J. & Hume, I.D. 1987. Passage of digesta markers in two species of arboreal folivorous marsupials – the great glider (*Petauroides volans*) and the brushtail possum (*Trichosurus vulpecula*). *Physiological Zoology* 60:103–113.

Foltz, D.W. & Schwagmeyer, P.L. 1989. Sperm competition in the thirteen-lined ground squirrel: differential fertilization success under field conditions. *American Naturalist* 133:257–265.

Ford, E.B. 1940. Polymorphism and taxonomy. In: Huxley, J., ed. *The New Systematics*, pp. 493–513. Clarendon Press, Oxford.

Fowler, C.W. 1987. A review of density dependence in populations of large mammals. In: Genoways, H.H., ed. *Current Mammalogy*, pp. 401–441. Plenum Press, New York.

Frankel, O.H. & Soulé, M.E. 1981. *Conservation and Evolution*. Cambridge University Press, Cambridge.

Franklin, I.R. 1980. Evolutionary change in small populations. In: Soulé, M.E. & Wilcox, B.A., eds. *Conservation Biology: an Evolutionary–Ecological Perspective*, pp. 135–149. Sinauer Associates, Sunderland, MA.

Franzmann, A.W. & LeResche, R.E. 1978. Alaskan moose blood studies with emphasis on condition evaluation. *Journal of Wildlife Management* 42:334–351.

Fraser, D. & Reardon, E. 1980. Attraction of wild ungulates to mineral-rich springs in central Canada. *Holarctic Ecology* 3:36–39.

Fryxell, J.M. & Sinclair, A.R.E. 1988a. Causes and consequences of migration by large herbivores. *Trends in Ecology and Evolution* 3:237–241.

Fryxell, J.M. & Sinclair, A.R.E. 1988b. Seasonal migration by white-eared kob in relation to resources. *African Journal of Ecology* 26:17–31.

Fryxell, J.M., Greever, J. & Sinclair, A.R.E. 1988. Why are migratory ungulates so abundant? *American Naturalist* 131:781–798.

Fullagar, P.J. 1985. The woodhens of Lord Howe Island. *Avicultural Magazine* 91:15–30.

Fuller, T.K. 1989. Population dynamics of wolves in north–central Minnesota. *Wildlife Monographs* 105:1–41.

Futuyma, D.J. 1986. *Evolutionary Biology*, 2nd edn. Sinauer Associates, Sunderland, MA.

Gasaway, W.C., Boertje, R.D., Grangaard, D.V., Kelleyhouse, D.G., Stephenson, R.O. & Larsen, D.G. 1992. The role of predation in limiting moose at low densities in Alaska and Yukon and implications for conservation. *Wildlife Monographs* 120:1–59.

Gates, C.C. & Hudson, R.J. 1981. Weight dynamics of wapiti in the boreal forest. *Acta Theriologica* 26:407–418.

Gates, C.E. 1979. Line transect and related issues. In: Cormack, R.M., Patil, G.P. & Robson, D.S., eds. *Sampling Biological Populations*. Statistical Ecology Series, Vol. 5, pp. 71–154. International Co-operative Publishing House, Burtonsville, MD.

Gates, C.E. 1980. LINETRAN, a general computer program for analyzing line-transect data. *Journal of Wildlife Management* 44:658–661.

Gates, C.E. 1986. LINETRAN *User's Guide*. Texas A and M University, College Station, Texas.

Gause, G.F. 1934. *The Struggle for Existence*. Williams and Wilkins, Baltimore, MD. (Reprinted 1964 by Hafner, New York.)

Gavin, T.A. & May, B. 1988. Taxonomic status and genetic purity of Columbian white-tailed deer. *Journal of Wildlife Management* 52:1–10.

Gibbs, H.L., Weatherhead, P.J., Boag, P.T., White, B.N., Tabak, L.M. & Hoysak, D.J. 1990. Realized reproductive success of polygynous red-winged blackbirds revealed by DNA markers. *Science (Washington)* 250:1394–1397.

Gill, R.B., Carpenter, L.H. & Bowden, D.C. 1983. Monitoring large animal populations: the Colorado experience. In: *Transactions of the 48th North American Wildlife and Natural Resources Conference*, pp. 330–341. Wildlife Management Institute, Washington, DC.

Gochfeld, M. 1980. Mechanisms and adaptive value of reproductive synchrony in colonial seabirds. In: Burger, J., Olla, B.L. & Winn, H.E., eds. *Behavior of Marine Animals: Current Perspectives in Research, Vol. 4, Marine birds*, pp. 207–270. Plenum, New York.

Golley, F.B. 1961. Energy values of ecological materials. *Ecology* 42:581–584.

Gordon, I.J. 1988. Facilitation of red deer grazing by cattle and its impact on red deer performance. *Journal of Applied Ecology* 25:1–10.

Grant, P.R. 1986. *Ecology and Evolution of Darwin's Finches*. Princeton University Press, Princeton, NJ.

Green, B. 1978. Estimation of food consumption in the dingo, *Canis familiaris dingo*, by means of ^{22}Na turnover. *Ecology* 59:207–210.

Green, G. & Brothers, N. 1989. Water and sodium turnover and estimated food consumption rates in free-living fairy prions (*Pachyptila turtur*) and common diving petrels (*Pelecanoides urinatrix*). *Physiological Zoology* 62:702–715.

Green, B., Anderson, J. & Whateley, T. 1984. Water and sodium turnover and estimated food consumption in free-living lions (*Panthera leo*) and spotted hyaenas (*Crocuta crocuta*). *Journal of Mammalogy* 65:593–599.

Grigg, G.C., Taplin, L.E., Green, B. & Harlow, P. 1986. Sodium and water fluxes in free-living *Crocodylus porosus* in marine and brackish conditions. *Physiological Zoology* 59:240–253.

Gullion, G.W. 1965. A critique concerning foreign game bird introductions. *Wilson Bulletin* 77:409–414.

Gunn, A., Shank, C. & McLean, B. 1991. The history, status and management of muskoxen on Banks Island. *Arctic* 44:188–195.

Guthrie, R.D. (ed.) 1990. *Frozen Fauna of the Mammoth Steppe: the Story of Blue Babe*. University of Chicago Press, Chicago.

Hallett, J.G. 1982. Habitat selection and the community matrix of a desert small-mammal fauna. *Ecology* 63:1400–1410.

Hamilton, W.D. 1971. Geometry of the selfish herd. *Journal of Theoretical Biology* 31:295–311.

Hanks, J. 1981. Characterization of population condition. In: Fowler, C.W. & Smith, T.D., eds. *Dynamics of Large Mammal Populations*, pp. 47–73. Wiley, New York.

Hansen, R.M., Mugambi, M.M. & Bauni, S.M. 1985. Diets and trophic ranking of ungulates of the northern Serengeti. *Journal of Wildlife Management* 49:823–829.

Hanson, W.R. 1967. Estimating the density of an animal population. *Journal of Research on the Lepidoptera* 6:203–247.

Hayne, D.W. 1949. An examination of the strip census method for estimating animal populations. *Journal of Wildlife Management* 13:145–157.

Heard, D.C. 1992. The effect of wolf predation and snow cover on musk-ox group size. *American Naturalist* 139:190–204.

Heard, D.C. & Williams, D. 1991. Wolf den distribution on migratory Barren Ground caribou ranges in the N.W.T. In: Butler, C.E. & Mahoney, S.P., eds. *Proceedings of 4th North American caribou workshop, St John's, Newfoundland*, pp. 249–250.

Henny, C.J., Byrd, M.A., Jacobs, J.A., McLain, P.D., Todd, M.R. & Halla, B.F. 1977. MidAtlantic coast osprey population: present numbers, productivity, pollutant contamination, and status. *Journal of Wildlife Management* 41:254–265.

Hik, D.S. & Jefferies, R.L. 1990. Increases in the net above-ground primary production of a salt-marsh forage grass: a test of the predictions of the herbivore-optimization model. *Journal of Ecology* 78:180–195.

Hofmann, R.R. 1973. *The Ruminant Stomach: Stomach Structure and Feeding Habits of East African Game Ruminants*. East African Monographs in Biology, Vol. 2. East African Literature Bureau, Nairobi.

Holling, C.S. 1959. The components of predation as revealed by a study of small-mammal predation of the European pine sawfly. *Canadian Entomologist* 91: 293–320.

Holt, R.D. 1977. Predation, apparent competition and the structure of prey communities. *Theoretical Population Biology* 12:197–229.

Holt, R.D. 1984. Spatial heterogeneity, indirect interactions, and the coexistence of prey species. *American Naturalist* 124:377–406.

Hope, J.H. 1972. Mammals of the Bass Strait islands. *Proceedings of the Royal Society of Victoria* 85:163–195.

Hornicke, H. & Björnhag, G. 1980. Coprophagy and related strategies for digesta utilization. In: Ruckebusch, Y. & Thivend, P., eds. *Digestive Physiology and Metabolism in Ruminants. Proceedings of 5th International Symposium on Ruminant Physiology*, Clermont Ferrand, France, pp. 707–730.

Houston, D.B. 1982. *The Northern Yellowstone Elk: Ecology and Management*. Macmillan, New York.

Howery, L.D. & Pfister, J.A. 1990. Dietary and fecal concentrations of nitrogen and phosphorus in penned white-tailed deer does. *Journal of Wildlife Management* 54:383–389.

Hudson, P.J. & Dobson, A.P. 1988. The ecology and control of parasites in gamebird populations. In: Hudson, P.J. & Rands, M.R.W., eds. *Ecology and Management of Gamebirds*, pp. 98–133. BSP Professional Books, Oxford.

Hudson, R.J., Drew, K.R. & Baskin, L.M. (eds) 1989. *Wildlife Production Systems*. Cambridge University Press, Cambridge.

Hume, I.D. & Warner, A.C.I. 1980. Evolution of microbial digestion in mammals. In: Ruckebusch, Y. & Thivend, P., eds. *Digestive Physiology and Metabolism in Ruminants. Proceedings 5th International Symposium on Ruminant Physiology*, Clermont Ferrand, France, pp. 669–684.

Hurlbert, S.H. 1984. Pseudoreplication and the design of ecological field experiments. *Ecological Monographs* 54:187–211.

Hutchinson, G.E. 1957. Concluding remarks. *Cold Spring Harbor Symposia on Quantitative Biology* 22:415–427.

Ims, R.A. 1990. On the adaptive value of reproductive synchrony as a predator-swamping strategy. *American Naturalist* 136:485–498.

Ims, R.A., Bondrup-Nielsen, S. & Stenseth, N.C. 1988. Temporal patterns of breeding events in small rodent populations. *Oikos* 53:229–234.

Jarman, P.J. 1973. The free water intake of impala in relation to the water content of their food. *East African Agricultural and Forestry Journal* 38:343–351.

Jarman, P.J. 1974. The social organisation of antelope in relation to their ecology. *Behaviour* 48:215–266.

Jarman, P.J. & Sinclair, A.R.E. 1979. Feeding strategy and the pattern of resource partitioning in ungulates. In: Sinclair, A.R.E. & Norton-Griffiths, M., eds. *Serengeti: Dynamics of an Ecosystem*, pp. 130–163. University of Chicago Press, Chicago.

Jefferies, R.L. & Gottlieb, L.D. 1983. Genetic variation within and between populations of the asexual plant *Puccinellia* X *phryganodes*. *Canadian Journal of Botany* 61:774–779.

Jeffreys, A.J., Wilson, V. & Thein, S.L. 1985. Hypervariable 'minisatellite' regions in human DNA. *Nature (London)* 314:67–73.

Jenks, J.A., Soper, R.B., Lochmiller, R.L. & Leslie, D.M., Jr. 1990. Effect of exposure on nitrogen and fiber characteristics of white-tailed deer feces. *Journal of Wildlife Management* 54:389–391.

Jewell, P.A., Holt, S. & Hart, D. (eds) 1981. *Problems in Management of Locally Abundant Wild Mammals*. Academic Press, New York.

Johnson, D.H., Krapu, G.L., Reinecke, K.J. & Jorde, D.G. 1985. An evaluation of condition indices for birds. *Journal of Wildlife Management* 49:569–575.

Jones, D.M. 1982. Conservation in relation to animal diseases in Africa and Asia. In: Edwards, M.A. & McDonnell, U., eds. *Animal Disease in Relation to Animal Conservation. Symposium of the Zoological Society of London*, no. 50, pp. 271–285. Academic Press, London.

Jones, M.M. 1980. Nocturnal loss of muscle protein from house sparrows (*Passer domesticus*). *Journal of Zoology (London)* 192:33–39.

Jones, W.T. 1987. Dispersal patterns in kangaroo rats (*Dipodomys spectabilis*). In: Chepko-Sade, B.D. & Halpin, Z.T., eds. *Mammalian Dispersal Patterns: The Effects of Social Structure on Population Genetics*, pp. 119–127. University of Chicago Press, Chicago.

Joubert, S.C.J. 1983. A monitoring programme for an extensive national park. In: Owen-Smith, R.N., ed. *Management of Large Mammals in African Conservation Areas*, pp. 201–212. HAUM Educational Publishers, Pretoria.

Källén, A., Arcuri, P. & Murray, J.D. 1985. A simple model for the spatial spread and control of rabies. *Journal of Theoretical Biology* 116:377–393.

Keith, L.B., Cary, J.R., Rongstad, O.J. & Brittingham, M.C. 1984. Demography and ecology of a declining snowshoe hare population. *Wildlife Monographs* 90:1–43.

Kelker, G.H. 1940. Estimating deer populations by a differential hunting loss in the sexes. *Proceedings of the Utah Academy of Sciences, Arts and Letters* 17:65–69.

Kelker, G.H. 1944. Sex ratio equations and formulas for determining wildlife populations. *Proceedings of the Utah Academy of Science, Arts and Letters* 20:189–198.

Kelsall, J.P. & Prescott, W. 1971. *Moose and Deer Behaviour in Snow in Fundy National Park, New Brunswick*. Canadian Wildlife Service Report Series no. 15.

Information Canada, Ottawa.

Kenagy, G.J. 1989. Daily and seasonal uses of energy stores in torpor and hibernation. In: Malan, A. & Canguilhem, B., eds. *Second International Symposium on Living in the Cold*, pp. 17–24. John Libbey Eurotext, London.

Kenagy, G.J., Sharbaugh, S.M. & Nagy, K.A. 1989. Annual cycle of energy and time expenditure in a golden-mantled ground squirrel population. *Oecologia* 78:269–282.

Kiff, L.F. 1989. DDE and the California condor *Gymnogyps californianus*: the end of a story? In: Meyburg, B.U. & Chancellor, R.D., eds. *Raptors in the Modern World*, pp. 477–480. World Working Group on Birds of Prey and Owls, Berlin.

King, D.R., Oliver, A.J. & Mead, R.J. 1981. *Bettongia* and fluoroacetate: a role for 1080 in fauna management. *Australian Wildlife Research* 8:529–536.

Kirkpatrick, R.L. 1980. Physiological indices in wildlife management. In: Schemnitz, S.D., ed. *Wildlife Management Techniques Manual*, 4th edn, pp. 99–112. The Wildlife Society, Washington, DC.

Kistner, T.P., Trainer, C.E. & Hartmann, N.A. 1980. A field technique for evaluating physical condition of deer. *Wildlife Society Bulletin* 8:11–17.

Kleiber, M. 1947. Body size and metabolic rate. *Physiological Reviews* 27:511–541.

Klein, D.R. 1968. The introduction, increase, and crash of reindeer on St Matthew Island. *Journal of Wildlife Management* 32:350–367.

Klein, D.R. & Olson, S.T. 1960. Natural mortality patterns of deer in southeast Alaska. *Journal of Wildlife Management* 24:80–88.

Knopf, F.L., Sedgwick, J.A. & Cannon, R.W. 1988. Guild structure of a riparian avifauna relative to seasonal cattle grazing. *Journal of Wildlife Management* 52:280–290.

Koford, C.B. 1953. *The California Condor*. Research Report 4, National Audubon Society, New York.

Korpimäki, E. & Norrdahl, K. 1991. Numerical and functional responses of kestrels, short-eared owls, and long-eared owls to vole densities. *Ecology* 72:814–826.

Krebs, C.J. 1985. *Ecology: The Experimental Analysis of Distribution and Abundance*, 3rd edn. Harper & Row, New York.

Krebs, C.J. 1989. *Ecological Methodology*. Harper & Row, New York.

Krebs, C.J., Gilbert, B.S., Boutin, S., Sinclair, A.R.E. & Smith, J.N.M. 1986. Population biology of snowshoe hares: I. Demography of food-supplemented populations in the southern Yukon, 1976–84. *Journal of Animal Ecology* 55:963–982.

Laake, J.L., Burnham, K.P. & Anderson, D.R. 1979. *User's manual for program TRANSECT*. Utah State University Press, Logan, Utah.

Lack, D. 1947. *Darwin's Finches*. Cambridge University Press, Cambridge.

Lamprey, H.F. 1963. Ecological separation of the large mammal species in the Tarangire game reserve, Tanganyika. *East African Wildlife Journal* 1:63–92.

Lavigne, D.M., Brooks, R.J., Rosen, D.A. & Galbraith, D.A. 1989. Cold, energetics and populations. In: Wange, L.C.H., ed. *Advances in Comparative and Environmental Physiology, Vol. 4, Animal Adaptation to Cold*, pp. 403–432. Springer, Berlin.

Laws, R.M. 1969. The Tsavo Research Project. *Journal of Reproduction and Fertility* (Suppl.) 6:495–531.

Leader-Williams, N. 1988. *Reindeer on South Georgia*. Cambridge University Press, Cambridge.

Leopold, A. 1933. *Game Management*. Charles Scribner's Sons, New York.

Leslie, D.M., Jr. & Starkey, E.E. 1985. Fecal indices to dietary quality of cervids in old-growth forests. *Journal of Wildlife Management* 49:142–146.

Leslie, D.M., Jr., Starkey, E.E. & Vavra, M. 1984. Elk and deer diets in old-growth forests in western Washington. *Journal of Wildlife Management* 48:762–775.

Lockyer, C. 1987. Evaluation of the role of fat reserves in relation to the ecology of North Atlantic fin and sei whales. In: Huntley, A.C., Costa, D.P., Worthy, G.A.J. & Castellini, M.A., eds. *Approaches to Marine Mammal Energetics*, pp. 183–203. Society for Marine Mammalogy, Special Publication no. 1. Lawrence, Kansas, Texas.

Lotka, A.J. 1925. *Elements of Physical Biology*. Williams & Wilkins, Baltimore, MD.

Lundberg, P. 1988. Functional response of a small mammalian herbivore: the disc equation revisited. *Journal of Animal Ecology* 57:999–1006.

MacArthur, R.H. 1957. On the relative abundance of bird species. *Proceedings of the National Academy of Sciences of the USA* 43:293–295.

MacArthur, R.H. 1958. Population ecology of some warblers of northeastern coniferous forests. *Ecology* 39:599–619.

MacArthur, R.H. 1960. On the relative abundance of species. *American Naturalist* 94:25–36.

MacArthur, R.H. & Levins, R. 1967. The limiting similarity, convergence and divergence of coexisting species. *American Naturalist* 101:377–385.

McCleery, R.H. & Perrins, C.M. 1985. Territory size, reproductive success and population dynamics in the great tit, *Parus major*. In: Sibly, R.M. & Smith, R.H., eds. *Behavioural Ecology: Ecological Consequences of Adaptive Behaviour*, pp. 353–373. The 25th Symposium of the British Ecological Society. Blackwell Scientific Publications, Oxford.

McCutchen, A.A. 1938. Preliminary results of wildlife census based on actual counts compared to previous estimates on national forests. *Transactions of the North American Wildlife Conference* 3:407–414.

McKilligan, N.G. 1984. The food and feeding ecology of the cattle egret, *Ardeola ibis*, when nesting in South-East Queensland. *Australian Wildlife Research* 11:133–144.

McLeod, M.N. 1974. Plant tannins – their role in forage quality. *Nutrition Abstracts and Reviews* 44:803–815.

MacLulich, D.A. 1937. *Fluctuations in the numbers of the varying hare (Lepus americanus)*. University of Toronto Press, Toronto.

McNaughton, S.J. 1976. Serengeti migratory wildebeest: facilitation of energy flow by grazing. *Science (Washington)* 191:92–94.

McNaughton, S.J. 1986. On plants and herbivores. *American Naturalist* 128:765–770.

Mace, G.M. 1979. *The Evolutionary Ecology of Small Mammals*. PhD thesis, University of Sussex, Brighton.

Magnusson, W.E., Caughley, G.J. & Grigg, G.C. 1978. A double-survey estimate of population size from incomplete counts. *Journal of Wildlife Management* 42:174–176.

Maloiy, G.M.O. 1973. The water metabolism of a small East African antelope: the dikdik. *Proceedings of the Royal Society of London. Series B: Biological Sciences* 184:167–178.

Malthus, T.R. 1798. *An essay on the principle of population as it affects the future improvement of society: with remarks on the speculation of Mr Godwin, M. Condoroet, and other writers*. Printed for J. Johnson, 1798. London.

Margules, C., Higgs, A.J. & Rafe, R.W. 1982. Modern biogeographical theory: are there any lessons for nature reserve design? *Biological Conservation* 24:115–128.

Marsh, R.E. 1988. Chemosterilants for rodent control. In: Prakash, I., ed. *Rodent Pest Management*, pp. 353–367. CRC Press, Boca Raton, FL.

Martin, P.S. & Klein, R.G. (eds) 1984. *Quaternary Extinctions: A Prehistoric Revolution*. University of Arizona Press, Tucson, Arizona.

May, R.M. 1972. Limit cycles in predator–prey communities. *Science (Washington)* 177:900–902.

Mech, L.D. & DelGiudice, G.D. 1985. Limitations of the marrow-fat technique as an indicator of body condition. *Wildlife Society Bulletin* 13:204–206.

Mendel, G. 1959. Experiments in plant-hybridization. In: Peters, J.A., ed. *Classic Papers in Genetics*, pp. 1–19. Prentice-Hall, Englewood Cliffs, NJ.

Messier, F. 1991. The significance of limiting and regulating factors on the demography of moose and white-tailed deer. *Journal of Animal Ecology* 60:377–393.

Messier, F. & Crete, M. 1985. Moose–wolf dynamics and the natural regulation of moose populations. *Oecologia* 65:503–512.

Messier, F., Huot, J., LeHenaff, D. & Luttich, S. 1988. Demography of the George River caribou herd: evidence of population regulation by forage exploitation and range expansion. *Arctic* 41:279–287.

Miller, F.L., Edmonds, E.J. & Gunn, A. 1982. Foraging behaviour of Peary caribou in response to springtime snow and ice conditions. *Canadian Wildlife Service Occasional Paper* 48:1–39.

Morrison, M.L., Marcot, B.G. & Mannan, R.W. 1992. *Wildlife–Habitat Relationships: Concepts and Applications*. University of Wisconsin Press, Wisconsin.

Moss, R. 1974. Winter diets, gut lengths, and interspecific competition in Alaskan ptarmigan. *Auk* 91:737–746.

Mould, E.D. & Robbins, C.T. 1981. Nitrogen metabolism in elk. *Journal of Wildlife Management* 45:323–334.

Mould, E.D. & Robbins, C.T. 1982. Digestive capabilities in elk compared to white-tailed deer. *Journal of Wildlife Management* 46:22–29.

Mountainspring, S. & Scott, J.M. 1985. Interspecific competition among Hawaiian forest birds. *Ecological Monographs* 55:219–239.

Mundy, P.J. & Ledger, J.A. 1976. Griffon vultures, carnivores and bones. *South African Journal of Science* 72:106–110.

Munger, J.C. & Brown, J.H. 1981. Competition in desert rodents: an experiment with semipermeable exclosures. *Science (Washington)* 211:510–512.

Murton, R.K., Isaacson, A.J. & Westwood, N.J. 1966. The relationships between wood-pigeons and their clover food supply and the mechanism of population control. *Journal of Applied Ecology* 3:55–93.

Myers, K. & Bults, H.G. 1977. Observations on changes in the quality of food eaten by the wild rabbit. *Australian Journal of Ecology* 2:215–229.

Nagy, K.A. 1980. CO_2 production in animals: analysis of potential errors in the doubly labeled water method. *American Journal of Physiology* 238:R466–R473.

Nagy, K.A. 1983. *The Doubly Labeled Water Method: A Guide to Its Use*. UCLA Publication no. 12-1417. University of California, Los Angeles, CA.

Nagy, K.A. 1989. Field bioenergetics: accuracy of models and methods. *Physiological Zoology* 62:237–252.

Nagy, K.A. & Peterson, C.C. 1988. *Scaling of Water Flux Rate in Animals*. University of California Publications in Zoology, Vol. 120. University of California: Los Angeles.

Nelson, M.E. & Mech, L.D. 1981. Deer social organization and wolf predation in northeastern Minnesota. *Wildlife Monographs* 77:1–53.

Nevo, E., Beiles, A. & Ben-Shlomo, R. 1984. The evolutionary significance of genetic diversity: ecological, demographic and life history correlates. *Lecture Notes in Biomathematics* 53:13–213.

Newton, I. 1972. *Finches*. Collins, London.

Norton, G.A. 1988. Philosophy, concepts and techniques. In: Norton, G.A. & Pech, R.P., eds. *Vertebrate Pest Management in Australia: a Decision Analysis/Systems Analysis Approach*, pp. 1–17. Project Report no. 5. CSIRO, Melbourne, Australia.

Norton, I.O. & Sclater, J.G. 1979. A model for the evolution of the Indian Ocean and the breakup of Gondwanaland. *Journal of Geophysical Research* 84:6803–6830.

Norton-Griffiths, M. 1978. Counting animals. In: Grimsdell, J.J.R., ed. *Counting Animals*, 2nd edn. Handbook no. 1. African Wildlife Leadership Foundation, Nairobi, Kenya. (A series of handbooks on techniques in African wildlife ecology.)

Noss, R.F. 1987. Corridors in real landscapes: a reply to Simberloff and Cox. *Conservation Biology* 1:159–164.

Noy-Meir, I. 1975. Stability of grazing systems: an application of predator–prey graphs. *Journal of Ecology* 63:459–481.

O'Brien, S.J., Nash, W.G., Wildt, D.E., Bush, M.E. & Benveniste, R.E. 1985a. A molecular solution to the riddle of the giant panda's phylogeny. *Nature (London)* 317:140–144.

O'Brien, S.J., Roelke, M.E., Marker, L., *et al.* 1985b. Genetic basis for species vulnerability in the cheetah. *Science (Washington)* 227:1428–1434.

O'Brien, S.J., Wildt, D.E. & Bush, M. 1986. The cheetah in genetic peril. *Scientific American* 254:68–76.

O'Brien, S.J., Wildt, D.E., Bush, M., *et al.* 1987. East African cheetahs: evidence for two population bottlenecks? *Proceedings of the National Academy of Sciences of the USA* 84:508–511.

O'Donoghue, M. 1991. *Reproduction, Juvenile Survival and Movements of Snowshoe Hares at a Cyclic Population Peak.* MSc thesis, University of British Columbia, Vancouver.

O'Kelly, J.C. 1973. Seasonal variations in the plasma lipids of genetically different types of cattle: steers on different diets. *Comparative Biochemistry and Physiology* 44:303–312.

Ojasti, J. 1983. Ungulates and large rodents of South America. In: Bourlière, F., ed. *Ecosystems of the World: 13. Tropical Savannas*, pp. 427–439. Elsevier Scientific, Amsterdam.

Orians, G.H. & Willson, M.F. 1964. Interspecific territories of birds. *Ecology* 45: 736–745.

Otis, D.L., Burnham, K.P., White, G.C. & Anderson, D.R. 1978. Statistical interference from capture data on closed animal populations. *Wildlife Monographs* 62:1–135.

Owen, R.B., Jr. & Reinecke, K.J. 1979. Bioenergetics of breeding dabbling ducks. In: Bookhout, T.A., ed. *Waterfowl and Wetlands – An Integrated Review*, pp. 71–93. Proceedings of a symposium at the 39th Midwestern Fish and Wildlife Conference, Madison, Wisconsin. The Wildlife Society, Washington, DC.

Owen-Smith, N. 1990. Demography of a large herbivore, the greater kudu *Tragelaphus strepsiceros*, in relation to rainfall. *Journal of Animal Ecology* 59: 893–913.

Owen-Smith, N. & Cooper, S.M. 1987. Palatability of woody plants to browsing ruminants in a South African savanna. *Ecology* 68:319–331.

Owen-Smith, N. & Cooper, S.M. 1989. Nutritional ecology of a browsing ruminant, the kudu (*Tragelaphus strepsiceros*), through the seasonal cycle. *Journal of Zoology (London)* 219:29–43.

Owen-Smith, R.N. (ed.) 1983. *Management of Large Mammals in African Conservation Areas.* HAUM Educational, Pretoria.

Parer, I. 1987. Factors influencing the distribution and abundance of rabbits (*Oryctolagus cuniculus*) in Queensland. *Proceedings of the Royal Society of Queensland* 98:73–82.

Parer, I., Conolly, D. & Sobey, W.R. 1985. Myxomatosis: the effects of annual introductions of an immunizing strain and a highly virulent strain of myxoma virus into rabbit populations at Urana, NSW. *Australian Wildlife Research* 12: 407–423.

Patterson, I.J. 1965. Timing and spacing of broods in the black-headed gull. *Ibis* 107:433–459.

Paulus, S.L. 1982. Gut morphology of gadwalls in Louisiana in winter. *Journal of Wildlife Management* 46:483–489.

Pease, J.L., Vowles, R.H. & Keith, L.B. 1979. Interaction of snowshoe hares and woody vegetation. *Journal of Wildlife Management* 43:43–60.

Pech, R.P. & McIlroy, J.C. 1990. A model of the velocity of advance of foot and mouth disease in feral pigs. *Journal of Applied Ecology* 27:635–650.

Pech, R.P., Sinclair, A.R.E., Newsome, A.E. & Catling, P.C. 1992. Limits to predator regulation of rabbits in Australia: evidence from predator-removal experiments. *Oecologia* 89:102–112.

Pehrsson, O. 1984. Relationship of food to spatial and temporal breeding strategies of mallards in Sweden. *Journal of Wildlife Management* 48:322–339.

Perrins, C.M. 1970. The timing of birds' breeding seasons. *Ibis* 112:242–255.

Peterson, M.J., Grant, W.E. & Davis, D.S. 1991. Bison-brucellosis management: simulation of alternative strategies. *Journal of Wildlife Management* 55:205–213.

Peterson, R.O. 1992. *Ecological Studies of Wolves on Isle Royale: Annual Report 1991–1992.* Michigan Technological University, Houghton, Michigan.

Pianka, E.R., Huey, R.B. & Lawlor, L.P. 1979. Niche segregation in desert lizards. In: Horn, D.J., Stairs, G.R. & Mitchell, R.D., eds. *Analysis of Ecological Systems*, pp. 67–116. Ohio State University Press, Columbus, OH.

Pimm, S.L. & Pimm, J.W. 1982. Resource use, competition and resource availability in Hawaiian honeycreepers. *Ecology* 63:1468–1480.

Pollock, K.H. 1974. *The Assumption of Equal Catchability in Tag–Recapture Experiments*. PhD dissertation, Cornell University, Ithaca, NY.

Pollock, K.H., Winterstein, S.R., Bunck, C.M. & Curtis, P.D. 1989. Survival analysis in telemetry studies: the staggered entry design. *Journal of Wildlife Management* 53:7–15.

Porter, R.N. 1977. *Wildlife Management Objectives and Practices for the Hluhluwe Game Reserve and the Northern Corridor*. Natal Parks Board (unpublished).

Pratt, H.D., Bruner, P.L. & Berrett, D.G. (eds) 1987. *A Field Guide to the Birds of Hawaii and the Tropical Pacific*. Princeton University Press, Princeton, NJ.

Provenza, F.D., Burritt, E.A., Clausen, T.P., Bryant, J.P., Reichardt, P.B. & Distel, R.A. 1990. Conditioned flavor aversion: a mechanism for goats to avoid condensed tannins in blackbrush. *American Naturalist* 136:810–828.

Quinn, T.W., Quinn, J.S., Cooke, F. & White, B.N. 1987. DNA marker analysis detects multiple maternity and paternity in single broods of the lesser snow goose. *Nature (London)* 326:392–394.

Raj, D. & Khamis, S.H. 1958. Some remarks on sampling with replacement. *Annals of Mathematical Statistics* 29:550–557.

Ralls, K., Ballou, J.D. & Templeton, A. 1988. Estimates of lethal equivalents and the cost of inbreeding in mammals. *Conservation Biology* 2:185–193.

Ransom, A.B. 1965. Kidney and marrow fat as indicators of white-tailed deer condition. *Journal of Wildlife Management* 29:397–399.

Rasmussen, D.I. & Doman, E.R. 1943. Census methods and their application in the management of mule deer. In: *Transactions of the 8th North American Wildlife Conference*, pp. 369–380. American Wildlife Institute, Washington, DC.

Redfield, J.A., Krebs, C.J. & Taitt, M.J. 1977. Competition between *Peromyscus maniculatus* and *Microtus townsendii* in grasslands of coastal British Columbia. *Journal of Animal Ecology* 46:607–616.

Redhead, T.D. 1982. *Reproduction, Growth and Population Dynamics of House Mice in Irrigated and Non-Irrigated Cereal Farms in New South Wales*. PhD thesis, Australian National University, Canberra.

Ringelman, J.K. & Szymczak, M.R. 1985. A physiological condition index for wintering mallards. *Journal of Wildlife Management* 49:564–568.

Robbins, C.T. 1983. *Wildlife Feeding and Nutrition*. Academic Press, New York.

Robbins, C.T., Hanley, T.A., Hagerman, A.E., *et al.* 1987. Role of tannins in defending plants against ruminants: reduction in protein availability. *Ecology* 68:98–107.

Robertshaw, D. & Taylor, C.R. 1969. A comparison of sweat gland activity in eight species of East African bovids. *Journal of Physiology* 203:135–143.

Robertson, G. 1987. Plant dynamics. In: Caughley, G., Shepherd, N. & Short, J., eds. *Kangaroos: Their Ecology and Management in the Sheep Rangelands of Australia*, pp. 50–68. Cambridge University Press, Cambridge.

Robertson, G., Short, J. & Wellard, G. 1987. The environment of the Australian sheep rangelands. In: Caughley, G., Shepherd, N. & Short, J., eds. *Kangaroos: Their Ecology and Management in the Sheep Rangelands of Australia*, pp. 14–34. Cambridge University Press, Cambridge.

Robertson, P.A. & Rosenberg, A.A. 1988. Harvesting gamebirds. In: Hudson, P.J. & Rands, M.R.W., eds. *Ecology and Management of Gamebirds*, pp. 177–201. BSP Professional Books, Oxford.

Robinette, W.L., Jones, D.A. & Loveless, C.M. 1974. Field tests of strip census methods. *Journal of Wildlife Management* 38:81–96.

Robinson, W.L. & Bolen, E.G. 1984. *Wildlife Ecology and Management*. Macmillan, New York.

Robson, D.S. & Whitlock, J.H. 1964. Estimation of a truncation point. *Biometrika* 51:33–39.

Rogers, L.L., Mech, L.D., Dawson, D.K., Peek, J.M. & Korb, M. 1980. Deer distribution in relation to wolf pack territory edges. *Journal of Wildlife Management* 44:253–258.

Rolley, R.E. & Keith, L.B. 1980. Moose population dynamics and winter habitat use at Rochester, Alberta, 1965–1979. *Canadian Field-Naturalist* 94:9–18.

Root, A. 1972. Fringe-eared oryx digging for tubers in the Tsavo National Park (East). *East African Wildlife Journal* 10:155–157.

Root, T. 1988. *Atlas of Wintering North American Birds: An Analysis of Christmas Bird Count Data.* University of Chicago Press, Chicago.

Rosenzweig, M.L. 1981. A theory of habitat selection. *Ecology* 62:327–335.

Routledge, R.D. 1981. The unreliability of population estimates from repeated, incomplete aerial surveys. *Journal of Wildlife Management* 45:997–1000.

Routledge, R.D. 1982. The method of bounded counts: when does it work? *Journal of Wildlife Management* 46:757–761.

Rowley, I. & Chapman, G. 1991. The breeding biology, food, social organisation, demography and conservation of the Major Mitchell or pink cockatoo, *Cacatua leadbeateri*, on the margin of the Western Australian wheatbelt. *Australian Journal of Zoology* 39:211–261.

Rubsamen, K., Heller, R., Lawrenz, H. & Engelhardt, W.V. 1979. Water and energy metabolism in the rock hyrax (*Procavia habessinica*). *Journal of Comparative Physiology B* 131:303–309.

Samuel, W.M., Pybus, M.J., Welch, D.A. & Wilke, C.J. 1992. Elk as a potential host for meningeal worm: implications for translocation. *Journal of Wildlife Management* 56:629–639.

Saunders, D.A. & Hobbs, R.J. (eds) 1991. *Nature Conservation: 2. The Role of Corridors.* Surrey Beatty & Sons, Chipping Norton, New South Wales.

Schaller, G.B. 1972. *The Serengeti Lion.* University of Chicago Press, Chicago.

Schaller, G.B., Jinchu, H., Wenski, P. & Jing, Z. 1985. *The Giant Pandas of Wolong.* University of Chicago Press, Chicago.

Schluter, D. 1988. The evolution of finch communities on islands and continents: Kenya vs. Galapagos. *Ecological Monographs* 58:229–249.

Schullery, P. 1984. *Mountain Time.* Nick Lyons Books, New York.

Schwagmeyer, P.L. & Woontner, S.J. 1985. Mating competition in an asocial ground squirrel, *Spermophilus tridecemlineatus. Behavioural Ecology and Sociobiology* 17:291–296.

Schwartz, C.C., Nagy, J.G. & Regelin, W.L. 1980. Juniper oil yield, terpenoid concentration, and antimicrobial effects on deer. *Journal of Wildlife Management* 44:107–113.

Scott, J.M., Mountainspring, S., Ramsey, F.L. & Kepler, C.B. 1986. *Forest Bird Communities of the Hawaiian Islands: Their Dynamics, Ecology, and Conservation.* Studies in Avian Biology no. 9, Cooper Ornithological Society, University of California, Los Angeles.

Seal, U.S., Thorne, E.T., Bogan, M.A. & Anderson, S.H. 1989. *Conservation Biology and the Black-Footed Ferret.* Yale University Press, New Haven, CT.

Seber, G.A.F. 1982. *The Estimation of Animal Abundance and Related Parameters,* 2nd edn. Macmillan, New York.

Seip, D.R. 1991. Predation and caribou populations. Proceedings of the Fifth North American Caribou Workshop, Yellowknife, Northwest Territories, Canada. *Rangifer* Special Issue no. 7:46–52.

Seip, D.R. 1992. Factors limiting woodland caribou populations and their interrelationships with wolves and moose in southeastern British Columbia. *Canadian Journal of Zoology* 70:1494–1503.

Semel, B. & Andersen, D.C. 1988. Vulnerability of acorn weevils (Coleoptera: Curculionidae) and attractiveness of weevils and infested *Quercus alba* acorns to *Peromyscus leucopus* and *Blarina brevicauda. American Midland Naturalist* 119:385–393.

Sen, A.R. 1982. A review of some important techniques in sampling wildlife. *Canadian Wildlife Service Occasional Paper* 49:1–15.

Shepherd, N. & Caughley, G. 1987. Options for management of kangaroos. In: Caughley, G., Shepherd, N. & Short, J., eds. *Kangaroos: Their Ecology and Management in the Sheep Rangelands of Australia*, pp. 188–219. Cambridge University Press, Cambridge.

Short, J. 1987. Factors affecting food intake of rangelands herbivores. In: Caughley, G., Shepherd, N. & Short, J., eds. *Kangaroos: Their Ecology and Management in the Sheep Rangelands of Australia*, pp. 84–99. Cambridge University Press, Cambridge.

Short, J., Bradshaw, S.D., Giles, J., Prince, R.I.T. & Wilson, G.R. 1992. Reintroduction of macropods (Marsupialia: Macropodoidea) in Australia – a review. *Biological Conservation* 62:189–204.

Sibly, R.M. 1981. Strategies of digestion and defecation. In: Townsend, C.R. & Calow, P., eds. *Physiological Ecology: An Evolutionary Approach to Resource Use*, pp. 109–139. Blackwell Scientific Publications, Oxford.

Simberloff, D. 1988. The contribution of population and community biology to conservation science. *Annual Review of Ecology and Systematics* 19:473–511.

Simberloff, D. & Cox, J. 1987. Consequences and costs of conservation corridors. *Conservation Biology* 1:63–71.

Sinclair, A.R.E. 1977. *The African Buffalo: A Study of the Resource Limitation of Populations*. University of Chicago Press, Chicago.

Sinclair, A.R.E. 1978. Factors affecting the food supply and breeding season of resident birds and movements of palaearctic migrants in a tropical African savannah. *Ibis* 120:480–497.

Sinclair, A.R.E. 1983. The adaptations of African ungulates and their effects on community function. In: Bourlière, F., ed. *Ecosystems of the World: 13. Tropical Savannas*, pp. 401–426. Elsevier Scientific, Amsterdam.

Sinclair, A.R.E. 1985. Does interspecific competition or predation shape the African ungulate community? *Journal of Animal Ecology* 54:899–918.

Sinclair, A.R.E. 1989. Population regulation in animals. In: Cherrett, J.M., ed. *Ecological Concepts: the Contribution of Ecology to An Understanding of the Natural World*, pp. 197–241. The 29th Symposium of the British Ecological Society. Blackwell Scientific Publications, Oxford.

Sinclair, A.R.E. & Duncan, P. 1972. Indices of condition in tropical ruminants. *East African Wildlife Journal* 10:143–149.

Sinclair, A.R.E. & Smith, J.N.M. 1984. Do plant secondary compounds determine feeding preferences of snowshoe hares? *Oecologia* 61:403–410.

Sinclair, A.R.E., Dublin, H. & Borner, M. 1985. Population regulation of Serengeti wildebeest: a test of the food hypothesis. *Oecologia* 65:266–268.

Sinclair, A.R.E., Krebs, C.J. & Smith, J.N.M. 1982. Diet quality and food limitation in herbivores: the case of the snowshoe hare. *Canadian Journal of Zoology* 60:889–897.

Sinclair, A.R.E., Krebs, C.J., Smith, J.N.M. & Boutin, S. 1988. Population biology of snowshoe hares: III. Nutrition, plant secondary compounds and food limitation. *Journal of Animal Ecology* 57:787–806.

Sinclair, A.R.E., Olsen, P.D. & Redhead, T.D. 1990. Can predators regulate small mammal populations?: evidence from house mouse outbreaks in Australia. *Oikos* 59:382–392.

Skogland, T. 1985. The effects of density-dependent resource limitations on the demography of wild reindeer. *Journal of Animal Ecology* 54:359–374.

Smith, J.N.M., Krebs, C.J., Sinclair, A.R.E. & Boonstra, R. 1988. Population biology of snowshoe hares: II. Interactions with winter food plants. *Journal of Animal Ecology* 57:269–286.

Smith, N.S. 1970. Appraisal of condition estimation methods for East African ungulates. *East African Wildlife Journal* 8:123–129.

Smuts, G.L. 1978. Interrelations between predators, prey, and their environment. *BioScience* 28:316–320.

Snell, G.P. & Hlavachick, B.D. 1980. Control of prairie dogs – the easy way. *Rangelands* 2:239–240.

Snyder, N.F.R. & Johnson, E.V. 1985. Photographic censusing of the 1982–1983 California condor population. *Condor* 87:1–13.

Snyder, N.F.R., Wiley, J.W. & Kepler, C.B. 1987. *The Parrots of Luquillo: Natural History and Conservation of the Puerto Rican Parrot*. Western Foundation of Vertebrate Zoology, Los Angeles, CA.

Solomon, M.E. 1949. The natural control of animal populations. *Journal of Animal Ecology* 18:1–35.

Southern, H.N. 1951. Change in status of the bridled guillemot after ten years. *Proceedings of the Zoological Society of London* 121:657–671.

Spinage, C.A. 1973. The role of photoperiodism in the seasonal breeding of tropical African ungulates. *Mammal Review* 3:71–84.

Spowart, R.A. & Thompson Hobbs, N. 1985. Effect of fire on diet overlap between mule deer and mountain sheep. *Journal of Wildlife Management* 49:942–946.

Spratt, D.M. 1990. The role of helminths in the biological control of mammals. *International Journal for Parasitology* 20:543–550.

Spratt, D.M. & Presidente, P.J.A. 1981. Prevalence of *Fasciola hepatica* infection in native mammals in southeastern Australia. *Australian Journal of Experimental Biology and Medical Science* 59:713–721.

Spray, C.J., Crick, H.Q.P. & Hart, A.D.M. 1987. Effects of aerial applications of fenitrothion on bird populations of a Scottish pine plantation. *Journal of Applied Ecology* 24:29–47.

Stafford, J. 1971. The heron population of England and Wales, 1928–1970. *Bird Study* 18:218–221.

Stanley Price, M.R. 1989. *Animal Re-introductions: The Arabian Oryx in Oman.* Cambridge University Press, Cambridge.

Stevens, G.C. 1989. The latitudinal gradient in geographical range: how so many species coexist in the tropics. *American Naturalist* 133:240–256.

Stonehouse, B. 1967. The general biology and thermal balances of penguins. In: Cragg, J.B., ed. *Advances in Ecological Research*, Vol. 4, pp. 131–196. Academic Press, London.

Sullivan, T.P. & Klenner, W. 1993. Influence of diversionary food on red squirrel populations and damage to crop trees in young lodgepole pine forest. *Ecological Applications* 3:708–718.

Swanson, G.A., Adomaitis, V.A., Lee, F.B., Serie, J.R. & Shoesmith, J.A. 1984. Limnological conditions influencing duckling use of saline lakes in south–central north Dakota. *Journal of Wildlife Management* 48:340–349.

Taber, R.D. 1956. Deer nutrition and population dynamics in the north coast range of California. *Transactions of the North American Wildlife Conference* 21:159–172.

Taitt, M.J. & Krebs, C.J. 1981. The effect of extra food on small rodent populations: II. Voles (*Microtus townsendii*). *Journal of Animal Ecology* 50:125–137.

Talbot, L.M. & Stewart, D.R.M. 1964. First wildlife census of the entire Serengeti–Mara region, East Africa. *Journal of Wildlife Management* 28:815–827.

Taylor, C.R. 1968a. The minimum water requirements of some East African bovids. In: Crawford, M.A., ed. *Comparative Nutrition of Wild Animals*, pp. 195–206. Symposia of the Zoological Society of London. Academic Press, London.

Taylor, C.R. 1968b. Hygroscopic food: a source of water for desert antelopes? *Nature (London)* 219:181–182.

Taylor, C.R. 1969. Metabolism, respiratory changes, and water balance of an antelope, the eland. *American Journal of Physiology* 217:317–320.

Taylor, C.R. 1970a. Strategies of temperature regulation: effect on evaporation in East African ungulates. *American Journal of Physiology* 219:1131–1135.

Taylor, C.R. 1970b. Dehydration and heat: effects on temperature regulation of East African ungulates. *American Journal of Physiology* 219:1136–1139.

Taylor, C.R. 1972. The desert gazelle: a paradox resolved. In: Maloiy, G.M.O., ed. *Comparative Physiology of Desert Animals*, pp. 215–227. Symposia of the Zoological Society of London, no. 31. Academic Press, London.

Taylor, C.R., Robertshaw, D. & Hofmann, R. 1969a. Thermal panting: a comparison of wildebeest and zebu cattle. *American Journal of Physiology* 217:907–910.

Taylor, C.R., Spinage, C.A. & Lyman, C.P. 1969b. Water relations of the waterbuck, an East African antelope. *American Journal of Physiology* 217:630–634.

Tedman, R. & Green, B. 1987. Water and sodium fluxes and lactational energetics in suckling pups of Weddell seals (*Leptonychotes weddelli*). *Journal of Zoology (London)* 212:29–42.

Telfer, E.S. 1970. Winter habitat selection by moose and white-tailed deer. *Journal of Wildlife Management* 34:553–559.

Temple, S.A. 1977. Plant–animal mutualism: coevolution with dodo leads to near extinction of plant. *Science (Washington)* 197:885–886.

Terborgh, J. 1989. *Where Have All the Birds Gone?: Essays on the Biology and Conservation of Birds that Migrate to the American Tropics.* Princeton University Press, Princeton, NJ.

Terborgh, J. 1992. Why American songbirds are vanishing. *Scientific American* 266:56–62.

Terborgh, J.W. & Janson, C.H. 1986. The socioecology of primate groups. *Annual Review of Ecology and Systematics* 17:111–135.

Thill, R.E. 1984. Deer and cattle diets on Louisiana pine-hardwood sites. *Journal of Wildlife Management* 48:788–798.

Thomas, V.G. 1988. Body condition, ovarian hierarchies, and their relation to egg formation in Anseriform and Galliform species. *Proceedings of the International Ornithological Congress* 19:353–363.

Thorne, E.T. & Williams, E.S. 1988. Disease and endangered species: the black-footed ferret as a recent example. *Conservation Biology* 2:66–74.

Trostel, K., Sinclair, A.R.E., Walters, C.J. & Krebs, C.J. 1987. Can predation cause the 10-year hare cycle? *Oecologia* 74:185–192.

Turner, J.W. & Kirkpatrick, J.F. 1991. New developments in feral horse contraception and their potential application to wildlife. *Wildlife Society Bulletin* 19:350–359.

Tyndale-Biscoe, C.H. 1989. Physiological strategies of Australian vertebrates. In: *CSIRO Division of Wildlife and Ecology Biennial Report 1986–1988*, p. 1. CSIRO, Melbourne.

Uhazy, L.S., Holmes, J.C. & Stelfox, J.G. 1973. Lungworms in the Rocky Mountain bighorn sheep of western Canada. *Canadian Journal of Zoology* 51:817–824.

Urquhart, D. & Farnell, R. 1986. *The Fortymile Herd.* Department of Renewable Resources, Whitehorse, Yukon.

Verme, L.J. & Holland, J.C. 1973. Reagent–dry assay of marrow fat in white-tailed deer. *Journal of Wildlife Management* 37:103–105.

Vesey-Fitzgerald, D.F. 1960. Grazing succession among East African game animals. *Journal of Mammalogy* 41:161–172.

Volterra, V. 1926. Variations and fluctuations of the numbers of individuals in animal species living together. In: Chapman, R.N., ed. *Animal Ecology*, pp. 409–448. McGraw-Hill, New York.

Walker, B.H., Ludwig, D., Holling, C.S. & Peterman, R.M. 1981. Stability of semi-arid savanna grazing systems. *Journal of Ecology* 69:473–498.

Walter, H. 1973. *Vegetation of the Earth in Relation to Climate and the Eco-physiological Conditions.* English Universities Press, London.

Walters, C. 1986. *Adaptive Management of Renewable Resources.* Macmillan, New York.

Ward, P. 1969. The annual cycle of the yellow-vented bulbul *Pycnonotus goiavier* in a humid equatorial environment. *Journal of Zoology (London)* 157:25–45.

Warner, R.E. 1968. The role of introduced diseases in the extinction of the endemic Hawaiian avifauna. *Condor* 70:101–120.

Watson, A. & Moss, R. 1971. Spacing as affected by territorial behavior, habitat, and nutrition in red grouse (*Lagopus l. scoticus*). In: Esser, A.H., ed. *Behavior and Environment: the Use of Space by Animals and Men*, pp. 92–111. Plenum Press, New York.

Weatherall, D.J. 1985. *The New Genetics and Clinical Practice*, 2nd edn. Oxford University Press, Oxford.

Wegener, A. 1924. *The Origins of Continents and Oceans.* Methuen, London.

Weir, J.S. 1972. Spatial distribution of elephants in an African National Park in relation to environmental sodium. *Oikos* 23:1–13.

Western, D. 1975. Water availability and its influence on the structure and dynamics of a savannah large mammal community. *East African Wildlife Journal* 13:265–286.

Westoby, M. 1989. Selective forces exerted by vertebrate herbivores on plants. *Trends in Ecology and Evolution* 4:115–117.

Wetton, J.H., Carter, R.E., Parkin, D.T. & Walters, D. 1987. Demographic study of a wild house sparrow population by DNA fingerprinting. *Nature (London)* 327: 147–149.

White, G.C., Anderson, D.R., Burnham, K.P. & Otis, D.L. 1982. *Capture–Recapture and Removal Methods for Sampling Closed Populations*. Report LA-8787-NERP, Los Alamos National Laboratory, Los Alamos, New Mexico.

White, R.G., Bunnell, F.L., Gaare, E., Skogland, T. & Hubert, B. 1981. Ungulates on arctic ranges. In: Bliss, L.C., Heal, O.W. & Moore, J.J., eds. *Tundra Ecosystems: A Comparative Analysis*, pp. 397–483. Cambridge University Press, Cambridge.

Whitehead, G.K. 1972. *Deer of the World*. Constable, London.

Whittaker, R.H. 1975. *Communities and Ecosystems*, 2nd edn. MacMillan, New York.

Whyte, R.J. & Bolen, E.G. 1985. Variation in mallard digestive organs during winter. *Journal of Wildlife Management* 49:1037–1040.

Wickstrom, M.L., Robbins, C.T., Hanley, T.A., Spalinger, D.E. & Parish, S.M. 1984. Food intake and foraging energetics of elk and mule deer. *Journal of Wildlife Management* 48:1285–1301.

Wiemeyer, S.N., Scott, J.M., Anderson, M.P., Bloom, P.H. & Stafford, C.J. 1988. Environmental contaminants in California condors. *Journal of Wildlife Management* 52:238–247.

Wiens, J.A. 1977. On competition and variable environments. *American Scientist* 65:590–597.

Wilbur, S.R., Carrier, W.D. & Borneman, J.C. 1974. Supplemental feeding program for California condors. *Journal of Wildlife Management* 38:343–346.

Wilcove, D.S. 1985. Nest predation in forest tracts and the decline of migratory songbirds. *Ecology* 66:1211–1214.

Wirtz, P. 1982. Territory holders, satellite males, and bachelor males in a high density population of waterbuck (*Kobus ellipsiprymnus*) and their association with conspecifics. *Zeitschrift für Tierpsychologie* 58:277–300.

Woodruff, D.S. 1989. The problems of conserving genes and species. In: Western, D. & Pearl, M.C., eds. *Conservation for the Twenty-First Century*, pp. 76–88. Oxford University Press, New York.

Wydeven, A.P. & Dahlgren, R.B. 1985. Ungulate habitat relationships in Wind Cave National Park. *Journal of Wildlife Management* 49:805–813.

Yuill, T.M. 1987. Diseases as components of mammalian ecosystems: mayhem and subtlety. *Canadian Journal of Zoology* 65:1061–1066.

Zar, J.H. 1984. *Biostatistical Analysis*, 2nd edn. Prentice-Hall, Englewood Cliffs, NJ.

Zeleny, L. 1976. *The Bluebird: How You Can Help Its Fight for Survival*. Indiana University Press, Bloomington.

Index

Page numbers in *italic* refer to figures. Page numbers in **bold** refer to tables.

325